Waste Paper in Early Modern England

Waste Paper in Early Modern England

Privy Tokens

ANNA REYNOLDS

OXFORD
UNIVERSITY PRESS

OXFORD
UNIVERSITY PRESS

Great Clarendon Street, Oxford, OX2 6DP,
United Kingdom

Oxford University Press is a department of the University of Oxford.
It furthers the University's objective of excellence in research, scholarship,
and education by publishing worldwide. Oxford is a registered trade mark of
Oxford University Press in the UK and in certain other countries

Published in the United States of America by Oxford University Press
198 Madison Avenue, New York, NY 10016, United States of America

British Library Cataloguing in Publication Data

Data available

Library of Congress Control Number: 2023942749

ISBN 9780198882701

DOI: 10.1093/oso/9780198882701.001.0001

Printed and bound by
CPI Group (UK) Ltd, Croydon, CR0 4YY

For Mum and Dad

Acknowledgements

First and foremost, I am immensely grateful to Helen Smith for her knowledge and support, which helped guide this project from its earliest stages. She has been and will continue to be an inspiration and a role model. I am indebted to many friends and colleagues for stimulating conversations and collaborations, and the encouragement that is invaluable to an early career scholar: in particular, to Matthew Augustine, Daniel Bellingradt, Danielle Clarke, Brian Cummings, Alex Davis, Megan Heffernan, Lorna Hutson, Mark Jenner, Ian Johnson, Kevin Killeen, Neil Rhodes, Hannah Ryley, and Adam Smyth. I am especially thankful to my colleagues at the University of St Andrews: their kindness and collegiality created a space in which I had the time and enthusiasm to complete this work.

The research for the project was made possible by an Arts and Humanities Research Council studentship and International Placement Scheme fellowship; the University of York's F. R. Leavis Fund, Graduate Research Support Fund, and a Humanities Research Centre Postgraduate Research Fellowship; and library fellowships from the Bibliographical Society, the Bibliographical Society of America, Durham University, the Harry Ransom Center, Marsh's Library, Princeton University, and the Renaissance Society of America. The book would have been far poorer without the patience, generosity, and expertise of the curators, librarians, and library assistants at Durham University Library, the Harry Ransom Center, the Huntington Library, Marsh's Library, the Plume Library, Princeton Library, and York Minster Library. Lara Haggerty, Helen Kemp, and Aaron Pratt deserve special mention. Thanks to Eleanor Collins, Karen Raith, and Jo Spillane at Oxford University Press for their expert editorial work, and to the anonymous readers for their insightful feedback.

To Marie Allitt, Doug Battersby, Sarah Cawthorne, Nik Gunn, Carla Suthren, and Jack Quin, thank you for your continuing comradeship and camaraderie. To Alex Alonso, thank you for your patient assistance with tangled ideas, unwieldy library ladders, and so much more. To Andy, thank you for your strength, and to Mum and Dad, though you won't ever read this book, it is indebted and dedicated to you.

An earlier version of Chapter 3 appeared in *The Journal of the Northern Renaissance*, 8 (2017) .

Contents

List of Figures and Illustrations

Note on Text and Abbreviations

In all direct quotations from early modern sources I have italicized expanded contractions and suspensions. I have not altered italicization or modernized 'u/v', 'i/j', and 'vv/w' in direct quotations, although I have silently replaced the long 's'.

The following abbreviations have been used:

OED	Oxford English Dictionary <www.oed.com>
EEBO	Early English Books Online <eebo.chadwyck.com>
MLN	Modern Language Notes
ELR	English Literary Renaissance
PMLA	Proceedings of the Modern Language Association
ESTC	English Short Title Catalogue <estc.bl.uk>

Introduction

Waste Matter

In 1652 Thomas Urquhart, a Scottish Royalist imprisoned by the recently instated commonwealth government, published *Ekskybalauron*, or *The Jewel*. In this macaronic text that is many things at once, Urquhart attempts to 'vindicat[e] [...] the honour' of his native Scotland, presently governed by the Presbyterian Covenanters so detested by the Royalists.[1] To achieve this, he catalogues Scottish heroes, focusing in particular on the Rabelaisian history of the 'Admirable Crichtoun'. At the same time, *Ekskybalauron* is an account of Urquhart's capture at the Battle of Worcester the previous year, as well as an advertisement for a 'Universal Language' of incalculable value to any nation, but which could only be completed if Urquhart was granted his freedom, his lands, and his property. In sum, *Eksykybalauron* is an exuberantly cavalier and so contradictory attempt to persuade Parliament of a Scottish Royalist's value to the commonwealth. Urquhart's account of his capture and imprisonment is of particular interest to historians of the book. In the 'Epistle Liminary', Urquhart offers a detailed account of the haphazard production of this composite text: Urquhart had thought the work complete, but because more and more gossip slandering his countrymen reached him daily, he 'was necessitated [...] to over-triple [his] diligence' and 'coop' himself up 'betwixt the case and the printing press' (A8r). Rapidly penning further evidence of his compatriots' reputability, Urquhart complains that he was unable to keep up with the 'nimble' compositor as they worked alongside one another. According to Urquhart, their interconnected acts of production were competitive: '[h]e and I striving thus who should compose fastest, he with his hand, and I with my brain [...] we would almost every foot so jump together in this joynt expedition, and so nearly overtake other in our intended course' (A8v).

This collaboration between the cognitive and the manual, the 'joint emulation betwixt the *theoretick* and *practical* part' (a1v), stops just short of crediting the unnamed 'workman' as a co-author of *Ekskybalauron*. But as a result, attention is drawn to the materials and manhandling that underpin Urquhart's own acts of textual production. In his haste, we are told, Urquhart was 'glad to tear off parcels

[1] Thomas Urquhart, *Ekskybalauron: or, The Discovery of a most Exquisite Jewel*, Wing U134 (London: Ja: Cottrel, to be sold by Rich. Baddeley, 1652), t.p. Further references to be given in the text in parentheses.

Waste Paper in Early Modern England. Anna Reynolds, Oxford University Press. © Anna Reynolds (2024).
DOI: 10.1093/oso/9780198882701.003.0001

of ten or twelve lines apeece' to give to his compositor (a1r). These lines were hastily scribbled 'upon the loose sheets of cording-quires' (a1r), the wrappings for reams of paper that, in Joseph Moxon's words, were the 'torn, wrinkled, stained, and otherwise naughty' sheets the papermaker had to hand, lying discarded in the printing house for use as waste.[2] These 'naughty' sheets are the materials of Urquhart's 'extemporanean' (a2r) textual generation, and when he 'minced & tore them', Urquhart continues, they 'look[ed] like pieces of waste paper, troublesome to get rallied, after such dispersive scatt[e]redness' (a1r). Because of this, he 'had not the leisure to read what [he] had written' (a1r), and so his description of these minced fragments is part-modesty topos and part-brag for having, 'in the space of fourteen working-daies, compleated this whole book' (a1v).

These torn quires are not the only foul papers that Urquhart describes in the preliminaries of this peculiar book. Behind the hastily torn cording quires lies another—perhaps sensationalized—narrative of rejected and repurposed pages. The origin story of *Ekskybalauron* begins a year earlier in the city of Worcester, where Urquhart had travelled with boxes containing 'sixscore & eight quires and a half' (2) of manuscripts prepared for print. Urquhart was in Worcester to fight on the side of the King in what would be the final battle of the Civil War. After the Royalists' defeat, Urquhart laments, his manuscripts were stolen and used as waste paper by the victorious Parliamentarian soldiers. Having 'plunder[ed]' his belongings, the soldiers initially cast away the papers 'as unfit for their use' (2), but, greedy for 'new booty', they turned back, 'apprehending how useful the paper might be unto them' (3). Handing out the sheets to their comrades 'for packeting up of Raisins, Figs, Dates, Almonds, Caraway, and other such-like dry Confections and other ware', they 'kindle[d] pipes of Tobacco with a great part thereof, and threw out all the remainder upon the streets, save so much as they deemed necessary for inferiour employments, and posteriour uses' (3). Discarded for a second time, 'those dispersedly-rejected bundles of paper' that had escaped the fate of toilet paper 'were gathered up by Grocers, Druggists, Chandlers, Pie-makers, or such as stood in need of any cartapaciatory utensil [for use as scrap writing paper], and put in present service, to the utter undoing of all the writing thereof, both in its matter and order' (3). But one lonely 'quinternion', or gathering of five sheets, we are told, was recovered by a goodly citizen (3). Stuck in the gutter amongst 'a heap of seven and twenty dead men, lying upon one another' (3), its rescuer recognized that it had been marked up for print and endeavoured, somewhat unbelievably, to 'preserve it' (5). These muddy sheets were 'but a parcel of the Preface' of 'the Grammar and Lexicon of an Universal Language' (5) which provide the basis for the first part of the *Ekskybalauron*: a list of 134 statements describing the promised

[2] Joseph Moxon, *Mechanick Exercises*, Vol. 1, Wing M3013 (London: for Joseph Moxon, 1677 [1678–1680]), 353. Moxon also details the various ways in which waste paper might be used in the printing house (199, 231, 300, 347, 352–354).

Language, minus a crucial bundle of explanatory pages that were among those pilfered (perhaps a little too conveniently) by the victorious soldiers.

The production of Urquhart's text is, then, framed by the material history of paper: its vulnerability and vagaries, its tendency toward 'dispersive scatt[e]redness' (a1r), and its potential transformation into waste. This waste paper is useful as a material for 'inferiour employments' (3) and as a surface to scribble on, but also as an idea. The physical experience of a plenitude of waste paper in the streets of Worcester and in London's printshops shapes Urquhart's acts of authorship and provides him with a narrative of composition, near complete loss, and partial reconstruction. This narrative might, after all, be wholly rhetorical, inspired by the author's wider knowledge of the potential biography of paper rather than actual events. We can't be sure if Urquhart's wasted manuscripts or his universal language ever truly existed. But despite Urquhart's status as an eccentric and potentially unreliable autobiographer, his attention to the fact that textual objects might readily become or recognizably be *like* waste paper is in keeping with a broader set of early modern, authorial preoccupations. Urquhart's book, like so many of the texts considered in the following chapters, moves in and out of a waste state: waste is at once, as Urquhart's introductory epistle makes clear, a practical reality and a textual fantasy. *Waste Paper in Early Modern England* is concerned with the sorts of objects that Urquhart described as being formed, handled, and discarded in the streets and shops of Worcester, and to have himself produced in a printing house in London. As such, it reveals that waste paper was commonly encountered and imagined stuff in early modern England. Examining textual renderings, like Urquhart's, of waste practices and imagery alongside the material evidence of waste extant in archives and libraries, this book makes a case for the powerful imaginative work that was generated by repurposed pages. In tracing the material and the tropic history of this category of object, we discover that the process of repurposing pages and the widespread sensitivity to the narratives that built up around them resulted in a solid metaphor, emblem, and potent thing to think with in early modern England.

*

Book historians have long been aware of the material practice of repurposing discarded texts. Nicholas Pickwoad and Neil R. Ker have described and catalogued fragments of medieval manuscripts pasted and stitched into early modern books, and David Pearson has outlined the regular use of both old books and the discards from the printshop in sixteenth- and seventeenth-century bindings.[3] Similarly,

[3] See, for instance, Nicholas Pickwoad, 'The Use of Fragments of Medieval Manuscripts in the Construction and Covering of Bindings on Printed Books', in *Interpreting and Collecting Fragments of Medieval Books*, ed. by Linda L. Brownrigg and Margaret M. Smith (London: The Red Gull Press, 2000), 1–20; N. R. Ker, *Fragments of Medieval Manuscripts Used as Pastedowns in Oxford Bindings: with a Survey of Oxford Binding, c.1515–1620* (Oxford: The Oxford Bibliographical Society, 1954;

social historians have noted the inevitability of paper re-use in the early modern household: Margaret Spufford describes the 'very real social need for lavatory paper', as well as 'twists of printed paper wrapped conically around spices', and, citing Spufford, David Cressy suggests that 'ephemeral publications' were regularly 'used as draught excluders', pie and box linings, kindling for fires and tobacco, spice wrappers, and toilet paper.[4] But despite the wealth of waste in our libraries and archives, scholars have only recently begun to consider the copiousness and imaginative heft of repurposed paper. Prior to this, scholarly reference to repurposing pages has typically been limited to brief asides and falls into one of two contradictory categories: it is either assumed that paper in the sixteenth and seventeenth centuries was too precious a commodity to waste, and so references to mustard stoppers and privy paper are understood to be rhetorical rather than literal in nature; or, that wasting paper was a mundane practice, and so does not warrant more than passing mention.[5] Most often, it has been elided with book destruction and reduced to a joke or 'quaint fear'.[6] Other references to waste paper are read as isolated instances, with its wider material and literary context overlooked. Scholarship on Thomas Nashe, for instance, as we will see in Chapter 4, has a tendency to portray waste paper as part of Nashe's singular representation of the material world, giving the impression that repurposed pages are weird and wonderful things from the depths of this author's imagination, rather than commonplace objects with a lengthy tropic history.

The first studies to consider the rhetorical potential of waste paper in depth share a focus on the eighteenth and nineteenth centuries. Ian Donaldson's lively essay on 'The Destruction of the Book', for instance, provides a wide-ranging overview of the 'apocalyptic yet bantering' trope from Horace to Byron, suggesting

repr. 2004); David Pearson, *English Bookbinding Styles, 1450–1800* (London: The British Library and Oak Knoll Press, 2005), 32–39, 147–154. On the recycling of parchment in the medieval period, see Hannah Ryley, *Re-Using Manuscripts in Late Medieval England: Repairing, Recycling, Sharing* (Woodbridge: Boydell and Brewer, 2022), and the work of Erik Kwakkel (blog entries such as 'X-Rays Expose a Hidden Medieval Library', *Medieval Books* (blog), December 18, 2015, last accessed August 19, 2020, https://medievalbooks.nl/category/fragments/).

4 Margaret Spufford, *Small Books and Pleasant Histories: Popular Fiction and Its Readership in Seventeenth-Century England* (Cambridge: Cambridge University Press, 1981), 48–50; David Cressy, 'Books as Totems in Seventeenth-Century England and New England', *The Journal of Library History* 21, no. 1 (1986): 93.

5 See, for instance, Mark Kurlansky's description of how, in nineteenth-century England, 'old books *started turning up* at groceries and cheese shops, where their pages were used for wrapping' (my emphasis), *Paper: Paging through History* (New York: W. W. Norton & Company, 2016), 238. Similarly, Lothar Müller's history of paper only mentions waste paper in relation to the 'exponential growth of the literary market' in the late eighteenth—and nineteenth centuries, *White Magic: The Age of Paper*, trans. by Jessica Spengler (Cambridge: Polity Press, 2014), 129–68.

6 See, for instance, Rowan Watson, 'Some Non-Textual Uses of Books', in *A Companion to the History of the Book*, ed. by Simon Eliot and Jonathan Rose (Oxford: Blackwell, 2007), 491; Wes Folkerth, 'Pietro Aretino, Thomas Nashe, and Early Modern Rhetorics of Public Address', in *Making Publics in Early Modern Europe: People, Things, Forms of Knowledge*, ed. by Bronwen Wilson and Paul Yachnin (New York: Routledge, 2010), 75.

that 'Augustan satire' is the primary site of waste paper play.[7] With a similar focus on the eighteenth century, scholars including Simon Werrett, William Noblett, Geoffrey Day and Amélie Junqua have outlined the multitude of ways in which paper was repurposed in shops and households.[8] Heather Tilley has traced networks of dust, buried wills, and waste paper in Charles Dickens' *Our Mutual Friend*, and, in her influential *How to Do Things with Books in Victorian Britain*, Leah Price makes a case for the development of a 'rejection history' of nineteenth-century literature, substituting 'handlership' for readership and expanding the definition of reading to encompass 'recycling'.[9] In the last few years, sustained attention has extended to an earlier stage in this 'rejection history', the period in which, as Chapter 1 demonstrates, waste paper began to circulate widely. In the wake of scholarship by Heidi Brayman Hackel, Leah Knight, and Wendy Wall that touches on the practice of repurposing books within the wider context of textual culture, Emily Butterworth and Adam Smyth have published the first in-depth examinations of the rhetoric of early modern waste.[10] Butterworth's 'Apothecaries' Cornets: Books as Waste Paper in the Renaissance' delves into the literary history of waste paper, highlighting the classical heritage of the trope and arguing that, in the works of French Renaissance writers such as Montaigne and Rabelais, the motif of repurposed pages 'conjoined literary and physical mortality'.[11] This motif is present in the work of the English authors examined in the present volume, and, as Butterworth explains, it provides a 'counter' to the 'early modern narrative of "posthumous extension"' or literary endurance.[12] *Waste Paper in Early Modern England* builds on Butterworth's article in its consideration of the classical lineage of the waste paper trope and its Renaissance reception in Chapter 2, and the way in

[7] Ian Donaldson, 'The Destruction of the Book', *Book History* 1, no. 1 (1998): 1–10.

[8] Simon Werrett, 'The Sociomateriality of Waste and Scrap Paper in Eighteenth-Century England', in *Working With Paper: Gendered Practices in the History of Knowledge*, ed. by Carla Bittel, Elaine Leong, and Christine Von Oertzen (Pittsburgh: University of Pittsburgh Press, 2019), 46–59; William Noblett, 'Cheese, Stolen Paper, and the London Book Trade, 1750–99', *Eighteenth-Century Life* 38, no. 3 (2014): 100–110; Geoffrey Day and Amélie Junqua, 'Á la recherche du papier perdu—Some Possible Circulations of Waste Paper in Eighteenth-Century England', in *Intermediality and the Circulation of Knowledge in the Eighteenth Century*, ed. by Peter Wagner and Frédéric Ogée (Trier: Wissenschaftlicher Verlag Trier, 2019), 203–218.

[9] Heather Tilley, 'Waste Matters: Charles Dickens' *Our Mutual Friend* and Nineteenth-Century Book Recycling', in *Book Destruction from the Medieval to the Contemporary*, ed. by Adam Smyth and Gill Partington (Basingstoke: Palgrave Macmillan, 2014), 152–171; Leah Price, *How to Do Things with Books in Victorian Britain* (Princeton: Princeton University Press, 2012), 7, 15, 256.

[10] See Heidi Brayman Hackel's work on paratexts in *Reading Material in Early Modern England: Print, Gender, and Literacy* (Cambridge: Cambridge University Press, 2005), 81; Leah Knight's consideration of the relationship between books and plants in *Of Books and Botany in Early Modern England: Sixteenth Century Plants and Print Culture* (Farnham: Ashgate, 2009), 125; and Wendy Wall's reading of a poem by Ben Jonson in *Recipes for Thought: Knowledge and Taste in the Early Modern Kitchen* (Philadelphia: University of Pennsylvania Press, 2015), 153–154; Emily Butterworth, 'Apothecaries' Cornets: Books as Waste Paper in the Renaissance', *MLN* 133, no. 4 (2018): 891–913; Adam Smyth, *Material Texts in Early Modern England* (Cambridge: Cambridge University Press, 2018), 137–174.

[11] Butterworth, 'Apothecaries' Cornets', 912.

[12] Ibid., 909, 912.

which the life cycle of a repurposed page might serve as a counternarrative to the widespread fantasy of literary immortality underpins the argument throughout.

Whereas Butterworth considers the literary representation of waste paper in French Renaissance literature, Smyth's chapter on waste in *Material Texts in Early Modern England* focuses on the nature of early modern readers' encounters with printed binding waste. In addition to exploring the potential for binding waste to 'invite or repay literary interpretation', signifying independently of the intentions of its author or binder, Smyth argues that waste paper presents a radical rebuttal to our modern-day understanding of books as being 'coherent, bound, unannotated, [and] "complete"'.[13] The present study is indebted to Smyth's compelling account of a textual culture centred on 'incorrigibly plural' and temporally multiple objects, but it differs from these previous studies of waste in two ways.[14] First is its scope: this book considers waste paper of all sorts (manuscript as well as printed); its use in spaces beyond the binder's shop; and its appearance in sixteenth- and seventeenth-century literature in English across the spectrum of devotional and controversial pamphlets, odes and elegies, antiquarian writing, prose narratives, and drama. Secondly, it tells the intertwined histories of waste paper's material *and* metaphorical lives, with Chapter 1 outlining the environmental pervasiveness of repurposed pages in the form of binding waste, food wrappers, tobacco lighters, box linings, textile padding, and toilet paper, and, the subsequent chapters demonstrating how waste paper served as a scaffold for imaginative thought.

Waste Paper in Early Modern England is not, however, a comprehensive history, limited as it is to the use and imagining of repurposed paper in a limited geographical area (England) and in a relatively narrow period (the sixteenth and seventeenth centuries). The focus on English texts allows for an in-depth consideration of the place of waste practices and waste tropes within a series of relatively unified religious, literary, bibliographical, and cultural developments. This book should, therefore, be understood as one of many possible case studies, rather than an account of a solely English phenomenon. Waste paper, as Butterworth and numerous book historians have shown, is both an idea and an object pervasive throughout early modern Europe. My focus on these specific centuries owes to their unique stage in the lengthy history of repurposing and imagining discarded texts. As Chapters 1 and 2 will demonstrate, these centuries saw an explosion of the employment of both waste paper and waste paper tropes, a consequence of the rapidly increasing circulation of paper objects (primarily, but not solely, printed books and administrative documents), and the increased exposure to classical descriptions of repurposing unwanted and unsatisfactory texts as a result of humanist scholarship. This is not to say that paper or parchment texts were not repurposed prior to this period, or that paper only gained meaning in the centuries

[13] Smyth, *Material Texts*, 169, 26.
[14] Ibid., 138.

covered by this study (as Orietta Da Rold has demonstrated, late medieval authors conceived of paper as 'a sophisticated object' that might be used 'for rhetorical effect').[15] Nor is it to deny the rich pre-history of representing books as fragile objects vulnerable to decay and destruction. Instead, this book is focused on one particular phase in this literary history, one that developed at the moment when waste paper and repurposed texts transitioned from being 'noticeable' (to borrow Da Rold's characterization of paper culture more broadly) to being 'ubiquitously present'.[16]

The growing interest in early modern waste paper follows on the heels of an explosion of interest in the history of paper more broadly, with numerous popular and scholarly publications overviewing its development and spread in recent years.[17] Orietta Da Rold, Caroline Fowler, and Joshua Calhoun have each sought to make the now 'invisible' or 'transparent' matter of paper visible, and to help reveal its forgotten cultural importance.[18] Da Rold's *Paper in Medieval England* corrects the widespread assumption that paper was adopted only hesitantly and suspiciously prior to the fifteenth century, instead demonstrating that '[p]aper reached maturity in the medieval period'.[19] Similarly revisionist, Fowler's *The Art of Paper* highlights the ways in which non-Western dominance and technological alternatives in the history of papermaking were collectively forgotten in the early modern period, erased by Eurocentric historiographies of the invention of paper.[20] At the same time, Fowler persuasively argues that paper came to be imagined as a blank and empty surface that served as a metaphor for both the human mind and, connectedly, indigenous lands conceived as empty and therefore ripe for colonial overwriting.[21]

Calhoun's intervention into this growing scholarly field shares Fowler's ambition to make modern readers aware of the ongoing damage wrought by the cultural narratives grounded, somewhat unexpectedly, in the mundane matter of paper. Critiquing the anthropocentrism that underpins the disciplines of book history and literary scholarship, Calhoun demonstrates that textual production and preservation is reliant on non-human as well as human participants. In contrast

[15] Orietta Da Rold, *Paper in Medieval England: From Pulp to Fictions* (Cambridge: Cambridge University Press, 2020), 4.

[16] Ibid., 213.

[17] Popular histories include Ian Sansom, *Paper: An Elegy* (London: Fourth Estate, 2012); Alexander Monro, *The Paper Trail: An Unexpected History of a Revolutionary Invention* (New York: Penguin Random House, 2014); Lothar Müller, *White Magic*; and Mark Kurlansky, *Paper: Paging through History*.

[18] Da Rold, *Paper in Medieval England*; Caroline Fowler, *The Art of Paper: From the Holy Land to the Americas* (New Haven: Yale University Press, 2019); Joshua Calhoun, *The Nature of the Page: Poetry, Papermaking, and the Ecology of Texts in Renaissance England* (Philadelphia: Pennsylvania University Press, 2020). Fowler explains how successful media become 'transparent, denied [their] own capability to convey a message' (2) and 'invisible' in their primary function as a vehicle for other information (30).

[19] Da Rold, *Paper in Medieval England*, 213, 179, 208.

[20] Fowler, *The Art of Paper*, 55–66.

[21] Ibid., 117–122.

to modern-day book users, he argues, early moderns were conscious that 'the very ecosystem that supplie[d] the material needed to give life to poetry on paper also thrive[d] on death and decay'.[22] Book history is not, then, concerned with an inevitable output of the unstoppable march of technological progress. Instead, like the 'urgency' of much early modern literature, it is rooted in environmental factors such as scarcity and corruptibility.[23] Indebted to the scholarship of Da Rold, Fowler, and Calhoun, this study understands paper to be a multifaceted and intellectually potent material. Early moderns were highly attuned to the history of paper, on both the grand scale of its technological development and on the microscopic level of a page's ecological make-up.

*

How do we account for such an intimate understanding of the life cycle of paper in early modern England? It is first important to rebut the myth that late medieval and early modern England were in any way paper-short.[24] In addition to the wealth of imports throughout this period, white and brown paper were produced domestically, with paper mills active in England in the 1490s and 1510s, and from the 1550s into the early 1600s.[25] The dominance of print in our histories of early modern books has led to the misconception that paper was a scarce and expensive commodity. It is, after all, an often repeated truism that paper made up a significant proportion of the costs of printing a book. This owes not, however, to the price of an individual sheet of paper, but to the large quantities of paper required for print runs. For most early moderns, paper was not purchased in bulk or at any great expense. Instead, white paper in the form of individual books and loose sheets, purchased from stationers' shops, was typically encountered alongside a range of cheaper brown and low-grade paper objects. Da Rold has shown that from the fourteenth century, paper was imported for purposes ranging from the 'medical and culinary' to 'wrapping goods', in addition to 'fulfilling writing needs'.[26] These non-textual functions only became more common in subsequent centuries, as Helen Smith has demonstrated, with paper regularly employed as poultices, prosthetics, bandages, toys, and mathematical instruments, as well as 'coffins' for

[22] Calhoun, *The Nature of the Page*, 127.

[23] Ibid., 144.

[24] For evidence to support this, see Da Rold, *Paper in Medieval England*, passim; Heather Wolfe, 'Was Early Modern Writing Paper Expensive?', *The Collation* (blog), February 13, 2018, last accessed August 20, 2020, https://collation.folger.edu/2018/02/writing-paper-expensive/; and Helen Smith, '"A Unique Instance of Art": The Proliferating Surfaces of Early Modern Paper', *Journal of the Northern Renaissance* 8 (2017) : last accessed August 19, 2020, http://northernrenaissance.org

[25] The standard historical overviews of papermaking are D. C. Coleman, *The British Paper Industry, 1495–1860* (Oxford: Clarendon Press, 1958), esp. 3-88; Richard Leslie Hills, *Papermaking in Britain, 1488–1988: A Short History* (London: Bloomsbury, 1988), esp. 1–12, 45–64; Dard Hunter, *Papermaking: The History and Technique of an Ancient Craft* (New York: Dover Publications, repr. 1978), esp. 114–120, 224–246; Alfred H. Shorter, *Paper Mills and Paper Makers in England, 1495–1800* (Hilversum: The Paper Publications Society, 1954), 19–50.

[26] Da Rold, *Paper in Medieval England*, 57.

baking in the oven and as filters for distilling liquids.[27] Elaine Leong has labelled these paper 'strategies' and 'technologies', and Smith argues that the ways in which paper was folded, stitched, and pinned made it 'both a practical and an intellectual resource'.[28] Though they strike a modern reader as somewhat peculiar, these diverse paper objects were cheap and widely available, and in their frequent handling, as Smith suggests, they served as 'modes of knowing' and things with which to 'materialis[e] problems' and 'extrapolat[e] ideas'.[29]

Early moderns were not only sensitive to the material affordances of these technologies, or what Da Rold terms the 'plasticity, porosity, and tensility' of paper.[30] They were also sensitive to the extended biographies of these white and brown sheets. Because competition over good quality linen rags was fierce, early moderns were aware of and often directly involved in the collection of these raw materials, particularly in papermaking regions such as the South East. In 1601, John Spilman's rag collectors were accused of aggressively 'begging at men's doors', and Heidi Craig has revealed the widespread presence of poor and typically female 'rag-gatherers' seeking cast-off textiles in gutters and 'dung-hills'.[31] Gabriel Plattes, in a 1644 husbandry manual, refers to the collection of rags and the repurposing of paper in the same breath. Urging his readers to practice thrift in gathering up paper scraps to sell to bookbinders, he praises an unnamed woman who 'reserved' her 'linen rags [...] and sold them yeerly to the Paper-Mils'.[32] In addition to making regular use of paper objects, then, early moderns might be involved in an early stage of paper's biography, just as they might enact its transformation into waste at later points in its life cycle.

Whether or not they were employed as rag-collectors or thrifty enough to gather them for the papermaker, it seems likely that the majority of early moderns were at least partially aware of how paper was made—how it was once a plant, transformed into clothing before being discarded as rags. In the words of Basilius Valentinus, 'the Matter of Flax [...] is pluck'd up' and 'putrefied and rotted in water', and then 'beaten, broken, peel'd, and last of all dress'd', then 'Cloth weaved of it', before being 'cut in pieces by Taylors', and 'when this Linnen is quite worn out, and torn, the old Rags are gathered together, and sent to the Paper-Mills, whereof they make Paper,

[27] Smith, '"A Unique Instance of Art"', para. 4.

[28] Elaine Leong, 'Papering the Household: Paper, Recipes, and Everyday Technologies in Early Modern England', in *Working With Paper: Gendered Practices in the History of Knowledge*, ed. by Carla Bittel, Elaine Leong, and Christine Von Oertzen (Pittsburgh: University of Pittsburgh Press, 2019), 32–45: 38; Smith, '"A Unique Instance of Art"', para. 3, 4, 43.

[29] Smith, '"A Unique Instance of Art"', paras. 3, 4, 43.

[30] Da Rold, *Paper in Medieval England*, 186.

[31] SP 12/279, f. 164 (National Archives), quoted in Smith, '"A Unique Instance of Art"', para. 23; Heidi Craig, 'English Rag-Women and Early Modern Paper Production', in *Women's Labour and the History of the Book in Early Modern England*, ed. by Valerie Wayne (London: Bloomsbury, 2020), 29–46: 33, 38.

[32] Gabriel Plattes, *The Profitable Intelligencer*, Wing P2414 ([London?]: for T.U., [1644]), A4r.

which is put unto divers uses'.[33] We certainly find evidence of this familiarity in the literature of the period, where it appears as a symbolic process imbued with allegorical significance. Joshua Calhoun has shown how Henry Vaughan meditates on the visibility of this material history in a page of the Bible. For Vaughan, Calhoun argues, 'matter [is] palimpsested in a Bible's paper', and 'the new text has not obliterated the old memories woven and worn into its constituent materials'.[34] By alluding to the histories of the multiple components that make up the material of the book, Vaughan creates an emblem for the corruption of earthly matter and the possibility of a religious 'restoration' after death.[35]

Vaughan's meditation is not an isolated one. A number of early modern poets make use of the metamorphic affordances of the life cycle of paper, reflecting on how, in the words of another religious thinker, 'the body when it rots in the grave, is as linnen worn to rags, and cast upon the dunghill: but at the resurrection it is like those Rags gathered up, and made into paper, which many times becomes gilt, and capable of noble and divine impressions'.[36] These symbolic narratives are not always spiritual in shape: decades earlier, in a panegyric to John Spilman's papermill, founded in 1588, Thomas Churchyard tells his readers how 'rags and shreds' are made 'to sweate. | Of whose thick froth, a creame or crudde should rise', which takes the 'shape' of a sheet of paper.[37] This process, in which 'drosse and rags' become 'Paper white and cleane', is 'comparde' to the process in which 'Mans secrete faults, and foule defects of minde, | must be reformde, like ragges in Paper mill'.[38] As Brandi Adams has argued, Churchyard uses paper to represent the importance of achieving a 'fairness' synonymous with 'economic and social value'.[39] In the literature of the period, Adams notes, the value of a woman is often figuratively entwined with her papery whiteness, whereas in Churchyard's verse the value of a man is conceived as paper-like in a different sense—in his capacity to transform, like rags, and achieve upward social mobility, and so is dependent on a socially performed rather than a literal paper-whiteness. Heather Wolfe's suggestion that the years in which Spilman set up his paper mill were also the years of a 'watershed' in paper literacy is, therefore, a persuasive one: in the following decades, poets such as John Taylor and Abraham Cowley found symbolic

[33] Basilius Valentinus, *Of Natural & Supernatural Things*, Wing B1020 (London: sold by Moses Pitt, 1671), 119–120.

[34] Calhoun, *The Nature of the Page*, 67–68.

[35] Ibid., 68.

[36] Henry Valentine, *Private Devotions, Digested into Six Letanies*, Wing 24576.3 (London: [M. Flesher] for John Marriot, [1635]), 256–257.

[37] Thomas Churchyard, *A Sparke of Frendship and Warme Goodwill*, STC 5257 (London: [T. Orwin], 1588), D2r.

[38] Churchyard, *Sparke of Frendship*, D3r–v.

[39] Brandi K. Adams, 'Fair/Foul', in *Shakespeare/Text: Contemporary Readings in Textual Studies, Editing, and Performance*, ed. by Claire M. L. Bourne (London: The Arden Shakespeare, 2021), 29–49: 36.

potential, like Vaughan and Churchyard, in the transformations of the materials that make up paper.[40]

Taylor, like Churchyard, uses the papermaking process to comment positively on the capacity for socio-economic transformation, and he does so to the point of factual inaccuracy. Describing how 'the linnen' and 'dirty socks' of a Tyburne slaue' might be 'transform'd' into expensive 'Paper-royall', while 'the torne shirt of a Lords or Kings' is 'pasht' into lowly 'Pot-paper', Taylor portrays a world in which one's social station is not determined by one's birth.[41] One might, he claims, 'by desert, by fauour and by chance' fall from honour or, conversely, from 'begg'ry' advance.[42] Referring to the difficulties that papermakers persistently had in sourcing the raw materials for their craft, Calhoun suggests that early moderns 'must have known the story of paper made from plants was not so simple, direct, or domestic as [Taylor's] commendatory lines suggest'.[43] It certainly seems probable that Taylor's readers were aware that 'dirty socks' and the rags collected by the hangman were unlikely to be of good enough quality to make paper of the best sort, whereas the Lord's or King's shirts were far more likely to constitute the clean white linen required in the production of 'Paper-royall'. Perhaps part of the humour in Taylor's extravagant praise of the papermaker's raw materials stems from such unrealistic flights of fancy: transformation from destitution to royalty was, after all, as impossible as making the whitest paper from the dirtiest rags. Regardless of whether Taylor intended his poetic license to serve as a social comment, he goes on to suggest a straightforward relationship between paper's history and his metamorphic social fantasy, and in doing so, makes explicit the relationship between things and metaphors. 'Thus', he concludes, 'are these tatters allegoricall | Tropes, types, and figure, of mans rise or fall'.[44] In Taylor's poem, the material biography of paper is the starting point for a series of fantastical transformations, and in these explanatory lines, we see the process through which objects accrete layers of meaning. Whether directly involved in rag-picking, observant of the material traces of papermaking embedded in the page, or simply knowledgeable of the papermaking process, by the end of the sixteenth century early moderns were able to draw on a range of allegories, tropes, and figures that had developed around the life cycle of paper.

[40] Heather Wolfe, 'Rethinking the Price, Quality, and Social Significance of Writing Paper in Early Modern England' (paper presented at the Paper, Pen and Ink: Manuscript Cultures in Early Modern England Seminar Series, London, April 14, 2014). Heather Wolfe and Henry Woudhuysen are in the process of researching Spilman's paper in detail. See 'An example of early modern English writing paper', *The Collation* (blog), February 4, 2014, last accessed August 19, 2020, http://collation.folger.edu/2014/02/an-example-of-early-modern-english-writing-paper/.

[41] *All the Workes of Iohn Taylor the Water Poet*, STC 23725 (London: I[ohn] B[eale], Elizabeth Allde, Bernard Alsop, and Thomas Fawcet for Iames Boler, 1630), 70.

[42] Ibid., 70.

[43] Calhoun, *The Nature of the Page*, 41.

[44] *All the Workes of Iohn Taylor*, 70.

The allegories, tropes, types, and figures explored in the chapters that follow are intertwined with the pre-life of paper as it transformed from flax to fabric to rags to sheets, but go beyond this, encompassing what we might term paper's afterlife—the later stages in its life cycle when paper has, in many instances, already served one purpose, and is turned to serve a second or even third function. Though inseparable from the wider cultural sensitivity to the biography of paper, the early modern preoccupation with waste is perhaps even more pervasive. This is in part a consequence of its conspicuousness: used in bindings or as wrappers, waste is far more visible than a fragment of flax or linen embedded in a page. The transition of text to waste is typically immediately apparent and not dependent on a pre-existing knowledge of a process such as papermaking. And so, just as the material *pre*-life of the page suggested a range of tropes and figures, the waste fate of paper prompted potent and widespread metaphorical work.

*

Alongside the recent turn toward the history of seemingly mundane objects such as paper, there has been surge of interest in rubbish, refuse, recycling, and waste.[45] It is no coincidence that we, at the turn of the twenty-first century, are increasingly preoccupied with the material constitution and disintegration of objects—textual and otherwise—through history. At first glance, the early modern preoccupation with textual loss and disintegration seems radically different to our own. Books, after all, appear to us as stable and secure objects. Even centuries-old books seem destined to survive indefinitely, stored in digital repositories such as *Early English Books Online* and what Calhoun wryly terms 'climate-controlled "archival temples"'.[46] At the same time, the development of new media and digital technologies has prompted a material consciousness more akin to the early centuries of paper use than the intervening centuries of paper's transparency and cultural invisibility. At risk of being superseded by media that gives the illusion of immateriality and permanence, the matter of the page is gaining a nostalgic solidity and heft within the cultural imagination.

The recent focus on the organic and multifaceted history of paper, along with the turn to histories of discarded stuff, is no doubt also a consequence of cultural anxiety surrounding the ongoing environmental crisis. We are becoming increasingly alert to the question of what will and what will not decompose over time, and the interconnectedness of our bodies and behaviours in a fragile and increasingly damaged ecosystem. Although I will turn to the entanglement of paper, human, and environment in Chapter 5, *Waste Paper in Early Modern England* is concerned

[45] Notable examples include Emily Cockayne, *Rummage: A History of the Things We Have Reused, Recycled, and Refused to Let Go* (London: Profile Books, 2020); Hannah Rose Shell, *Shoddy: From Devil's Dust to the Renaissance of Rags* (Chicago: University of Chicago Press, 2020); Lisa Woollett, *Rag and Bone: A History of What We've Thrown Away* (London: John Murray Press, 2021).

[46] Calhoun, *Nature of the Page*, 139.

primarily with the immediate and imaginatively potent afterlife of texts in the form of waste in the sixteenth and seventeenth centuries. While many of these materials remain extant and continue to very gradually decay in our libraries and archives, my focus is primarily on how they were handled and imagined in the years and decades immediately following their transformation into waste. Thanks to the work of Adam Smyth and Jeffrey Todd Knight, it is no longer taken for granted that books were understood and experienced as stable objects in the early modern period.[47] What is more, we are becoming increasingly attuned to the importance of loss and decay to the negotiation and representation of textual culture in these centuries. As Smyth writes, the 'history of the early modern book is also a history of loss and destruction: of print as impermanent; of literature as something that needs to be forgotten; of books that burn; of reading as an act of throwing away'.[48] Counter-intuitively, that which no longer survives and that which is imagined as being on the brink of disintegration and decomposition is as present and influential in the early modern imagination as that which is whole and carefully preserved. To fully understand early modern literature and textual culture, it is crucial that we work to recover and reassemble this material and figurative landscape of textual fragmentation and decay. Waste paper, we will discover, is a particularly potent and pervasive manifestation of bibliographic vulnerability, and so looms large in the early modern imagination.

To study the use and understanding of waste fragments also means working towards an enriched and altered phenomenology of early modern reading. In his influential work on the subject, Bruce Smith suggests that though attempts to reconstruct 'the kinesthetic knowledge that comes with reading' might look 'presumptuous', 'it is perfectly possible' to 'insinuate' oneself into the subject position of an early modern reader, so long as we recognize that we are 'intruder[s]' when we do so.[49] What Smith terms historical phenomenology underpins much recent work on the history of reading, centred as it is on the assumption that knowledge and thought are embedded in the embodied experience of the world—specifically, that the embodied experience of reading books mutually constitutes the way that books, in Andrew Stauffer's phrase, 'exist for us, in the world'.[50] This volume is grounded in the related assumption that haptic encounters with waste fragments shaped the way in which all early modern books were read and conceived, with the idea or rhetorical construct of the book indebted to the everyday rhythms of textual reuse as well as more conventional acts of reading complete texts. We can 'intrude' on, or hypothetically inhabit, the subjective experience of an early modern reader

[47] See Jeffrey Todd Knight, *Bound to Read: Compilations, Collections and the Making of Renaissance Literature* (Philadelphia: University of Pennsylvania Press, 2013) and Adam Smyth, *Material Texts*.

[48] Smyth, *Material Texts*, 55.

[49] Bruce R. Smith, *Phenomenal Shakespeare* (Oxford: Wiley-Blackwell, 2010), xvi, 36.

[50] Andrew Stauffer, *Book Traces: Nineteenth-Century Readers and the Future of the Library* (Philadelphia: University of Pennsylvania Press, 2021), 124.

by compiling material and representational evidence of textual encounters, and in doing so it is crucial that we include the less obviously textual but commonplace encounters with waste fragments. In this period, the more recognizable actions of turning a book's pages and opening and closing its covers were part of a wider textual kinaesthetics: the tearing, cutting, folding, crumpling, burning, wiping, and unwrapping that took place within grocers' shops, kitchens, and privies, and that therefore need to be incorporated within our histories of reading.

This broadened phenomenology of reading does more than deepen our understanding of the material make-up of books and the mechanics of reading in this earlier period. As Sarah Wall-Randell states, a fixation with 'materiality' in isolation results in the elision of 'the multiple imaginative and affective qualities' bound up in the 'interaction between the individual reader and the text'.[51] Texts are at once 'profoundly physical and powerfully metaphysical' entities, and so the idea of the book is as important for a history of reading and an understanding of literature as actual books themselves.[52] Wall-Randell's work is particularly pertinent to this study because of her emphasis on the importance of 'dematerializing, losing the wholeness and heft, the certainty and understandability' of the book as an idea in the early modern period.[53] As Pauline Reid suggests in her work on early modern readers' visual interactions with texts, we need to consider the 'perceptual' and 'epistemic' as well as the material dimensions of 'fragmentation, destruction, and recyclability'.[54] These result, Reid concludes, in a cultural understanding of the book as a 'vibrant, unsettled, and unsettling medium'.[55] As well as attending to the material history of waste paper, therefore, *Waste Paper in Early Modern England* interrogates the vibrant and unsettling effect of waste fragments on the early modern imagination, and in doing so provides further evidence for Wall-Randell's claim that books persistently serve as phenomenological vehicles, or 'metaphorical processes' through which 'such immaterial concepts as consciousness, identity, and the meaning of lived experience [could] be manifested and considered'.[56] The early modern idea of the book, and so lived experience more broadly, always verges on and is haunted by the processes of fragmentation and waste.

My understanding of how the embodied experience of waste paper transforms into waste-paper metaphors, and how these in turn structure an understanding of self, book, and world, is indebted to the work of Christopher Tilley. Tilley, a leading proponent of a phenomenological approach to archaeology, has theorized the relationship between metaphor and material culture: 'structures of bodily experience', he writes, 'work their way up into abstract meanings [and] embodied

[51] Sarah Wall-Randell, *The Immaterial Book: Reading and Romance in Early Modern England* (Ann Arbor: University of Michigan Press, 2013), 3.

[52] Ibid., 79.

[53] Ibid., 2.

[54] Pauline Reid, *Reading by Design: The Visual Interfaces of the English Renaissance Book* (Toronto: University of Toronto Press, 2019), 5–6.

[55] Ibid., 22.

[56] Wall-Randell, *The Immaterial Book*, 14.

imagination'.[57] The bodily experience of repeatedly entering and exiting a house, for instance, works its way up into a range of spatio-temporal metaphors and so shapes our fundamental perception of 'dwelling in the world'.[58] As with this fundamental example of the house, the actions that we carry out and the objects and material structures that we encounter again and again accrete 'layers of metaphorical meaning like the rings of an onion', with these meanings 'becom[ing] attached to the artefact'.[59] The relationship between self, culture, and metaphor is, then, mutually constitutive. Although these meanings are 'encode[d]' into objects that, Tilley notes, 'would themselves have no meaning', the accretion of these imaginative layers mean that objects become 'active rather than passive' in relation to the selves that make use of them.[60] In other words, because we have imbued objects with meaning over decades and centuries of use, they become eloquent things capable of articulating a range of metaphors and meanings. Waste paper is one particularly potent example of what Tilley terms solid or concrete metaphors.

Throughout, I use the term kinaesthetics to refer to the 'structures of bodily experience' that constitute waste paper use.[61] I do so to make explicit the importance of the movement of body and page in conjunction with one another in the creation and use of waste. I also refer frequently to the *haptic* dimensions of waste paper use—the mode of experiencing the world that encompasses the kinaesthetic alongside the sense of touch and the body's sense of itself within space. As Laura Marks notes in her 'sensuous theory' of touch, the haptic is typically imagined in opposition to the optic, a consequence of both the cultural prioritization of sight dating as far back as Aristotle and the more recent disembodiment of sight in the post-Enlightenment era.[62] In this study, I emphasize the haptic at the expense of the optic not only in response to its relative neglect in cultural history, but because early modern representations of encounters with waste firmly ground them in the kinaesthetic and the tactile. This is in contrast to the more common representations of reading non-waste, or whole, texts, in which hapticity is often obscured. In centring the haptic in the experience of repurposed paper, we are also able to account for its imaginative importance. As Abbie Garrington suggests, '[t]he haptic is intimately connected to the constitution of the self', functioning as it does as the 'carnal border between self and world'.[63] Waste paper, so frequently handled, is one vehicle through which early moderns imagined the fate of texts more broadly, and negotiated their own position within the wider world. Waste paper's

[57] Christopher Tilley, *Metaphor and Material Culture* (Oxford: Blackwell, 1999), 35. Tilley is indebted to George Lakoff and Mark Johnson, *Metaphors We Live By* (Chicago: University of Chicago Press, 1980).

[58] Ibid., 36, 40–48.

[59] Ibid., 266, 71, 75.

[60] Ibid., 76.

[61] Ibid., 75.

[62] Laura Marks, *Touch: Sensuous Theory and Multisensory Media* (Minneapolis: University of Minnesota Press, 2002), xiii.

[63] Abbie Garrington, *Haptic Modernism: Touch and the Tactile in Modernist Writing* (Edinburgh: Edinburgh University Press, 2013), 17.

onion-like layers of meaning meant that it was, in these centuries, an object capable of both shaping thought and serving as a vehicle through which to comprehend fundamental aspects of human experience.

*

Some readers of this book might be sceptical regarding these claims for the intellectual potency of small fragments of paper. As Adam Smyth notes in his essay 'What We Talk about When We Talk about Scissors', scholarship on material culture has a 'tendency to select a single object or textual moment (like a scribbled shopping list in the binding of a book) and to argue that an entire culture was somehow embodied in this little scrap'.[64] It is important to emphasize, however, that while waste paper often circulates in 'little scrap[s]', it is not a single object but a pervasive presence in the early modern domestic, urban, administrative, and commercial environment, and one that was frequently described in texts of the period. Writing a decade earlier, Jonathan Gil Harris expressed a similar concern to Smyth regarding the scholarly representation of objects as '"everyday" emblems of cultural formations'.[65] Harris is suspicious of such criticism's treatment of objects as 'docile and unchanging' symbols of grand historical developments, particularly in combination with the simultaneous understanding of these objects as strange and alluring.[66] This contradiction, Harris argues, repeats the errors of an 'old' and uncritical antiquarianism, which 'fastens encyclopedically on strange objects as the positive residua [...] of a reassuring, synchronically conceived totality'.[67] Representing a 'freeze-frame of a historical moment', objects are too often slotted into the nooks of a critical *Wunderkammer* and so are understood as a synecdoche of a paradoxically stable yet exotic cultural whole.[68]

This book might at times seem guilty of what Harris labels 'old' antiquarianism. As we will see in Chapter 3, waste paper was (and continues to be) gathered and pinned in place by 'old'-fashioned, though not necessarily uncritical, antiquarians and bibliographical collectors.[69] Furthermore, waste paper will be revealed as being both unusual and everyday—the scholarly contradiction that Harris complains of—unusual to us as modern-day book users, but an everyday object in the sixteenth and seventeenth centuries. Waste paper also serves an emblematic function (another critical term that Harris is suspicious of), though it is emblematic

[64] Adam Smyth, 'What We Talk about When We Talk about Scissors', in *Ornamentalism: The Art of Renaissance Accessories*, ed. by Bella Mirabella (Ann Arbor: University of Michigan Press, 2011), 295.

[65] Jonathan Gil Harris, 'The New New Historicism's Wunderkammer of Objects', *European Journal of English Studies* 4, no. 2 (2000): 111–123, 115.

[66] Ibid., 114–115.

[67] Ibid., 116.

[68] Ibid., 123.

[69] I have considered this stage in the history of waste in more detail in 'What Do Texts and Insects Have in Common?; or Ephemerality before Ephemera', in *Practices of Ephemera in Early Modern England*, ed. by Callan Davies, Hannah Lilley, and Catherine Richardson (Abingdon: Routledge, 2023), 25–43.

of far more than one cultural moment or formation, and despite suggesting that it carries an unexpected imaginative potency and importance, *Waste Paper in Early Modern England* does not claim that this extends to the 'totality' of the early modern period. Instead, we will go on to see that waste has the capacity to figure for multiple ideas and processes. Waste paper figures so variously for the same reason that waste (when understood as at once an idea and a thing) cannot be slotted comfortably into pre-existing cultural histories or critical *Wunderkammern*. Waste paper is too lively and multifarious a category of thing to signify singly, and an account of it must encompass the breadth of its contradictory and slippery histories. An object and a metaphor, waste is at once a useful object and a piece of rubbish; a valuable curio and a scrap of paper; something mundane and something deeply meaningful; something ephemeral and something enduring, and so something lost and absent, as well as present and persisting. Where waste falls on this spectrum depends on the object, the time, and the onlooker or handler: which individual instance of waste is under scrutiny? What moment in its history is being considered? And who is the one grasping and engaging with the scrap, be it physically or imaginatively? Waste might mean many different things depending on these circumstances. This book traces these historical and imagined fluctuations in the biography of waste, and so uncovers a shifting landscape of lively slips and scraps of repurposed paper and the mark that they left on the early modern imagination.

A study of waste paper is, then, a study of a 'thing-in-motion', and like so many studies of material culture, *Waste Paper in Early Modern England* is informed by the methodological framework outlined by Arjun Appadurai in his 1986 *The Social Life of Things*.[70] Appadurai offers an antidote to the historical materialist understanding of 'the world of things as inert and mute, set in motion and animated, indeed knowable, only by persons and their words'.[71] Appadurai argues instead that matter is not irrevocably commodified, reified, and fetishized within the capitalist system. Instead, 'the term "commodity" is used [...] to refer to things that are at a certain *phase* in their careers and in a particular *context*, meet the requirements of commodity candidacy'.[72] By studying the 'total trajectory' of an object 'from production, through exchange/distribution, to consumption', we discover that far from inert and mute, objects are enlivened and endowed with meaning by 'human transactions, attributions, and motivations', and so, in turn, 'illuminate their human and social context'.[73] Appadurai distinguishes between two sorts of trajectories in the life of an object—between the 'cultural biography' of an

[70] Arjun Appadurai, 'Introduction: Commodities and the Politics of Value', in *The Social Life of Things: Commodities in Cultural Perspective*, ed. by Arjun Appadurai (Cambridge: Cambridge University Press, 1986), 3–63: 5.

[71] Ibid., 4.

[72] Ibid., 16.

[73] Ibid., 13, 5.

individual object as it moves through different hands, contexts, and uses, and a 'social history of things' that encompasses the 'larger ebb and flow' of a whole category of object.[74] These two histories are of course interconnected, with shifts in the biographies of individual objects leading to shifts in the category's broader social history, and the social history constraining the ways in which individual objects might be used and understood. *Waste Paper in Early Modern England* tells both sorts of histories, outlining a social history pieced together from the cultural biographies of individual instances of waste extant in archives and libraries alongside the complex web of waste paper imagery that spreads through sixteenth- and seventeenth-century literature. The latter reveals both the nature of individual encounters with waste as well as the 'total trajectory' of this category of thing and its capacity to illuminate aspects of early modern culture.

Throughout, I refer to waste paper as an object, a thing, and a form of stuff. In doing so, I make explicit the ways in which its biography is constituted of numerous phases, and the way that it serves its users as a particularly potent vehicle for understanding related categories of things (such as books more broadly), as well as its user's own position in an ecology of diverse human and non-human matter. Bill Brown's theory of things, indebted as it is to Heidegger's broken hammer, is particularly useful when accounting for the effect of waste paper. Objects 'asser[t] themselves as things', Brown writes, when 'they stop working for us', and so disrupt our typical relationship with the objects around us.[75] Waste paper can, therefore, be usefully conceived as a 'thing' in so much as it is an object that has stopped working as a complete text (though it is important to note that it continues to 'work' for us in other ways), and so behaves differently to other non-wasted objects.[76] Jane Bennett extends Brown's theorization of things into the sphere of contemporary politics and environmentalism: what she terms the 'thing-power' of the material world causes a complete breakdown in the distinction between subject and object, a rewriting or rejection of Brown's understanding of a more subtle shift.[77] I return to Bennett's political ecology in Chapter 5, but it is worth noting here that both Brown's and Bennett's theorization of things move away from the economic contexts and rhythms so central to Appadurai's framework. Brown's and Bennett's theories account for when objects become things, and so do not consider the previous stages or potential future relapses in their life cycle. As I will outline in Chapter 1, waste paper remains a commodity—what Appadurai defines as 'any thing intended for exchange'—even after it has become a thing.

[74] Ibid., 34–36.
[75] Bill Brown, 'Thing Theory', *Critical Inquiry* 28, no. 1 (2001): 1–22, 4.
[76] Ibid., 4.
[77] Jane Bennett, *Vibrant Matter: A Political Ecology of Things* (Durham: Duke University Press, 2010), esp. 1–19.

Maurizia Boscagli's *Stuff Theory* provides a useful model in negotiating the mutually possible definition of waste as a thing (disrupting or dissolving the relation between subject and object) and a commodity (a stage in the life cycle of an object in which the hierarchical relation between subject and object remains relatively stable and secure). Although not strictly 'stuff' as Boscagli defines it, waste paper shares stuff's status as something that has shed its 'commodity glamour without yet being quite cast aside', existing 'brazenly as neither one thing nor the other: not quite saleable, and certainly not garbage, not monumental or important [...] but still bearing traces of a past, of desire, of life, and of the interactions between subject and object that formed them and wore them out'.[78] As we will see, early modern waste paper *is* saleable, but otherwise Boscagli's characterization of stuff maps readily onto repurposed paper. Distinguishing the 'liminal' and 'hybrid' nature of stuff from Brown's assertive and Bennett's powerful things, Boscagli argues that, on a spectrum of stable object to disruptive thing, stuff sits somewhere between the two.[79] Worn and used, stuff continues to exist and circulate within a capitalist system of commodification, though in a markedly different way to shiny, new, and individuated objects. As it persists in the sphere of use and retains a level of value (however fractional), waste paper enacts a subtler pressure on the subject–object binary than a broken and so wholly conspicuous thing. The fact that waste remains functional and saleable, however, means it exerts this subtle pressure far more frequently and persistently than fully broken and disruptive things, which, like Heidegger's broken hammer, are typically encountered less often. Whenever an early modern paused to dwell on the simultaneously discarded and useful stuff of a repurposed page, we might imagine their understanding of their own place in the world shifting ever so slightly more toward the hybrid and the liminal.

It is the liminal status of waste paper that prompts its widespread invocation in early modern literature and makes it particularly magnetic in terms of its accumulated layers of meaning. In addition to theories of things and stuff, theories of waste can help us to understand this effect. As David Halpern suggests, waste is an especially captivating category of thing because it is both an answer to and a question about 'materiality, time, and value'.[80] Repurposed paper, like other waste stuff, is particularly unruly and unsettling because of the way it prompts us to dwell on its past lives at the same time as imagining its future use and further transformation, and the way that this biography might blur with that of other categories of things,

[78] Maurizia Boscagli, *Stuff Theory: Everyday Objects, Radical Materialism* (London: Bloomsbury, 2014), 11, 6.
[79] Ibid., 14.
[80] David M. Halperin, 'Out of Australia', in *Culture and Waste: The Creation and Destruction of Value*, ed. by Gay Hawkins and Stephen Muecke (Lanham: Rowman and Littlefield, 2003), 4.

including ourselves. It is important to approach waste theory with some caution: a number of its proponents have a tendency to ignore historical difference, describing waste as 'not reducible to a particular historical moment', or confessing to a desire to 'bypass chronology' and allow waste artefacts and literary waste 'from different cultures, time periods, and genres' to 'cuddle up' with one another.[81] As we will see in Chapter 1, early modern conceptualizations of waste are not straightforwardly mappable onto our own understanding of rubbish, trash, junk, garbage, and excrement. With this proviso in mind, two useful tenets emerge from the field of waste studies: first, that the history of waste is inextricable from constantly 'shifting ind[ices] of value', akin to Appadurai's 'things-in-motion'; and second, that waste has a peculiar relationship to temporality.[82] To 'call something "waste"', as Suzanne Raitt writes, 'is to invoke its history', gesturing 'back to the productive economies that generated [it]' and so waste, as numerous critics note, 'haunts' the present, and makes us dwell on the subject of time itself.[83]

The fluctuating worth of waste is comprehensively theorized in Michael Thompson's 1979 *Rubbish Theory*, in which he argues that rubbish is socially defined as the zero-degree of value which in turn determines the relative value of other things.[84] Thompson describes the two normative categories of objects: the 'transient', in which things, like cars, progressively decrease in value, and the 'durable', in which things, such as antiques, have permanent or increasing value and can, seemingly, never become rubbish.[85] Rubbish is a 'covert' third category of thing which draws attention to the instability of these seemingly fixed systems of value. Instead of disappearing once it has zero value, rubbish often persists 'in a timeless and valueless limbo' and so might, at some point, be discovered and abruptly transform from a transient to a durable thing—from a piece of trash to an antique.[86] This is certainly the case in relation to a significant subset of waste paper that has been extracted, collected, and preserved by antiquarians, as we will see in Chapter 3, and so has been 'magically transmogrifi[ed]' (to borrow John Frow's phrase) into a bibliographic relic or curio.[87]

[81] William Viney, *Waste: A Philosophy of Things* (London: Bloomsbury, 2015), 15; Susan Signe Morrison, *The Literature of Waste: Material Ecopoetics and Ethical Matter* (London: Palgrave Macmillan, 2015), 7.

[82] Valerie Allen, cited in Morrison, *Literature of Waste*, 8.

[83] Suzanne Raitt, 'Psychic Waste: Freud, Fechner, and the Principle of Constancy', in *Culture and Waste: The Creation and Destruction of Value*, ed. by Gay Hawkins and Stephen Muecke (Lanham: Rowman and Littlefield, 2003), 73. 'Haunting' recurs throughout Hawkins and Muecke, *Culture and Waste*, 4, 40, 46, 121; Morrison, *Literature of Waste*, 11; John Scanlan, *On Garbage* (London: Reaktion, 2005), 172; and Viney, *Waste*, 49.

[84] Michael Thompson, *Rubbish Theory: The Creation and Destruction of Value* (London: Pluto Press, first published 1979, new ed. 2017), 27.

[85] Ibid., 27.

[86] Ibid., 27.

[87] John Frow, 'Invidious Distinction: Waste, Difference, and Classy Stuff', in *Culture and Waste: The Creation and Destruction of Value*, ed. by Gay Hawkins and Stephen Muecke (Lanham: Rowman and Littlefield, 2003), 35.

We might, then, describe the range of meanings accrued by waste, like waste itself, as 'a product of time'.[88] In addition to revealing the socially constructed nature of value, waste makes visible the process of becoming and ending—how, as John Scanlan writes, time 'is always running matter down' into various states of damage, decomposition, and decay.[89] The temporality of waste is not, however, straightforwardly linear. In addition to marking the trajectory of entropy, waste can erupt into the present and 'stay with us', as Gay Hawkins and Stephen Muecke argue, destabilizing our usual sense of time's relentless march into the future.[90] Michael Taussig suggests that waste inhabits an 'Other Time', evoking a 'dreamy, other-worldly feeling' because, in persisting beyond its appointed juncture, it 'telescope[s]' time.[91] The idea of waste as being somehow telescopic is central to William Viney's nuanced philosophy of discarded matter. Drawing, like Brown, on Heidegger, Viney argues that all objects progress through two states of 'material being'.[92] Objects first inhabit a 'use-time', in which we conceive of them as subsisting within the linear progression of time, with a beginning and an end and diminishing potential as they gradually break down or wear away.[93] Once discarded, such objects enter the phase of 'waste-time': not used up but no longer in use, these objects are suspended in inaction.[94]

Waste paper conforms to Viney's theory of 'waste-time', although with some important caveats. Though discarded, it is not a 'thing without use'. It has been laid aside, moving from its primary function (as text) to a secondary one (as binding waste, lining, or wrapper) where it still serves a variety of functions. Despite these differences, Viney's theorization of the imaginative impact of waste is useful. Waste paper is both a remainder and a reminder of the past which, when encountered, interrupts the present. Binding waste, for instance, is often visually distinct from the text that it binds: darker and more stained with dirt, it bears visible traces of a past life in its folds, stitchmarks, and annotations. It might also be obviously 'out of date', marked as such by bibliographical clues, like the dates found in the running headers of an old almanac, or the rubricated letters of a Catholic mass book circulating as a wrapper in post-Reformation England. Encountered years, decades, or centuries after their production, these fragments linger, temporally out of joint. They therefore prompt their users to think about both the pastness carried into the present and the ways that the fragment—like all other texts, waste and non-waste—might be used in the future. The memory of these 'extrinsic' temporalities is encoded, to borrow the phrasing of Jonathan Gil Harris and Anna Neill, in the

88 Hawkins and Muecke, 'Introduction', xiv.
89 Scanlan, *On Garbage*, 34.
90 Hawkins and Muecke, 'Introduction', xiv.
91 Michael Taussig, 'Miasma', in *Culture and Waste: The Creation and Destruction of Value*, ed. by Gay Hawkins and Stephen Muecke (Lanham: Rowman and Littlefield, 2003), 15.
92 Viney, *Waste*, 10.
93 Ibid., 6–9.
94 Ibid., 9–14.

physical form of the waste object as it 'rupture[s]' and 'tear[s]' the 'timeless apparition' of the present.[95] Harris expands on this process in his influential *Untimely Matter in the Time of Shakespeare* (2009), where he argues that objects and materials in the early modern world were often 'polytemporal', with elements surviving from older times and so 'collat[ing] many different moments'.[96] These objects were and often remain 'untimely matter', or palimpsests that 'cross temporal borders'.[97] Encountering these palimpsests provokes one of three modes of cognition: in certain contexts, older elements might be superseded by the present, demonstrating that the past has been transcended or overcome, and that time moves inexorably forward; these older elements might also explode into the present, enacting a 'living agency' or matter-based haunting; finally, they might offer a 'conjunction' of 'the combined activity of all its polychronic components'.[98]

Waste paper is one such untimely object and has the potential to trigger a comparable range of responses and imaginings. It might demonstrate the progression of time and the supersession of the past by the present when, for instance, old editions of a book are wasted when new ones are available, or when remnants of the Catholic, liturgical past, wrap the sermons and theological treatises of the Protestant present. But repurposed pages might also provoke an unwelcome consciousness of temporality. As we will see in the chapters that follow, these waste pages often reminded their users of the capacity for other books and other bodies to similarly turn to waste, or the lingering of things that we might expect to be dead and buried. Finally, the rupturing of the present and the conjunction of different temporal moments might cause a thrill or frisson that can be traced in the literary record. John Aubrey, whom we will meet in Chapter 3, was an antiquarian certainly sensitive to the polychronicity of waste, and Thomas Nashe, whose diverse works are the focus of Chapter 4, finds a striking degree of pleasure and figurative fecundity in the kinaesthetics and multitemporal matter of repurposed pages.

As William Viney observes, waste is arresting in its tendency to prompt narrativization. It compels us, he writes, 'to deal with th[e] gaps between what an object is, what its status as waste suggest it might have been and what it might yet be', and so we feel the urge to supplement its material shape, traces, marks, and absences with stories about its past.[99] We also shape narratives about its future: will it be used again, and if so in what capacity, and by whom? When will it finally waste away? We are surprised, after all, that it hasn't already. This urge to narrativize and explain waste is visible in the writings of John Donne and Thomas

[95] Jonathan Gil Harris and Anna Neill, 'Hollywood's Pacific Junk: The Wreckage of Colonial History in *Six Days and Seven Night* and *Rapa Nui*', in Hawkins and Muecke, *Culture and Waste*, 87–92.
[96] Jonathan Gil Harris, *Untimely Matter in the Time of Shakespeare* (Philadelphia: University of Pennsylvania Press, 2009), 4.
[97] Ibid., 2.
[98] Ibid., 29–30, 91–94, 143–146.
[99] Viney, *Waste*, 11.

Middleton (to whom I turn in Chapters 2 and 5), as well as those of Nashe and Aubrey. Waste paper invites us to dig through layers of use and accreted meanings, to 'unravel' its biographical thread and stretch out its 'telescopic' time, and to unfold its 'crumpl[ed]' archive.[100] Once they had stopped working in one sense and so decreased in economic value, early modern texts gained a potent figurative worth as they crossed over into the liminal category of disruptive things and stuff. Unexpectedly, then, at this later stage in their life cycle, repurposed pages might signify an even richer set of meanings than those contained within the words printed or written on their fibres.

*

This study will perform two stages, or levels, of uncrumpling. On one hand, it tells the material history of waste paper in early modern England, unfolding both the cultural biography of individual fragments of waste paper and the social history of waste paper as a category of object as it is revealed in the Stationers' Company archives, in probate inventories, and in the holdings of rare books libraries. On the other hand, it uncovers the narratives and layers of meaning that early moderns folded around waste paper, pursuing the multitude of waste paper commonplaces in addition to extended episodes of waste paper play in literary texts. We might describe the relationship between the material and the metaphorical histories of waste paper in the same terms as Urquhart characterized his collaboration with the compositor in the printing shop: as the 'joint emulation betwixt the *theoretick* and *practical* part'.[101] Waste paper commonplaces and figurative waste paper play are a consequence of its material history and the haptics of its handling. At the same time, material encounters with and uses for waste are in part determined by its cultural history—how it was understood and perceived—and alterations in how waste was imagined in turn altered its practical use.

Chapter 1 tells the material history of repurposing texts, noting the antecedents of waste paper in repurposed papyrus and earlier parchment and charting its rapid rise in the sixteenth century. It begins by defining waste and hypothesizing why, in the middle of the sixteenth century, this particular term came to be applied to discarded and repurposed texts. Considering two early references to this category of object—a 1551 probate inventory entry and a 1566 Privy Council ordinance—it considers how the etymological valences of waste might help us to understand how early moderns conceived of repurposing and repurposed pages. The chapter then goes on to examine the evidence found in extant binding waste of patterns in the sorts of texts that were transformed into waste; the moment in its life cycle that

[100] See Tilley, *Metaphor and Material Culture*, 266 regarding these onion-like layers; on waste's unravelled threads and telescopic effect see Viney, *Waste*, 11, 1; on 'uncrumpling' see Patricia Yaeger, 'Trash as Archive, Trash as Enlightenment', in *Culture and Waste: The Creation and Destruction of Value*, ed. by Gay Hawkins and Stephen Muecke (Lanham: Rowman and Littlefield, 2003), 114.

[101] Urquhart, *Ekskybalauron*, A1v.

a book might be deemed waste; and the various ways in which waste paper was used in early modern England. It demonstrates that rather than being the most common or significant manifestation of repurposed paper in the sixteenth and seventeenth centuries, binding waste is simply the form of waste paper most likely to have survived. Because of this, it is a crucial source of information regarding waste paper more broadly, and it can be supplemented with the evidence available in Stationers' Company Records, still life paintings, and texts ranging from dictionaries and recipe books to craft and trade manuals. From these sources, we glean enough information to estimate the going rate of a sheet of repurposed paper, and to discover the need for revised bibliographical categories of waste. Chapter 1, then, reveals the precise ways in which waste paper permeated the environment inhabited by the writers who form the focus of the rest of the book.

After recounting its material history, this study goes on to demonstrate the ways in which waste's legible biography provided the structural apparatus for conceiving of a range of chronologies. These chronologies encompass the ephemerality of all texts; the passage of time and the persistence of its remnants; the rhythms of creating and consuming literature; and the decay and renewal of all organic matter. Central to each chapter is the way in which the dispersal of a book's pages came to signify the eventual fate of other matter; in particular, the disintegration of books more broadly and of the human body. Furthermore, within the eschatological worldview widespread in post-Reformation England, waste paper's capacity to model dispersal and decay might even extend to the disintegration of the world itself. The polychronic temporality of waste, however, means that repurposed pages might also serve to complicate such linear narratives of entropy and ending. Interspersed throughout these meditations on corporeal and textual loss, we find accounts of resurrection, of material survival and unexpected endurance, of wholesome and curative consumption, and of the intermingled reconstitution of human and non-human matter. These non-linear narratives of the fate of books, bodies, and the wider world are grounded in the vagaries of repurposed pages and are a cultural response to the capacity of discarded sheets and fragments to unexpectedly endure and fluctuate in and out of the status of waste.

Chapter 2 traces the lineage of the waste paper commonplace. It begins with its earliest appearance in the poetry of Catullus, Horace, Persius, and Martial, before moving on to its reception in the sixteenth and seventeenth centuries. As a result of the labour of humanist scholars who edited and published the works of classical writers, the trope re-emerged in the same decades that books, paper, and waste began to proliferate. Furthermore, books were becoming cheaper and smaller, and there was a growing sense of their flimsiness and ephemerality. It felt increasingly unlikely, within this environment, that a new publication would endure. Classical descriptions of papyri providing togas for fish and wrappers for pepper chimed with an everyday material practice and articulated an overriding sense of textual vulnerability. The phrase '*Pipere & Scombris*' emerged as a shorthand for this practice, and the metaphor manifested itself most frequently as either a commonplace insult or modesty trope, grounded in wish-fulfilment regarding another's book or anxiety concerning one's own. The chapter concludes by attending to the underlying sentiment of these commonplaces—their concern with the survival of material and intellectual things in time—as it appears in John Donne's odes and elegies on waste paper, dissolution, and death. Donne's poetry encapsulates a counter-narrative to Horace's perennial poetic monuments—monuments that underly the literary culture excavated throughout the chapter. In Donne, we find an eschatological narrative of entropy and decay, in which waste paper emblematizes the disintegration of all things. The emphasis is, though, on fragmentation rather than loss. Donne's faith in the reconstitution of the decayed body at the Resurrection is neatly materialized in the survival of discarded texts in slips and scraps that can still be read, though 'in peeces', and that might, theoretically, be pieced back together.[102]

In Chapter 3, the focus shifts from the tropic history of waste paper to its material presence in post-Reformation England. It focuses on the sudden influx of waste manuscripts in the decades following the dissolution of the monastic libraries and examines the work of two antiquarians who were alert to the histories of these fragmentary objects. Bibliographical writings by John Bale and John Aubrey provide evidence for the thoughtful apprehension of repurposed manuscripts, and reveal the ways in which waste parchment, like waste paper, might provide a vehicle for understanding both the decay of time and the world and the prolonged and partial survival of seemingly flimsy objects. The first half of the chapter considers the waste that haunts *The Laboryouse Journey & serche of John Leylande, for Englandes Antiquities*, a text heavily edited by John Bale and printed in 1549. Sensitive to the material histories of the fragments he sought to catalogue in binders' and stationers' shops, Bale's description of these objects demonstrates their bifurcating value in the decades after the dissolution, as well as their figurative potential. They become, for Bale, an emblem of his eschatological understanding of time. The chapter then outlines the telescopic effect of waste in Aubrey's late seventeenth-century meditation on the fates of monastic manuscripts. Aubrey's 'Digression', an autobiographical addition to his *History of Wiltshire*, reveals that these waste objects continued to narrate stories of dissolution and Reformation to attentive handlers. Detached by a century and a half from these events, however, they also provoke Aubrey to dwell on the fragments' relationship to history and to time more broadly. In the 'Digression' we encounter a snapshot of the decades in which waste manuscripts were perceived as both cheap wrappers and as aesthetically valued artefacts, in addition to witnessing Aubrey's self-reflexive appraisal of waste pages as the props for his antiquarian thought and identity.

[102] '*Incipit Joannes Donne*' in Thomas Coryate, *Coryats Crudities*, STC 5808 (London: W[illiam] S[tansby for the author], 1611), D3v.

Chapters 4 and 5 turn to two authors who inhabited the landscapes populated by vast quantities of printed, manuscript, paper, and parchment waste. Thomas Nashe and Thomas Middleton, like all early moderns living in urban areas and near dissolved monasteries, would have regularly encountered waste in grocers' shops, privies, and the bindings of books. As participants in London's vibrant literary culture, they would have come across waste paper in the printers', binders', and booksellers' shops of St Paul's. Furthermore, as a result of their grammar school and university educations, they would have been exposed to the waste paper commonplace in the works of classical poets, as well as in the sermons and controversial pamphlets of their contemporaries. The final two chapters will demonstrate that, like Bale and Aubrey, Nashe and Middleton paid particularly close attention to the narratives contained within repurposed pages, and that, like Donne, Nashe and Middleton repeated, experimented with, and outdid conventional waste paper insults and modesty tropes.

Chapter 4 examines the centrality of waste paper to Thomas Nashe's literary output. Many of Nashe's references to waste paper are insults, aimed predominantly at his nemesis Gabriel Harvey, and a small number are modesty tropes directed toward aristocratic patrons. Nashe, however, goes on to radically redefine these conventions, developing an intricate poetics of waste that, after the summer of 1593, underpins the worldview that characterizes his works. Nashe's preoccupation with the wasteful consumption of the materials of writing is prompted in part by his involvement in the Martin Marprelate controversy, which in turn spurred his decade-long feud with Harvey. In subsequent years, Nashe revised the dominant sense—widespread in contemporary pamphleteering—that waste paper emblematized the widespread waste of ink, paper, and talent. In the plague-ridden summer of 1593, perhaps residing in a city brimming with noxious air, plague bills, and plague dead, Nashe produced two texts that transformed waste paper from a symbol of literary prodigality and impotence to an emblem of wholesome consumption and an antidote for disease. It is, according to Nashe, miserliness and hoarding that, in 1593, provoked and materialized the plague. The generous gift and prodigal consumption of texts as waste forms the opposite of this miserly kinetics, and so serves Nashe as a gestural cure for such sinfulness. Writers such as Harvey are, according to Nashe, sinful, textual misers. Harvey's mistaken belief (ventriloquized by Nashe) that his texts might survive indefinitely in print means that he only 'la[ys] waste paper in pickle', and so attempts to preserve decaying matter that should instead be generously (and therefore safely) dispersed. As for most authors considered in this study, then, the dismantling of unwanted texts prompts Nashe to think on the disintegration of other books as well as bodies. This dissolution, though, provokes little anxiety, and instead models a kinetics that Nashe encourages his readers to celebrate and imitate in all areas of life.

The final chapter turns to an early seventeenth-century writer who, indebted to Nashe, shared the older writer's preoccupation with discarded books. But whereas

Nashe was primarily interested in the wasting of his own works and those of his enemies, Thomas Middleton's fascination lies in a specific sub-category of waste: the almanacs that, discarded at the end of the year or categorized as waste when unsold in the autumn months, were likely the most commonly repurposed text in the period. The chapter uses Jane Bennett's *Vibrant Matter* as a framework for understanding the science of almanacs, grounded as they are in the interconnection of the human and the non-human, as well as in the capacity of the little books to communicate particularly potently once discarded. In his plays and masques, Middleton frequently stages almanac–human hybrids. Perhaps costumed in actual, torn up almanacs, these hybrids are Middleton's response to the disturbing capacity of out-of-date almanacs to forecast the inevitable disintegration of the human body back into the homologous matter of the world. Unlike the majority of the writers considered in this study, waste paper does not prompt Middleton to meditate on religious chronologies of eschatology, resurrection, or Nashe's unorthodox pattern of godly dissolution. In these religious meditations, and in the preceding chapters of this book, the biography of waste is typically substituted for the biography of the human, serving as a vehicle through which to imagine the impermanence of the writer along with his text, or the fate of the human body within God's decaying world. Old almanacs, however, decentre the human and suggest that their fate is identical, rather than simply analogous, to that of discarded and dismembered texts. They reiterate that the human will eventually become indistinguishable from the non-human world. In the final section of Chapter 5 and in the Coda, *Waste Paper in Early Modern England* asks if there is anything that we might learn from the widespread early modern sensitivity to discarded matter that has been the focus of this volume. Although we no longer cohabit with our waste, and though our waste is materially distinct from the organic pages of old books, there is perhaps a benefit to cultivating an awareness of the full biographies of the things that we throw away.

Each of these chapters draws on waste objects that I have encountered in my archival research, either through systematic surveys of collections or through fortuitous discoveries.[103] Considering these extant objects alongside the accounts of them in John Donne, John Bale, John Aubrey, Thomas Nashe, Thomas Middleton, and an array of more obscure writers, enables us to trace the layers of thought that grew up around waste paper in the early modern period. These texts reveal how waste structured an early modern understanding of the passage of time and the fragmentation of things in time, as well as how waste narrated the fluctuating

[103] These surveys will form part of the online database of printed waste paper that I am creating with Adam Smyth and Megan Heffernan. They include the binding waste recorded in the catalogues of the Huntington Library, the Harry Ransom Center, Marsh's Library, Princeton Library, and the Cosin and Bamburgh Libraries at Durham University, and full surveys of the bindings in Innerpeffray Library, Perthshire, and the Plume Library, Maldon (the second of which is ongoing). Bindings and other objects that contain waste referenced in this volume are listed in the Appendix.

value of objects and their capacity to unexpectedly survive. Repurposed pages are one stage in the long life cycle of texts, which progress through the stages of flax, linen, and rags before transforming into paper, text, and then (in many instances) waste, before, on occasion, becoming historical artefacts 'rescued' from their waste fate, and so catalogued and valued in their own right.[104] *Waste Paper in Early Modern England* suggests that we should endeavour to become sensitive, like our early modern counterparts, to the palimpsestic nature of the things around us, attending to their capacity to shape our thought, and so, perhaps, to prompt us to inhabit the world differently.

[104] Sean Coughlan, '"Incredibly rare" William Caxton print discovered'. *BBC*, May 9, 2017, last accessed August 18, 2020, www.bbc.co.uk/news/education-39846929.

1
The Material History of Waste Paper

The practice of repurposing texts significantly pre-dates the emergence of the term 'waste paper' in the sixteenth century. Papyrus documents, for instance, were used to bandage Egyptian mummies, and, as we will see in Chapter 2, classical Roman writers such as Horace and Martial describe wrapping spices and fish in the papyrus on which bad poetry was written.[1] Throughout the intervening centuries scribes, bookbinders, and tailors reused unwanted and redundant manuscripts, either scraping away their ink and reusing them as a writing substrate, or cutting and stitching them into new shapes and structures.[2] These manuscripts were predominantly composed of parchment, but as the work of Orietta Da Rold has demonstrated, low-grade paper was used as a wrapping material as early as the fifteenth century, and so it seems likely that a small quantity of redundant texts written on paper would have been put to similar use.[3] Paper manuscripts were repurposed in the same way as parchment ones, and misprints and proof sheets were used as waste from the earliest days of print.[4] As the sixteenth century progressed, increasing volumes of waste material became widely available. The subject of Chapter 3, huge quantities of parchment manuscripts flooded the English market for waste after the dissolution of the monastic libraries in the 1530s and 1540s and, at much the same time, printshops, bureaucratic institutions, and households consumed paper at ever-increasing rates.[5] A surfeit of discarded sheets and an intensification of waste practices was the inevitable result.

This chapter charts the emergence of the term 'waste paper' and the extraordinary proliferation of the material in the sixteenth and seventeenth centuries,

[1] Pallab Ghosh, 'Scan technique reveals secret writing in mummy cases'. *BBC*, December 31, 2017, last accessed July 20, 2021, https://www.bbc.co.uk/news/science-environment-42357259.

[2] Ryley, *Re-Using Manuscripts*; Erik Kwakkel, 'Destroying Medieval Books—and Why That's Useful', *Medieval Books* (blog), October 31, 2014, last accessed August 19, 2020, https://medievalbooks.nl/2014/10/31/destroying-medieval-books-and-why-thats-useful/

[3] Da Rold, *Paper in Medieval England*, 56.

[4] Ryley, *Re-Using Manuscripts*. On the use of incunabula as binding waste, see Lotte Hellinga, *Incunabula in Transit: People and Trade* (Leiden: Brill, 2018), 208–218 and Oliver Dunke and Falk Eisermann, 'Fortschritt oder Fidibus? Zur Bestimmung, Bewahrung und Bedeutung von Inkunabelfragmenten', in *Fragment und Makulatur: Überlieferungsstörungen und Forschungsbedarf bei Kulturgut in Archiven und Bibliotheken*, ed. by Hanns Peter Neuheuser and Wolfgang Schmitz (Wiesbaden: [Harrassowitz], 2015), 281–308.

[5] See *Archives and Information in the Early Modern World*, ed. by Kate Peters, Alexandra Walsham, and Liesbeth Corens (Oxford: Oxford University Press, 2018) and Randall C. Head, *Making Archives in Early Modern Europe: Proof, Information, and Political Record-Keeping, 1400–1700* (Cambridge: Cambridge University Press, 2019).

Waste Paper in Early Modern England. Anna Reynolds, Oxford University Press. © Anna Reynolds (2024).
DOI: 10.1093/oso/9780198882701.003.0002

and in doing so sets the scene for the literary history that is told in the following chapters. The majority of extant waste paper survives within bookbindings, with particularly high proportions in collections that have not undergone programmes of repair and rebinding in the course of the eighteenth, nineteenth, and earlier twentieth centuries. Much waste has been removed and either discarded or sold as loose fragments in antiquarian markets and, in some cases, librarians have stored waste fragments from bindings in archival envelopes and folders, with the practice of recording the relationship between fragment and host binding only becoming common in the later twentieth century.[6] Though the material presence of early modern waste paper is now limited almost exclusively to libraries and archives, binding fragments do not tell the full story of this category of object. Instead, they reveal the unlikely survival of more typically ephemeral slips and scraps. Vast quantities of waste paper would have served 'base ends' and 'inferiour employments', as various writers phrased it, scattered across early modern kitchens, shops, libraries, and privies.[7] Waste paper, therefore, populated almost every kind of urban and domestic space. Although bindings provide vital evidence regarding the chronological and geographical patterns of waste paper production and use, we need to keep in mind the fragility and vagaries of larger proportions of repurposed pages. After exploring the origins and etymology of the term 'waste paper', this chapter situates the conclusions we can draw from the wealth of extant binding waste alongside a small number of less likely material survivals, as well as the waste paper visible in the early modern textual and visual record.

1 Defining Waste

What, then, is waste paper? Waste paper—also 'wastepaper', 'waste-paper', or 'wast paper'—is defined by the *OED* as (1) 'Paper cast aside as spoiled, superfluous, or useless for its original purpose. Also *fig*.' and (2), 'Blank or unused paper'.[8] The second definition of 'blank or unused paper' cites a 1691 usage and is described as obsolete, though as we will see, the idea of waste paper as blank, useable space has a much longer and richer early modern heritage than the *OED* allows. For the purposes of this study, waste paper is defined more narrowly: it is paper that bears text, either handwritten or printed, that has been reused for a non-textual purpose and so is no longer intended, primarily, to be read. Within the parameters of

[6] See, for instance, the folder of Huth Fragments housed at the Henry E. Huntington Library, San Marino, and the boxes of waste fragments that previously wrapped pamphlets at the Plume Library, Maldon.

[7] Richard Younge, *Englands unthankfulness striving with Gods goodness, for the victory*, Wing Y152 ([London: J. Bell for James Crump, 1643?]), 30; Urquhart, *Ekskybalauron*, 4.

[8] 'Waste-paper', *OED*. The Latin 'maculatur', meaning 'soiled' or 'stained', is synonymous with waste paper and was current in the sixteenth and seventeenth centuries. It was used less frequently in England than on the Continent.

the *OED*'s definition, blank paper might also be categorized as waste, such as the cording quires—like those described by Thomas Urquhart—that, already damaged, were used by the papermaker to protect the outer-edges of a ream of paper as it was transported from paper mill to paper sellers' or printers' shops. These inferior sheets of writing or printing paper were, therefore, 'spoiled' and so 'useless for [their] original purpose'.[9] Early modern references to waste paper, however, refer more specifically to repurposed *textual* objects, perhaps because repurposed *blank* paper is difficult to identify as such, or because it had its own, more precise nomenclature, such as 'cording quires'. It seems likely that repurposed blank sheets did not capture the imagination in the same way as wasted texts, and so go relatively unmentioned in the literature of the period.

Although not comprehensive, the *OED*'s two definitions go some way to highlighting the complexities and ambiguities that characterize waste paper as a category of thing. It is at once an object (a wrapper or fragment) and a literary trope ('*figurative*').[10] It is often 'superfluous' to requirement, suggesting the underappreciated plenitude and excess of paper in the sixteenth and seventeenth centuries, but its presence also indicates a frugality and thriftiness more typically associated with scarcity.[11] Waste paper is 'cast aside', torn up and reshaped because it is unwanted or considered irrelevant, but coterminous with the act of discarding is that of gathering up, setting aside, and immediate or eventual reuse.[12] Often branded 'useless' because it has either been spoiled or 'used up', waste paper is nonetheless useful, serving a multitude of non-textual functions.[13] But the *OED* omits one important dimension of waste: it might often be used, once again, for its 'original purpose', reinstated as reading material or 'cartapaciatory utensil' (Urquhart's term for scrap writing paper), and so again serve as a surface to be written on or a text to be read.[14]

It is also worth noting that, transformed into waste, a paper, parchment, or papyrus sheet remains materially the same. Though its shape and size has altered as it has been torn, cut, folded, and crumpled, it is still made of paper, parchment, or papyrus. The transformation is, then, phenomenological, and the term 'waste' denotes a reclassification of an object and a stage in the life cycle of a text rather than a more radical conversion of matter, of the sort that occurs in the conversion of flax to linen or rags to paper.[15] This is why the term 'repurposed' is more accurate than 'recycling' when describing waste practices, although neither was current in the early modern period. 'Repurposing' is useful because it indicates conversion 'for a different purpose or for use in a different way', suggesting continuity

[9] See pp. 1–3 above; 'Waste-paper', *OED*.
[10] Ibid.
[11] Ibid.
[12] Ibid.
[13] Ibid.
[14] Urquhart, *Ekskybalauron*, 3.
[15] There are some exceptions: waste parchment boiled to make size and waste paper pulped to make pulpboard are reconstituted and reformed, rather than simply dismantled and used differently.

between the object before and after its re-categorization.[16] Although 'recycling' is sometimes used in this sense, it carries the connotation of a more substantial transformation. Recycled objects 'return to a previous stage of a cyclic process', often melted down or chemically treated to form radically altered and reusable products.[17] Waste paper is, therefore, not always a permanent state or the end-point of a text's biography. As William Viney writes of waste more broadly, '[t]here is not a single or uniform trajectory that sees objects move from use to waste but a rich series of comings and goings, a crisscrossing in and out of the projective time of use and the slack, indeterminate and disoriented time of waste'.[18] Waste paper might fluctuate between the status of text and waste depending on the nature of its use and the perspective of its user. For contemporary scholars, particularly medievalists and book historians, an extant fragment might be valued more highly as a remnant of an absent textual whole than as a scrap of binding waste. Attending to its textual contents or making use of its blank space, a waste fragment might similarly slip back into the status of text for its early modern user.

We might ask why, in the middle of the sixteenth century, waste paper came to be labelled as *waste*, and why the process of dismantling and repurposing texts was considered an act of *wasting*. How do these early modern waste objects compare to the sorts of matter we categorize as waste today? Twenty-first-century waste comes in many forms, ranging from the bodily waste we excrete to the rubbish or garbage that ends up in landfill and refuses to disappear, or the chemical or nuclear waste that, like the growing quantities of non-degradable plastics, leaks toxic matter into our oceans and air. Waste today is what is left behind, what has been used up and spat out by industrial processes or daily acts of life, and it threatens, haunts, pollutes, and persists. The central questions concerning waste in the industrialized West concern *how* it can be disposed of, *where* it can be disposed of, and, ultimately, *if* it can ever be disposed of fully. Waste disposal was not a topic of such anxiety in the early modern period. Certainly, bodily excrement and other potentially infective matter had to be gotten rid of, particularly in densely populated urban areas, but bodily fluids and expulsions were not termed *waste* in the sixteenth and seventeenth centuries.[19] Though we as modern-day readers might label a work such as John Harington's *Metamorphosis of Ajax* a discourse on waste

[16] 'Repurpose', *OED*.

[17] 'Recycle', I.1.a, *OED*.

[18] Viney, *Waste*, 60.

[19] It is difficult to prove that a word did not mean something at a particular time, but the works of Thomas Nashe provide compelling evidence. Nashe, a reveller in excrement and wasted stuff, never uses the term 'waste' to refer to bodily expulsions. Waste, when used as a noun, typically refers to a specific set of objects, such as waste soap ashes, waste cloth, and—most often—waste paper. Waste ground and wasteland is also commonly referred to, as will be discussed below. None of these waste objects are expelled from the human body, although they often come into contact with it. Scholarship that describes excrement in medieval and early modern literature as waste does so anachronistically, using a later valence of the term. I am not certain when waste and excrement do become synonymous, but it is not before the end of the seventeenth century.

disposal, concerned as it is with the technology of 'jakes', or toilets, this is not a term that would have been applied by its early readers. Instead, as the scholarship of Ayeesha Mukherjee and Frances Dolan has demonstrated, discourse surrounding waste centred on its potential productivity, rather than on its grossness and toxicity, and writers voiced anxieties regarding the wasting of waste, rather than its unwanted presence. How, they asked, might by-products such as soap ash and the shavings of animal skins be put to use and so enrich other useful matter, such as soil and manure?[20] In this and other ways, waste paper defies many of our modern-day assumptions regarding waste stuff. Repurposed pages are not toxic or difficult to dispose of, and are far from being 'refuse matter', 'unserviceable [...] by-products' or 'material [...] so damaged as to be useless or unsaleable'.[21] Although a by-product of a process—printing, reading, writing, and book ownership—waste paper was in fact extremely serviceable.

With this in mind, then, what was it that made waste paper *waste*? The earliest instance of the term I have come across dates from 1551, when 'bookes for waste paper' were valued at 2 d. in a Cambridge probate inventory, and the term was used sporadically before it became commonplace in literary texts from the 1590s onward.[22] We can assume that wasting in the sense of squandering was central to the coinage of the term: to waste, as is made clear by a 1574 dictionary, was a morally loaded act, meaning 'to consume, spend, or wast riotously'.[23] As Lorna Hutson has demonstrated, the end of the sixteenth century saw a rise in anxiety regarding the consumption and spending of materials and resources, and to be wasteful was considered a grave sin within the religious and economic climate of Elizabethan England.[24] A sheet of paper was more valuable when blank than when repurposed, and so a page bearing text used for a non-textual purposes might be understood as 'wasted' paper.

The verb 'waste' can also indicate an entropic trajectory, denoting when something has worn out, decayed, or wasted away over time.[25] This sense of the word suggests the material condition of 'olde broken bokes and other trasshe', often described in probate inventories, that might be dismantled and used as scrap, and indicates how early moderns perceived of waste paper as being at a later stage in

[20] Ayeesha Mukherjee, *Penury into Plenty: Dearth and the Making of Knowledge in Early Modern England* (London: Routledge, 2015), esp. 71–88 and Frances E. Dolan, *Digging the Past: How and Why to Imagine Seventeenth-Century Agriculture* (Philadelphia: University of Pennsylvania Press, 2020), esp. pp. 26–32.

[21] 'Waste', (*OED*, III.11.a).

[22] See E. S. Leedham-Green, *Books in Cambridge Inventories: Book-lists from Vice-Chancellor's Court Probate Inventories in the Tudor and Stuart Periods*, vol. 2 (Cambridge: Cambridge University Press, 1986), 816–20.

[23] John Baret, *An Aluearie or Triple Dictionarie*, STC 1410 (London: by Henry Denham, 1574).

[24] Lorna Hutson, *Nashe in Context* (Oxford: Clarendon Press, 1989), pp. 15–37.

[25] See the entry for 'Consumer', meaning 'to consume, to waste, to weare away', in Claudius Hollyband, *A Dictionarie French and English*, STC 6737 (London: T[homas] O[rwin] for Thomas Woodcock, 1593).

the life cycle of a text.[26] In addition to this gradual wearing away, 'to waste' hints at violent destruction—a form of 'laying waste' that correlates to the act of cutting, tearing, and reshaping a sheet of paper.[27] This violence is suggested by the early use of the term 'waste paper' in the 1566 Privy Council *Ordinaunces decreed for reformation of diuers disorders in pryntyng and utteryng of Bookes*, which were intended to punish both the importation of Continental Catholic publications and the violation of patents. The *Ordinaunce* instructs wardens that 'all bookes to be so forfayted, shalbe brought into the Stationers hall in London' to be 'destroyed or made waste paper' at the discretion of the Company.[28] Here, transformation into waste is treated as synonymous with a book's destruction. Although the processes have distinct material results—in one an object persists, in the other it doesn't—the effect is in this instance considered the same: a Catholic publication, fragmented and dismantled, is perceived as no longer dangerous either because, circulating in units of a sheet at most, its meaning is no longer comprehensible, or because it has been broken and so is demarcated as worthless. This is in keeping with the wider practices of iconoclasm in the sixteenth and seventeenth centuries, in which religious objects were only partially defaced or, less symbolically, used for a different function, and so might also be described as acts of wasting.[29]

The proportion of waste paper originating from censored books would, though, have been insignificant in comparison to the waste paper that emerged as a by-product of the printing process and the trade in unwanted, old books. It is useful, though, to conceive of waste paper as a surface or space that has been laid waste—as waste ground or a wasteland—either depopulated and emptied out, or wild and uncultivated.[30] Wastelands make visible the etymological root of 'waste' in the Latin *vāstus*, meaning either 'a des[e]rt or solitary place' or an immense and extensive 'void'.[31] This lies behind the *OED*'s secondary definition of waste paper as 'blank or unused', as well as the 'waste blanks' of Shakespeare's Sonnet 77 and early modern 'waste books'—notebooks with printed forms and blank space for drafting accounts which became popular in the seventeenth century.[32] Waste paper has

[26] Leedham-Green, *Books in Cambridge Inventories*, 817.

[27] See '[T]o Waste: Depopulate, dilapidate, divast', in Henry Cockeram, *The English Dictionarie*, STC 5461 (London: [Eliot's Court Press] for Edmund Weauer, 1623) and 'to Waste, destroy, or spoyle', in Baret, *An Aluearie*.

[28] *Ordinaunces decreed for reformation of diuers disorders in pryntyng and utteryng of Bookes*, STC 7754.8 ([London: s.n., 1566]), broadside. For a detailed account of this ordinance, see David J. Harvey, *The Law Emprynted and Englysshed: The Printing Press as an Agent of Change in Law and Legal Culture 1475–1642* (London: Bloomsbury, 2015), 52–53, 69.

[29] See *The Archaeology of Reformation, 1480–1580*, ed. by David Gaimster and Roberta Gilchrist (Abingdon: Routledge, 2003), especially Margaret Aston, 'Public Worship and Iconoclasm', 9–28 and Sarah Tarlow, 'Reformation and Transformation: What Happened to Catholic Things in a Protestant World?', 108–121.

[30] On the history of wastelands as a specific category of space, see Vittoria di Palma, *Wasteland: A History* (New Haven: Yale, 2014).

[31] William Camden, *Remaines, of a Greater Worke*, STC 4521 (London: George Eld, 1605), 103.

[32] Shakespeare's Sonnet 77.9–10: 'Look What thy Memory Cannot Contain | Commit to These Waste Blanks', in *The Oxford Shakespeare: Complete Works*, 2nd ed., ed. by John Jowett et al. (Oxford:

much in common with these wastelands, waste blanks, and waste books. Addressing young learners in his 1612 *Ludus Literarius*, John Brinsley instructs them to practice their letters 'in some voyde place of their book, or some wastepaper', blurring the distinction between 'voyde', *vastus*, and waste.[33] Much extant waste paper, like a book's flyleaves and margins, contains pen trials. But, in the case of waste, these trials often trace or imitate text imprinted or inscribed on the page (see Figs. 1.1 and 1.2). Waste spaces and surfaces seem to invite intrusion and further use. They brim with potential, resembling the 'wastfull wilderness' and 'desert waste' of literary texts such as *The Faerie Queene*, as well as the scriptural wastelands of the biblical books of Isaiah, Jeremiah, and Ezekiel. The wildernesses of such romances and religious histories are filled with noise, movement, and surplus, and so the snippets of text inserted into the 'voyde place[s]' of a waste sheet (to borrow Brinsley's phrase) are akin to the 'wild beasts', 'doleful creatures', 'owls', and dancing 'satyrs' that occupy the spaces laid waste in the Old Testament, as well as the thickly populated wildernesses of poems such as Edmund Spenser's.[34]

Furthermore, waste paper, like the biblical wastelands in the books of the latter prophets, might be redeemed, repopulated and redefined as non-waste. In the book of Isaiah, a post-waste future is promised: 'For the Lord shall comfort Zion: he will comfort all her waste places; and he will make her wilderness like Eden, and her desert like the garden of the Lord; joy and gladness shall be found therein, thanksgiving, and the voice of melody'.[35] In a similar fashion, a waste sheet might be reinstated as a textual object by any user who cares to read it. The pen trials visible in Figures 1.1 and 1.2 suggest the frequency with which the textual contents of a waste sheet came into focus, and fictional accounts of unwrapping and reading waste hint at the possibility such events. Edmund Gayton, for instance, recounts a fantastical biography of a repurposed book in his 1654 *Pleasant notes upon Don Quixot*. He claims that Cervantes's text, along with Rabelais' *Gargantua and Pantagruel*, was 'preserv'd in that famous and wonderfull hollow tooth of Gargantua'.[36] The giant, in Gayton's version of the story, thriftily stopped up his capacious tooth with 'wast paper' consisting of 'all the Pamphlets then extant, all Romances, English, Spanish, French, and throughout the world'.[37] This results in a fortuitous form of textual endurance, not wholly dissimilar to that of Urquhart's

Clarendon Press, 2005). 'Waste book' can refer to any 'day book' for rough account-keeping: see 'Waste Book' in Peter Beal, *A Dictionary of Manuscript Terminology, 1450–2000* (Oxford: Oxford University Press, 2011).

33 John Brinsley, *Ludus Literarius*, STC 3768 (London: [Humphrey Lownes] for Thomas Man, 1612), 37–39.

34 Edmund Spenser, *The Faerie Queene*, ed. by A. C. Hamilton (London: Routledge, 2013), Book I, Canto i. verse 32; Book I, Canto ii, verse 42; Isaiah 13:1, KJV.

35 Isaiah 51:3, KJV.

36 Edmund Gayton, *Pleasant Notes Upon Don Quixot*, Wing G415 (London: by William Hunt, 1654), 33.

37 Gayton, *Pleasant Notes*, 33.

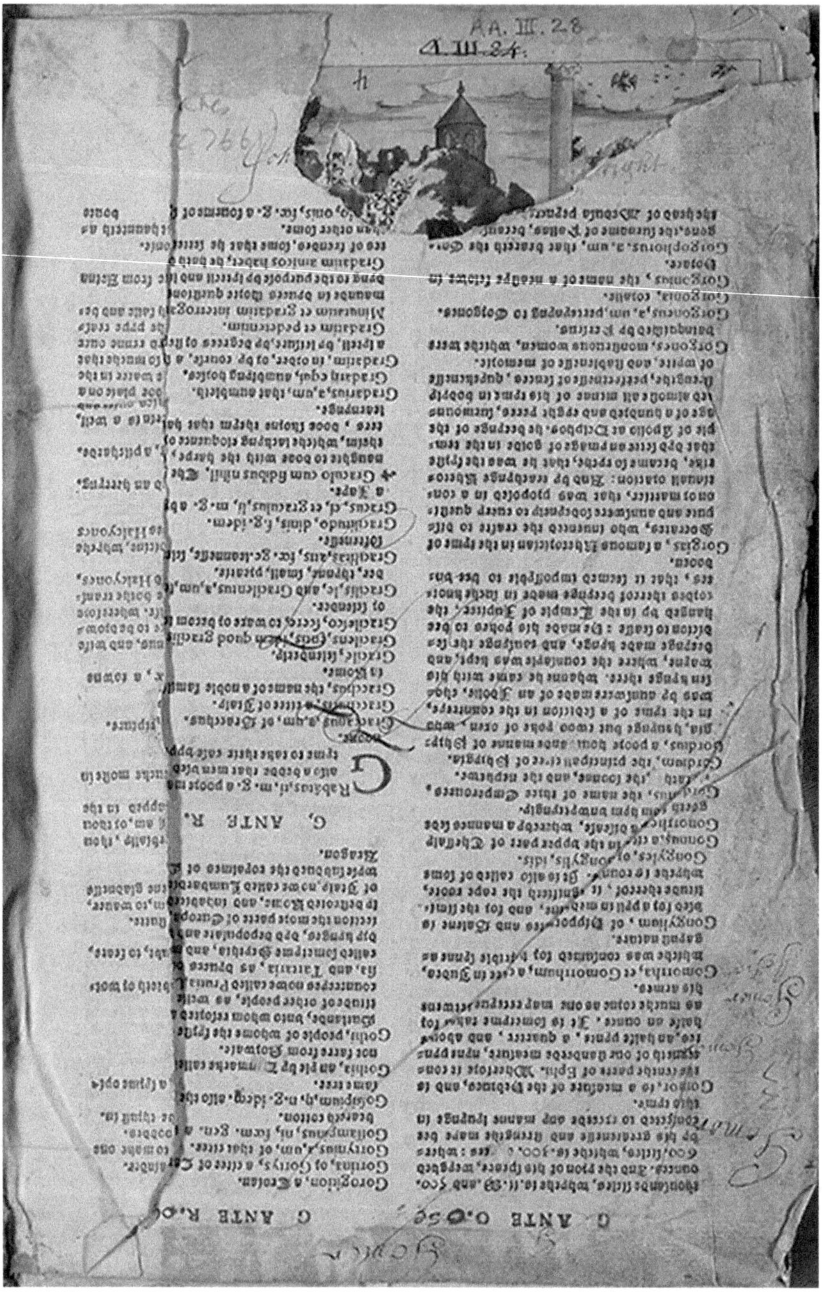

Fig. 1.1 An annotated sheet of Thomas Elyot's *Bibliotheca* Eliotae, STC 7661 (London: Thomas Bethelet, 1548) used upside down as a flyleaf in Robert Burton's *The Anatomy of Melancholy*, STC 4161 (Oxford: John Lichfield for Henry Cripps, 1628), Cosin AA.3.28. 'Gomor' has been copied multiple times in the margins.

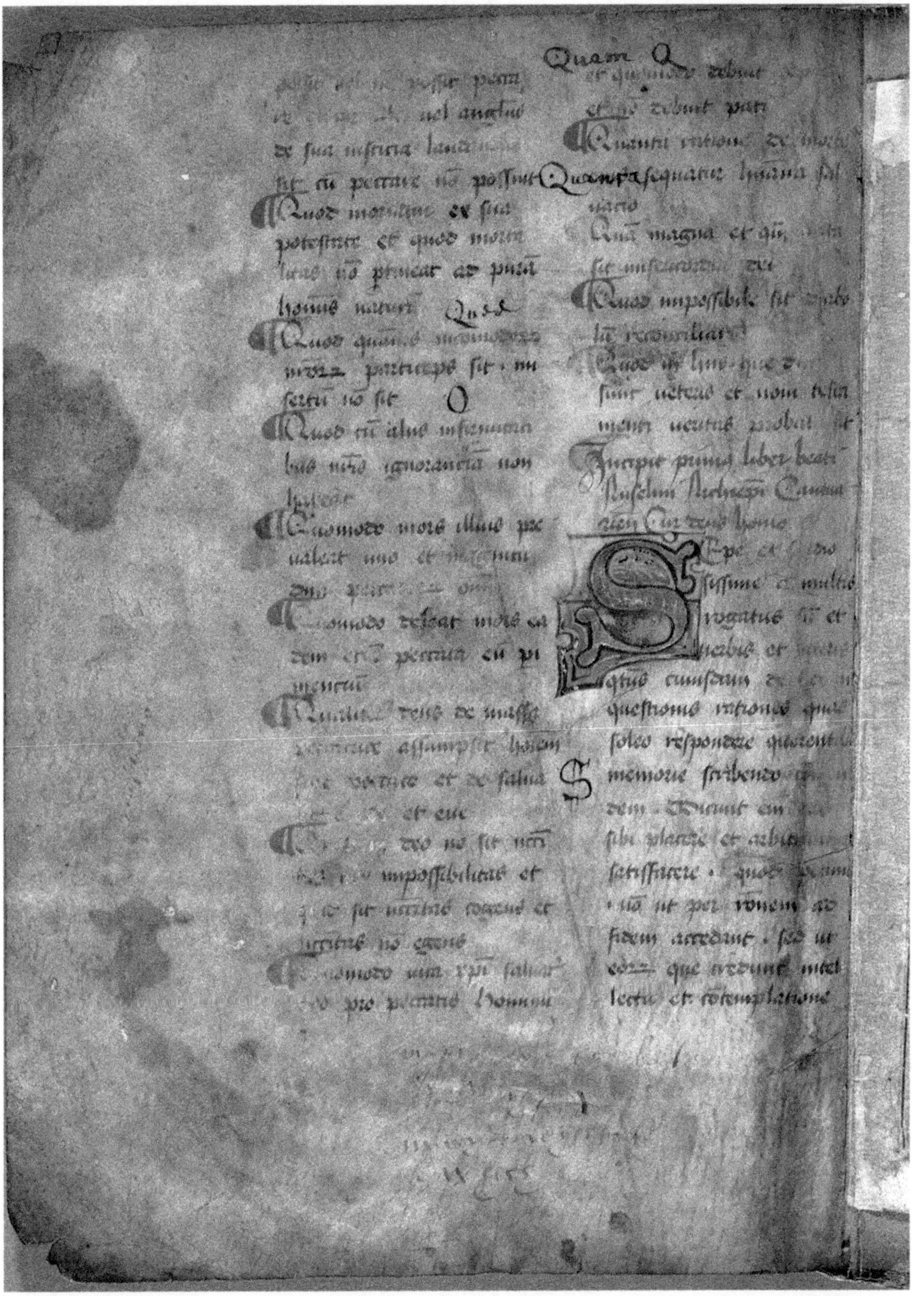

Fig. 1.2 A fifteenth-century manuscript of St. Anselm used as a wrapper for Thomas Morley's *Canzonets. Or Little Short Songs to Three Voyces*, STC 18121 ([London]: Tho: Est, [1593]), 13101v.1, with the illuminated 'S' and 'Q' copied in black ink in the blank spaces of the page. The Huntington Library, San Marino, California.

Eksykybalauron recounted above.[38] Similarly, though less fantastically, William Lilly claims that a prophecy regarding Charles II was, in 1651, 'found by chance in a wast Paper, wherein was wrapped a pair of gloves'.[39] Like that of a biblical wasteland, therefore, the biography of a sheet of repurposed paper is neither straightforward nor linear. Though discarded books were said to 'li[e] [as] wast-paper, and [be] clean forgotten' or to 'dye in obscurity', these textual objects might have a lengthy and unpredictable afterlife.[40] The varied fates of waste paper in kitchens, privies, grocers' shops, and printing houses meant that it was at once ephemeral and enduring stuff, conserving other matter—foodstuffs, medicines, the pages of the book it bound—but sometimes itself being preserved. Much extant waste remains the sole survivor of a lost textual whole, and so owes its endurance, counter-intuitively, to its use as waste.

The term 'waste', then, is apt for these sheets and fragments that have been rejected, dismantled, and discarded but that might also, in the same motion, have been reformed and preserved. Variously squandered, consumed, destroyed, and worn out, waste paper straddles many seemingly stable binaries and so proffers expansive meanings. Having been laid aside as surplus or superfluous, it might be gathered up for future use as wrapping paper, as binding waste, or, by antiquarians and other interested readers, as historical curios and documents. Even as it is cut up or torn apart, waste paper is remade, stitched into the bindings of other books. Even as it is emptied of its textual function, it might be filled up with new matter such as spices, roast meat, pies, tobacco, excrement, fresh ink, and additional text. Finally, like a wasteland that has been violently 'laid waste', waste paper continues to brim with textual noise, and so carries the possibility of future rereading, repopulation, and resurrection. Having considered the reasons that repurposed paper might have been labelled waste and the ways in which such pages might slip in and out of this category, we will now turn to the contexts in which a text might be transformed into waste in the sixteenth and seventeenth centuries.

2 Becoming Waste

Two of the earliest uses of the term 'waste paper', both quoted in the section above, indicate two key ways in which a printed or manuscript sheet might be repurposed. The first is the description of 'bookes for waste paper', valued at 2 d., in a 1551

[38] See pp. 1–3 above.

[39] William Lilly, *Monarchy or No Monarchy in England*, Wing L2228 (London: for Humfrey Blunden, 1651), 68. See also the historical examples of the sheets recovered by Anthony Wood, p. 48 below.

[40] John Everard, *The Gospel-Treasury Opened*, Wing E3531 (London: by John Owsley for Rapha Harford, 1657), B6r; Thomas Dekker, *Lanthorne and Candle-Light*, STC 6486 (London: [by Edward Allde] for John Busby, 1609), F4v.

Cambridge probate inventory;[41] the second is the 1566 Privy Council ordinance that instructs the Stationers' Company's wardens to gather books that either violate patents or have been printed or imported illegally, and to bring them 'into the Stationers hall in London' to be 'destroyed or made waste paper'.[42] The first suggests a widespread but haphazard market for 'olde broken bokes and other trasshe', sold for use as waste either after their owners' death or when no longer wanted, and the latter reveals an organized trade in rejected printed books contingent on the stationers' trade more broadly.[43]

It is likely that the 'broken bokes' recorded in probate inventories encompassed handwritten commonplace and account books as well as printed volumes, and perhaps also included collections of loose papers if present in significant enough quantities to be recorded in an inventory.[44] Loose leaves of draft accounts and letters were certainly used around the household and are present in bookbindings particularly frequently from the end of the sixteenth century.[45] They might also have been gathered and sold by thrifty householders, as Gabriel Plattes advises in a 1644 treatise:

> waste paper of all sorts, either white, or brown, or written, or printed [...] will make good passe-board, the white is worth three farthings a pound, and the other an half penny a pound to make brown passe-board good to cover books, and all other things where the colour is hidden in the work, and therefore worthy to be reserved, for in some houses it is of very considerable value.[46]

The presence of shop accounts in a small number of seventeenth-century bindings suggests that merchants and shopkeepers also gathered and traded their waste slips and scraps, though the relatively small quantities suggest that this 'written' waste might have been more often pulped to form pasteboard, as Plattes describes, or put

[41] Leedham-Green, *Books in Cambridge Inventories*, 816. See also the 'certayn olde broken bokes and trash' valued at 4 d. in 1558 (817) and the 'other wast bookes' valued at 4 d. in 1605 (820).

[42] *Ordinaunces decreed*, broadside.

[43] Leedham-Green, *Books in Cambridge Inventories*, 817.

[44] The diary of Anthony Wood suggests the frequency with which collections of papers were sold as waste upon their owners' deaths. See *The Life and Times of Anthony Wood*, ed. by Andrew Clark, 5 vols (Oxford: Clarendon Press, 1891-95), I. 150, 429, 476. Elizabeth Yale outlines these waste practices in *Sociable Knowledge: Natural History and the Nation in Early Modern Britain* (Philadelphia: University of Pennsylvania Press, 2016), 205–248.

[45] Inventories and accounts have been used in the bindings of Harry Ransom Center DA 330 B12 1629, D 158 F95 1647, and BV 4615 B647 1635, as well as Durham University Library Cosin BB.5.27. Drafts of letters and legal documents, as well as scraps of scholarly notes, are also common.

[46] Gabriel Plattes, *The Profitable Intelligencer*, Wing P2414 ([London?]: for T.U., [1644]), A4r. By 'houses', Plattes most likely means the binders' shops that made pasteboards for their own bindings. See 'pulp board' in the Ligatus Project's 'Language of Bindings Thesaurus', https://www.ligatus.org.uk/lob/concept/1528.

to more ephemeral uses in the kitchen, grocer's shop, printer's shop, or privy.[47] It is also worth noting that the related trade in waste parchment likely resembled the gathering and sale of old books at probate and the loose sheets that lay around shops and households. We know that monastic libraries were often sold in bulk as waste after their dissolution, and extant binding waste reveals that a steady stream of contemporary parchment documents made their way into binders' shops in the late sixteenth and seventeenth centuries.[48] These leases, deeds, wills, suits, and indentures, often dated within a decade of their reuse, suggest that legal institutions sold large quantities of old and redundant documents, capitalizing on the need for durable and flexible scrap parchment as the supply of monastic waste dwindled.

The Privy Council's *Ordinaunces decreed for reformation of diuers disorders in pryntyng and utteryng of Bookes* suggests that, by 1566, waste paper was a recognized facet of the book trade, and that the Stationers' Company exercised some control over its use and sale. The 'Selling and reprinting of books, waste paper, [and] buying and delivering paper' were among the 'Duties of the Warehouse Keeper' listed by the Stationers' Company when it renamed and outlined the role in 1687.[49] The waste in the Company warehouse was made up of seized publications and the misprinted and surplus sheets that were a by-product of printing the English Stock—the bestsellers that the Company held lucrative patents for, such as prayer books, almanacs, primers, and certain school books.[50] Printers of the English Stock were supplied with paper from the Company warehouse, but only paid in full when 'the number of copies and the waste overrun sheets were added and found to equal the paper' they had been issued.[51] If partners across multiple printing houses were working on the same text, 'wast' was distributed between them 'to make vp those [sheets] that be vnperfect'.[52] There are records of the Company selling waste paper (composed of unlicensed books and the misprints and surplus from the aforementioned English Stock) in 1602, 1663/4, and 1664/5, as well as the purchase of a number of reams of waste in 1621.[53] It is likely that the profits of these sales were shared internally, particularly for the benefit

[47] The binding waste in Harry Ransom Center BV 4817 B45 1657, for instance, is an account dated a decade prior to the book's publication and lists various fabrics, suggesting that it is a fragment from the rough accounts of a seventeenth-century clothier.

[48] On the sale of medieval libraries by the pound, see p. 97 below.

[49] *A Chronology and Calendar of Documents Relating to the London Book Trade, 1641–1700*, vol. 3, ed. by D. F. McKenzie and Maureen Bell (Oxford: Oxford University Press, 2005), 25.

[50] See the entries describing the seizing, or attempted seizing, of books for use as waste on 28 June 1658; 23 November 1661; 5 July 1669; and 7 December 1685 (*A Chronology and Calendar of Documents*, ed. by McKenzie and Bell). On the history of the English Stock, see Cyprian Blagden, 'The English Stock of the Stationers' Company: An Account of Its Origins', *The Library*, series 5, 10:3 (1955): 163–185.

[51] *Records of the Court of the Stationers' Company, 1576–1602, from Register B*, ed. by W. W. Greg and E. Boswell (London: The Bibliographical Society, 1930), x, 200.

[52] Ibid., lxvi, 16.

[53] Ibid., 86; Cyprian Blagden, *The Stationers' Company: A History, 1403–1959* (London: George Allen and Unwin, 1960), 185, cited in Juliet Fleming, 'Damask Papers', in *The Elizabethan Top Ten: Defining Print Popularity in Early Modern England*, ed. by Andy Kesson and Emma Smith (Farnham: Ashgate, 2013), 190.

of the Company's needier members: in 1592, the Company stipulated that the 'Cordinge quieres of the paper' given to English Stock printers should be sold to the wealthiest of the partners, and the money 'thereof commyng' to be shared between them all.[54] Towards the end of the seventeenth century, waste was more often damasked for use as decorated wallpaper, box linings, and flyleaves, and so was sold for a larger profit.[55] But the practice of supporting poorer members of the Company with the profits from its sale continued, as is revealed by the records surrounding the seizure of unlicensed and seditious books in one Benjamin Harris's shop in 1685.[56] Harris claimed to have sold them only as 'wast paper', and 'prayed' the copies seized by the wardens 'be Damasked for the benefitt of the Poore'. Ironically, a 1686 entry describes how, on account of 'his indigence & poverty', the profit made from damasking Harris's waste paper was 'paid unto the said Benjamin Harris'.[57]

Waste paper resulting from publications that weren't part of the English Stock would have been the property of the individual that had supplied the paper—usually the printer or publisher. This waste would have been used in-house, as frisket sheets in the printshop, or sold privately by the bookseller, as Benjamin Harris claimed to have done in 1685, and as Elizabeth Calvert similarly informed the Stationers' Company in 1661.[58] Calvert's husband and business partner, Giles Calvert, had been imprisoned for printing a politically suspect book.[59] Elizabeth was instructed by the Company to 'make up all the waste sheets' of the book in question 'and send them to Secretary Nicholas'.[60] She 'made them up' as requested, 'but', she claims, 'as they were not sent for, [she] sold them for waste paper'.[61] As well as providing the Stationers' Company with a practical means of censoring illicit books, as the 1566 *Ordinaunces* indicates, waste paper practices seem to have provided seditious printers with an excuse for concealing and destroying evidence. Benjamin Harris's and the Calvert's testimony reveals the routine production and sale of waste in the printing process. They attempt (or are accused of attempting) to pass prohibited texts off as some of the many sheets of waste paper that emerged from the printing press: the misprints, proofs, trials, cancellations, and surplus that are typically termed 'printers' waste' by bibliographers and book historians.[62]

[54] *Records of the Court of the Stationer's Company*, ed. by Greg and Boswell, 48.

[55] See Fleming, 'Damask Papers', 179–191.

[56] 7 December 1685 in *A Chronology and Calendar of Documents*, ed. by McKenzie and Bell.

[57] Ibid., 1 March 1686.

[58] On frisket sheets, which might be reused a second time and so preserved as binding waste, see Smyth, *Material Texts*, 142–143 and Elizabeth Upper, 'Red Frisket Sheets, c. 1490–1700: The Earliest Artefacts of Colour Printing in the West', *Papers of the Bibliographical Society of America* 108, no. 4 (2014): 477–522 and its regularly updated census.

[59] 23 November 1661 in *A Chronology and Calendar of Documents*, ed. by McKenzie and Bell.

[60] Ibid., 23 November 1661.

[61] Ibid., 23 November 1661.

[62] Henry Bradshaw was the first bibliographer to categorize waste, and his taxonomy of 'printers'' and 'binders'' waste is still widely used (*List of the Founts of Type and Woodcut Devices Used by Printers in Holland in the Fifteenth Century* (London: Macmillan, 1871), 7). For a discussion of the limitations of these terms, see Joseph Dane, *The Myth of Print Culture: Essays on Evidence, Textuality, and*

In addition to printers' waste and the haphazard trade in old and unwanted books, there is a third route through which a text commonly came to be categorized and sold as waste. Stationers' Company records hint at the vast quantities of books that went unsold in the period. In addition to poor sales caused by unpopularity, this was also on occasion the result of the infringement of patents and the flooding of particular markets. In 1636, for instance, William Stansby continued to print the Middlesbrough Psalms after being ordered 'to the Contrary' by the Company.[63] He was fined for the 'loss' of sales, and so we catch a glimpse of a portion of the English Stock going unsold and an influx of sheets of psalms into the waste paper market for that year.[64] Similarly in 1666, one Francis Mawborne of York was penalized for infringing on the English Stock when he sold 4000 almanacs in addition to an unspecified quantity of sheet calendars.[65] As a result, several thousands of licensed almanacs went unsold and were 'turnd only into wast paper'.[66] In contrast to printers' waste, it can be difficult to identify waste that has been taken from unsold books because it comes in a variety of forms: these books typically remained in sheets, sitting in stacks of paper on the bookseller's shelves until determined unsaleable. Some would have been folded, and smaller numbers stitched into a book-block or temporary binding before sale. It is often difficult to distinguish, therefore, between this semi-bound, unsold waste and pages that have been recovered from old, unwanted books. Recovered waste can sometimes be identified by annotations and signs of use that date from prior to the repurposing of the text, and, more commonly, might bear traces of paste in its gutters as well as sewing holes revealing that it had once been bound. The quantity of extant binding waste that bears no evidence of folding, cutting, stitching, or sewing, however, suggests that unsold books made up the majority of early modern waste paper.

The waste paper in the Stationers' Company records is described, in large part, because it relates to criminal activities—unlicensed publications and the investigation of clandestine activity. The trade in repurposed pages was, however, an everyday business activity in urban early modern environments, and was so commonplace as to have gone largely unrecorded. Waste paper, whether recovered from old books, made up of loose slips and scraps of draft letters and accounts, composed of unusable sheets from a printshop, or books that had gone unsold, was most likely purchased in quantities too small to have been worth noting, or by traders, such as binders, whose accounts do not typically survive. This makes it difficult to pinpoint the going rate of waste paper with certainty. Furthermore,

Bibliographical Method (Toronto: Toronto University Press, 2003), 61–62 and Anna Reynolds, '"Worthy to be Reserved": Bookbindings and the Waste Paper Trade in Early Modern England and Scotland', in *The Paper Trade in Early Modern Europe* (Leiden: Brill, 2021), 346–355.

[63] 28 April 1636 in *A Chronology and Calendar of Documents*, ed. by McKenzie and Bell.

[64] Ibid., 28 April 1636.

[65] Ibid., 7 August 1666.

[66] Ibid., 7 August 1666.

its price would have varied depending on the quantity and quality of the waste, and the available surface area useable for wrapping other objects. The waste paper market would have also fluctuated depending on the amount of waste available in the time and place in which the sale took place. The small number of records that relate the price of waste paper per quire or ream do, however, allow us to draw tentative conclusions. The Stationers' Company received £3 10 s. for forty reams of 'wast paper' at 21 d. a ream, or just over 1 d. a quire, in 1602, and, in 1621, the Company paid Thomas Gubbin 22 d. a ream, again just over 1 d. a quire, 'for so many Reames of Accedences as the Company had of him being wast paper'.[67] In a 1626 text, William Vaughan complains of religious texts selling for 'two pence a quire' and, in a similar vein, Michael Sparke laments, in 1652, the 'misery of the whole Company of Stationers', damaged by popish booksellers who had stolen the custom of 'poor honest stationers'.[68] Sparke claims that good honest books 'which cost 10 or 12 s. a Ream' are now 'turned to wast paper' and 'sold at 2 s. 6 d'., or 1.5 d. a quire.[69] Despite the polemic context of Vaughan's and Sparke's complaints, the prices they cite correlate roughly to those of the Stationers' Company: in all of these four instances, a quire of waste paper was valued between 1 and 2 d.

The 'bookes for waste paper', valued at 2 d. and 4 d. in sixteenth- and early seventeenth-century probate inventories do not specify quantity, but these seem to have sold for significantly less than 1 or 2 d. a quire, as we can assume that the books contained more than two and four quires of paper respectively.[70] This is most likely because the Stationers' Company records, like Vaughan and Sparke, refer to either printers' or unsold waste in sheets, whereas probate inventories refer to recovered waste—pages that would have been removed from a binding and so have had a significantly smaller surface area for use as waste. These smaller pages would have been less useful for wrapping and binding purposes, and more suited for pulping into boards or binding only smaller format books. We might tentatively, then, compare the value of waste to plain white paper. As Heather Wolfe has shown, the average cost of writing paper in the early modern period was approximately 4 d. a quire.[71] James Daybell quotes the 1604 *Rates of Merchandize*, which assesses ordinary paper at 2 s. 6 d. a ream (1.5 d. a quire) for taxation purposes, but

[67] *Records of the Court of the Stationers' Company*, ed. by W. W. Greg and E. Boswell, 86; *Records of the Court of the Stationers' Company, 1602–1640*, ed. by William A. Jackson (London: The Bibliographical Society, 1957), 134. For a discussion of these records in the context of the most expensive category of waste paper, damasked sheets, see Fleming, 'Damask Papers', 179–191.

[68] William Vaughan, *The Golden Fleece*, STC 24609 (London: William Stansby, Miles Flesher [and another] for Frances Williams, 1626), C3v–4r, cited in Ben Higgins, '"The Book-sellers Shop": Browsing, Reading, and Buying in Early Modern England', in *The Oxford Handbook to the History of the Early Modern Book*, ed. by Adam Smyth (Oxford: Oxford University Press, forthcoming); Michael Sparke, *A Second Beacon Fired by Scintilla*, Wing S4818BA (London: for the author, 1652), B1r.

[69] Sparke, *A Second Beacon*, B1r.

[70] Leedham-Green, *Books in Cambridge Inventories*, 817–820.

[71] Heather Wolfe, 'Was Early Modern Writing Paper Expensive?' *The Collation* (blog), February 12, 2018, last accessed August 19, 2020, https://collation.folger.edu/2018/02/writing-paper-expensive/.

Daybell goes on to say that paper, purchased in smaller amounts, cost between 2 and 5 d. a quire.[72] It is striking that, at 1–2 d. a quire, waste paper remained a relatively valuable commodity, costing as much as half the price of writing paper purchased in small quantities.

The evidence for both a commercial trade in unsold books and a more sporadic trade in unwanted books and papers counters widespread assumptions regarding the worthlessness of waste. It also helps us to answer the question of what sorts of texts were turned to waste, which captivates early modern writers and more recent book historians alike. The answer is, in fact, quite straightforward: texts that were printed in large quantities were more likely to be used as waste, simply because of the sheer volume of sheets emitting from the presses. This is certainly the case in relation to printers' waste, as more printed sheets meant a greater likelihood of misprints and surplus. It also holds true for unsold waste: though unpopular books would have remained unsold and so been categorized as waste, the waste fate is not a straightforward marker of unpopularity. This is because, even if all the sheets from an entire print run went unsold, they would still be outnumbered by the surplus and unsold sheets of steady sellers such as the Bible and the Book of Common Prayer, which were periodically replaced by newer editions. Ephemerality would certainly have been a factor in the likelihood of a book being used as waste. Almanacs, as we will see in Chapter 5, were only relevant for one year, and so would have been classified as unsold and used as waste far more frequently than other sorts of texts.

Patterns in the sorts of books that were most likely to be turned to waste at a further point in their history, with their pages cut and removed after circulating for a period, are more difficult to decipher. This is partly because smaller quantities of recovered waste are extant and identifiable than unsold waste, but also because the classification of a book as waste by its owner would have been determined in part by personal judgement of its relevance and quality. There is, however, one broad category of book that was dismantled and used as binding waste with frequency in the early modern period: large-format Latin texts, such as editions of Canon Law, theological commentaries, and philosophical texts, in particular editions of Aristotle. These were likely transformed to waste decades after they were first printed, either after the death of their owner or when replaced by a newer edition. This would have been a particularly prominent part of the waste paper trade in university towns such as Oxford and Cambridge. Large-format books, particularly folios, were also more likely to be dismantled and used as waste in comparison to smaller-format books because of the surface area of each recovered fragment. Format, however, does not determine the usefulness of printers' or unsold waste that remained unbound, as an unfolded and uncut sheet of an octavo or duodecimo

[72] James Daybell, *The Material Letter in Early Modern England: Manuscript Letters and the Culture and Practices of Letter-Writing, 1512–1635* (Basingstoke: Palgrave Macmillan, 2012), 34.

book is the same size as an unfolded and uncut folio sheet. Similarly, the smallest fragments remained useful for the bookbinders who specialized in pulping waste paper to make boards, but these texts are only fleetingly visible in the form of flecks of type within the coarse surface of a board, and so cannot usually be identified.

Most waste paper, then, is a by-product of the book trade. It was either categorized as waste within a printer's or bookseller's shop, or, in the case of old books, manuscripts, and papers, was 'laid aside' and recovered as waste after a period of use.[73] Profits from the sale of waste paper went either to the Stationers' Company members, individual booksellers and publishers, or private individuals. We will now turn to the sorts of individuals who purchased this waste paper, and what they did with it.

3 Using Waste

Because of both the nature of book historical research and the survival of waste primarily in bookbindings, the best-known uses of waste paper are those relating to the book trade. In addition to providing crucial evidence for seventeenth-century printing practices, Joseph Moxon's *Mechanick Exercises* outlines the multiple ways in which waste paper was used in early modern printing houses. The waste that printers used as frisket sheets, for 'papering up' type, and for protecting and wrapping freshly printed pages would most likely have originated from the shop's own misprints, proofs, and surplus.[74] Bookbinders, however, would have had to source large quantities of waste from outside of their own shops, and they most likely did so through their connections with booksellers located nearby, who often had large quantities of unsold books on their hands. Once purchased from the bookseller or perhaps traded for services rendered, the binder would have cut, folded, stitched, and sewn the waste sheets to form guards, spine linings, pastedowns, flyleaves, and wrappers, as well as layering them to form laminated boards or pulping them to form pulpboards.[75]

A small number of extant objects, however, demonstrate that the demand for waste paper extended far beyond the book trade. Significant quantities of waste

[73] William Troughton, *Saints in England Under a Cloud*, Wing T2319 (London: Mathew Simmons for Hannah Allen, 1648), *5v; Anon., *Bradshaw's ghost*, Wing B4165 ([London: s.n.], 1659), 11.

[74] Moxon, *Mechanick Exercises*, 300, 303, 335, 353–354. For further detail on the waste practices described by Moxon, see Reynolds, 'Waste, Offcuts, Remains, Reuse: The History of the Book in Pieces', *The Oxford Handbook to the History of the Early Modern Book*, ed. by Adam Smyth (Oxford: Oxford University Press, 2023), 365–384.

[75] Binding waste is well documented by book historians. The majority of this scholarship, however, focuses on medieval manuscript and early printed waste at the expense of the binding practices that were commonplace in the late sixteenth and seventeenth centuries. The work of David Pearson and Nicholas Pickwoad are important exceptions: see David Pearson, *Oxford Bookbinding 1500–1640. Including A Supplement to Neil Ker's Pastedowns in Oxford Bindings* (Oxford: Oxford Bibliographical Society, 2000), 139–141 and *English Bookbinding Styles 1450–1800* (London: The British Library, 2005), 20–23, 32–34, 37–39, 151–154; Nicholas Pickwoad, 'Onward and Downward: How Binders

paper also survives lining seventeenth-century boxes. These range from the small, cheap, and portable archival boxes, typically lined with sheets of unsold books, and used to store documents and records, to the more luxurious boxes designed for use in domestic households.[76] An early seventeenth-century elm trunk and a late seventeenth-century oak box, both held at the Victoria and Albert Museum, contain waste, while printed papers dated 1597 line a box of trenchers held at Shakespeare's Birthplace Trust.[77] Extant waste paper can also be found lining textiles, such as the pair of early seventeenth-century gloves, held at the Metropolitan Museum of Art, that contain scraps of pattern books and receipts.[78]

Like the fragments and sheets used in bindings, these examples of waste paper have survived because they have been sheltered by another material, be it leather, wood, or fabric, and because the object itself (box, book, or gloves) has been considered valuable enough to store and conserve by a series of owners. A significant proportion of waste paper was used in very different ways, and, unsheltered by another material, does not survive. Much sixteenth- and seventeenth-century waste paper use inverted the structure of binding waste and box lining, with waste paper often forming highly disposable wrappers, 'cornets', and 'coffins' for other, typically perishable, materials. Because these waste wrappers were ephemeral by design, we must rely on textual and pictorial evidence to establish the nature and extent of their use. It is striking that, though many writers refer to books that are 'scarce worthy to be wast paper [*sic*] for the Binder to put before this to shelter it', or 'waste paper to defend this Book from the injury of its covers', far more refer, like Urquhart, to waste paper used more variously by 'Apothecaries [...] Glovers, Cookes, and Bakers', as well as grocers and chandlers.[79] In fact, dictionary definitions refer almost exclusively to waste paper in relation to wrapping such commodities. John Higgins's 1585 dictionary references 'waste paper [...] wherein occupiers wrap their severall wares' and John Florio translates 'incartocciato' ('put

Coped with the Printing Press before 1800', in *A Millenium of the Book: Production, Design and Illustration in Manuscript and Print, 900–1900*, ed. by Robin Myers and Michael Harris (Winchester: St Paul's Bibliographies, 1994), 61–106.

[76] Significant numbers of archival boxes in the National Archives at Kew Gardens are lined with sheets of unsold books. See, for instance, the lining of C 107/110.

[77] Victoria and Albert Museum, W.17:1–1910 and W.51–1926; Shakespeare's Birthplace Trust, Stratford-upon-Avon, SBT: 1992-4. My thanks to Claire Canavan for bringing these boxes to my attention.

[78] The Metropolitan Museum of Art, New York, 28.220.7,.8. See Stefan Hanß, 'Zooming into History: An interview with Cristina Balloffet Carr on Examining Early Modern Textiles Under the Microscope', *Materialized Identities* (blog) July 4, 2017, last accessed August 19, 2020, http://www.materializedidentities.com/single-post/2017/07/04/Zooming-into-History-An-Interview-with-Cristina-Balloffet-Carr-on-Microscoping-Early-Modern-Textiles. My thanks to Rebecca Unsworth for bringing these gloves to my attention.

[79] *The Works of George Swinnock*, Wing S6264 (London: J.B. for Tho. Parkhurst, 1665), 504; Nathanael Culverwel, *An Elegant and Learned Discourse of the Light of Nature*, Wing 7569 (London: T. R[atcliffe] and E. M[ottershed] for John Rothwell, 1652), A1v; Vaughan, *The Golden Fleece*, 8; Henry Jeanes, *A Second Part of The Mixture of Scholasticall Divinity*, Wing J508 (Oxford: H. Hall [and A. Lichfield] for Thomas Robinson, 1660), N1v; Wawrzyniec Goslicki, *The Sage Senator Delineated*, Wing G2027 (London: Cottrel for Sam. Speed, 1660), 207. On Urquhart, see pp. 2–3 above.

into a paper bag or cornet') as 'rouled up in a piece of waste paper' and 'cartoccio' (a 'paper bag or cornet') as 'a piece of waste paper to put any thing in'.[80] Richard Perceval translates 'Stráça, as Papél de stráça', or *charta emporetica*, as 'waste paper to winde up ware in, browne paper', and Philemon Holland renders 'Emporetica' as 'merchant Paper or shop-paper [...] as wast Paper for sarplers [a coarse wrapper or sackcloth] to wrap and pack up wares in: also for coffins or coronets to lap spice and fruits in'.[81]

Spices, in particular pepper, are the wares most frequently referred to in relation to these waste paper cornets, but it seems likely that all manner of foods, medicines, and goods sold by apothecaries, grocers, glovers, and chandlers were stored, sold, and transported in twists and parcels of repurposed pages.[82] A range of seventeenth-century still life paintings, particularly those of Willem Claeszoon Heda and Pieter Claesz, provide further evidence for this facet of waste paper use. A pattern emerges in these visual representations: spices such as pepper are rolled neatly in a cornet of black and red printed text, in columns and with a border, as in Heda's *Still Life with a Gilt Cup* (1635, Rijksmuseum) and *Breakfast with a Lobster* (1648, The Hermitage), and in Claesz's *Still Life with a Turkey Pie* (1627, Rijksmuseum; see Fig. C.1). The colours and patterns of these cornets suggests that the waste is a fragment of the rubricated calendar of an almanac. Other still life paintings suggest that tobacco was also wrapped in waste paper, though in a slightly different way. In Heda's *Still Life with Glasses and Tobacco* (1633, Museum of Fine Arts, Boston) and *Tobacco Still Life* (1637, Los Angeles County Museum of Art), the dried tobacco leaves sit at the centre of a crumpled scrap of paper, rather than a conical cornet, bearing black text. It is difficult to tell if these paintings depict printed or manuscript waste, but they do suggest that practices for wrapping goods with waste differed depending on the substance in question. Whereas expensive spices were rolled neatly within pages of unsold almanacs, tobacco may have been folded within whatever scraps of paper were to hand. It is possible that, because tobacco would have been consumed more quickly and in larger quantities than spices, its wrappers were even more disposable, and so they were simply crumpled rather than neatly rolled. This might also explain why waste paper is often described as 'dr[ying] and inflam[ing] Tobacco', rather than wrapping it, seemingly at odds with the visual record.[83] Perhaps it served the dual function of both

[80] 'Segestria' in Hadrianus Junius, *The Nomenclator*, trans. by John Higgins, STC 14860 (London: for Ralph Newberie and Henrie Denham, 1585); Incartocciato' in John Florio, *A Worlde of Wordes*, Wing 11098 (London: Arnold Hatfield for Edw. Blunt, 1598).

[81] Richard Perceval, *A Dictionarie in Spanish and English*, STC 19620 (London: Edm. Bolifant, 1599); Pliny the Elder, *The Historie of the World*, trans. by Philemon Holland, STC 20030 (London: Adam Islip, 1634), 392–393.

[82] See Lilly, *Monarchy or No Monarchy*, 68 and Francis Anthony, *The Apologies, or Defence of a Verity Heretofore Published Concerning a Medicine Called Aurum Potabile*, STC 667 (London: Iohn Legett, 1616), 108.

[83] Thomas Dekker, *Nevves from Graues-End*, STC 12199 (London: T[homas] C[reede] for Thomas Archer, 1604), B1r.

wrapper and kindling: in David Teniers the Younger's *Boors Drinking* (1635–40, National Galleries of Scotland), one of the boors is in the process of lighting his pipe, and the loosely crumpled paper on the table in front of him contains what might be remnants of tobacco, the edges of which glow as if alight.

The textual record suggests that, in addition to wrappers, cornets, and binding waste, repurposed paper stopped mustard pots (the jars in the top right-hand corner of Tenier's *Boors Drinking* have, for instance, been stopped with scraps of paper), as well as being widely used in the kitchen to line 'pye-bottoms' and prevent 'Roast-meat' from burning.[84] It was also used to line jars containing medicines and distilled liquids, and serving as filters in the process of distillation. Moyse Charas, for instance, describes how one might distil vinegar 'by means of a Cornet or horn of wast paper, moisten'd in water'.[85] Waste paper was also, as we have already seen, set aside for scrap paper on which to write out mathematical calculations or practice handwriting.[86] Although these waste filters, linings, stoppers, and scraps inevitably do not survive, the wide range of recipes and manuals that instruct readers to use paper (whether white, brown, or waste) demonstrate that repurposed pages were used throughout the domestic and commercial environment, and so were not just a figment of the early modern imagination.[87] The most notorious function of waste paper, and that which is hardest to find material evidence for, is its use 'about *refusely occasions*', as '*bumm-fodder*', and for 'inferiour employments, and posteriour uses'.[88] According to William Cornwallis, who claims to keep 'printed Bawdery' in his 'privy', this was the use 'that waste paper is most subject too'.[89] Early modern writers revelled in the scatological potential of using old and unwanted books as toilet paper, and one extant document does suggest that the act was not a literary invention. The library of Anthony Wood contains a 1675 sheet bearing a note, in Wood's hand, that reads, 'This I found in Dr Lowers privy house 24. May 1675 in Bow Street, Lond[on]'.[90] Wood evidently acquired this document, perhaps hanging from a hook or sitting in a stack of papers, in his friend's toilet. Wood was, in fact, a connoisseur of waste: another printed sheet in his library reads, in a handwritten addition, 'Tobacco wrapt up in this paper in the beginning of May 1693'.[91]

[84] John Hall, *A Serious Epistle to Mr. William Prynne*, Wing H359A (London: for John Place, 1649), 32.

[85] Moyse Charas, *The Royal Pharmacopoeia*, Wing C2040 (London: for John Starkey, 1678), 54.

[86] Brinsley, *Ludus Literarius*, 37–39; Robert Chamberlain, *Chamberlain's Arithmetick*, Wing C1813 (London: for John Clark, 1679), 58–59, 66.

[87] See Smith, '"A Unique Instance of Art"'; Leong, 'Papering the Household', 32–45.

[88] Samuel Fisher, *Rusticus ad academicos*, Wing F1056 (London: for Robert Wilson, 1660), 135; Alexander Brome, *Bumm-fodder, or Waste-Paper Proper to Wipe the Nation's Rump With, or Your Own*, Wing B4864A (London: s.n., 1660); Urquhart, *Ekskybalauron*, 3.

[89] William Cornwallis, *Essayes*, STC 5775 ([London]: [S. Stafford and R. Read] for Edmund Mattes, 1600–1601), I7r.

[90] Nicolas K. Kiessling, *The Library of Anthony Wood* (Oxford: Oxford Bibliographical Society, 2002), xix.

[91] Kiessling, *The Library of Anthony Wood*, xix.

As these diverse uses show, it is important that we look beyond bindings when we think about early modern waste paper. Far from being limited to libraries, printers' and binders' shops, and bureaucratic spaces, waste paper pervaded every area of domestic and urban life in the sixteenth and seventeenth centuries, entering kitchens, toilets, laboratories, all manner of shops, and any space in which food was eaten and tobacco smoked. The environment was, to borrow Urquhart's phrase, one of 'such dispersive scatt[e]redness'.[92] Waste paper 'save[d] the charge of brown paper', as John Hall writes, and so was purchased by box-makers as well as binders, and gathered, saved, and purchased by housewives, tailors, apothecaries, spicers, glovers, grocers, cooks, and merchants.[93] It was used for any purpose brown paper might be put to, as Perceval's dictionary entry for 'Stráça, as Papél de stráça' suggests, with waste and brown paper presented as synonymous.[94] Both were, after all, cheap and readily accessible forms of a light and relatively durable material, good for folding and crumpling around other things.

Brown and waste paper do, though, look very different from one another: brown paper is of a low-grade, rough and unsized, and usually a murky shade of grey, blue, or brown. Waste paper, in contrast, typically has smoother and whiter fibres and contains lines of text that make visible its previous life as a book, manuscript, or document. Waste paper has, essentially and materially, changed very little since the life stage in which it was categorized as a text; it is now a fragment or incomplete part of a text, moved and encountered elsewhere. It is no wonder, then, that the scraps of misprinted, unsold, and unwanted texts that permeated the early modern environment took on a rich imaginative life. Waste paper makes explicit its past life—its dismantling, dispersal, and rejection—just as it suggests the vulnerability of all other texts to a similar fate. The material history of waste paper provides evidence for physical and intellectual interactions with repurposed sheets. Dictionary definitions, among other textual and pictorial sources, reveal that it was understood as a malleable, fragmentary object, valued for its physical properties. This haptic knowledge was not limited to binders but was shared by early modern housewives, apothecaries, cooks, tailors, and grocers. Because of its capacity to catch alight quickly it served as kindling, but its ability to withstand the heat of the oven also made it a frugal way to cover and protect meat roasting in the oven. Its capacity to fold, unfold, and retain its given shape meant that it might usefully wrap other objects, such as spices and tobacco, and its capacity to absorb greases, oils, and excrement, but also to safely peel away when lining pies and jars and wiping arses, contributed to its versatility. This, then, is the material and imaginative environment that early modern writers inhabited: like so many of their contemporaries, authors participated in the waste paper trade, encountering and using it

[92] Uruquhart, *Ekskybalauron*, A6r.
[93] Hall, *A Serious Epistle*, 32.
[94] Perceval, *A Dictionarie in Spanish and English.*

in their daily lives and, as they so often noted, contributing to its production in their very acts of composition.

4 Waste Endings

This study of waste paper is limited to the sixteenth and seventeenth centuries. This is not because pages stopped being repurposed after this date, but instead because of the significant shift in the physical use and, consequently, imaginative life of waste paper at the end of the seventeenth century. Waste paper is still used today, though it is typically pulped and so invisible to its users. Two instances that have drawn comment in recent decades are the practice of wrapping fish and chips in old newspapers, proscribed by hygiene regulations in the 1970s, and when, in 2003, millions of old novels were pulped and mixed with the tarmac of the M6 motorway.[95] Light-hearted news pieces reveal striking similarities between modern and early modern waste paper use and the haptic forms of knowledge and imaginative thought that they prompted. In a 2017 BBC podcast, fish and chip shop owner Ahmet Ziyaeddin mused on what it meant to read yesterday's news imprinted on the surface of one's food, as well as the way that ink and paper might affect its taste.[96] Strikingly, these musings echo scholarship on the relationship between text and food in early modern England, recalling, in particular, Wendy Wall's work on housewives' marchpane food displays—what she terms 'ephemeral commonplace books bearing letters, ink stains, phrases, and sentences'.[97] Similarly, Richard Beal, project manager of the M6 toll road, may have been unaware of the literary trope he was invoking when he gave an interview to the BBC in 2003. Beal punned on the reputed 'slushiness' of the Mills and Boon novels that made up a portion of the road's pulped fiction, explaining that their 'no-slushiness'—their capacity to absorb moisture and prolong the integrity of the road surface—was what made the books so attractive to his company.[98] In addition to assessing the material qualities of repurposed paper, like so many waste paper users before him, Beal revelled in the rhetorical opportunities afforded by the stuff. In his interview, his claim to have 'use[d] copies of Mills & Boon books not as a statement about what we think of the writing, but because it is so absorbent', carries a whiff of apophasis, raising as it does the ghost of the classical commonplace (that we will meet in Chapter 2) in which bad books are seemingly inevitably turned to waste.

[95] See Emily Thomas (host), 'What happened to newspaper wrapped fish and chips?' The Food Chain (podcast), November 1, 2017, last accessed August 13, 2020, https://www.bbc.co.uk/programmes/p05lq6yq and 'M6 Toll built with pulped fiction', *BBC News*, December 18, 2003, last accessed August 13, 2020, http://news.bbc.co.uk/1/hi/england/west_midlands/3330245.stm.
[96] Emily Thomas (host), 'What happened to newspaper wrapped fish and chips?'
[97] Wall, *Recipes for Thought*, 154.
[98] *BBC News*, 'M6 Toll built with pulped fiction'.

As these recent interviews suggest, there are broad similarities in the way that paper has been wasted and the way that waste paper is conceived across the centuries: waste paper appears, for instance, in the work of Alexander Pope, Jonathan Swift, and Laurence Sterne, as well as in the prints of William Hogarth, and later in the novels of Charles Dickens and James Joyce.[99] But the nuances of the metaphorical life of waste, alongside the intricacies of its material use, are dependent on specific historical conditions, and the subsequent chapters will explore the particularities of repurposed paper in its sixteenth- and seventeenth-century contexts. These contexts altered significantly in the final decades of the seventeenth century when, after two and a half centuries of reliance on imports from Italy, France, and the Low Countries, the growth of a domestic paper trade meant that white and brown paper became more readily available in England than ever before.[100] With this, the aesthetic sensibilities regarding waste shifted. Whereas earlier binders and bookbuyers were apparently unconcerned by the prominence of waste fragments in their bindings, it became common practice to hide waste beneath white or marbled pastedowns from the 1670s onwards. As Juliet Fleming has shown, the marbling and damasking of paper (often waste) became increasingly popular in these decades.[101] The use of white paper flyleaves paired with thinner waste paper guards was a common structure from the turn of the seventeenth century, but the size of the fragments of waste used in bindings shrank through the decades, until in the eighteenth century waste was largely reduced to spine linings only detectable when the binding fell or was taken apart. This shift in binding practice is perhaps indicative of a broader shift in taste and an altered attitude toward waste materials. Perhaps the torn and uneven edges of rejected pages, bearing printed or handwritten ink, were now considered too messy or too unseemly to frame a newly purchased work. Late seventeenth- and eighteenth-century buyers seem to have preferred neat and uniform bindings, like Samuel Pepys who, in addition to purchasing books 'for the love of the binding', had volumes rebound according

[99] See, for instance 'Sandy's Ghost', in *The Poetical Works of Alexander Pope* (London: Macmillan & Co. Ltd, 1961), 473–474; Jonathan Swift, 'Advice to the Grub-Street Verse Writers', in *The Complete Poems* (Harmondsworth: Penguin, 1983); Laurence Sterne, *A Sentimental Journey* (Harmondsworth: Penguin, 2005), 97; William Hogarth, *The Bad Taste of the Town*, 1724, engraving on paper, 170×242 mm., London, Victoria and Albert Museum; Charles Dickens, *Bleak House* (Oxford: Oxford University Press, 1996), 586; James Joyce, *Ulysses* (London: The Bodley Head, 2008), 57. This is a far from exhaustive list of waste paper in post-seventeenth-century literature. For the place of waste paper in later literature and book culture, see Donaldson, 'The Destruction of the Book', 1–10; Tilley, 'Waste Matters', 152–171; and Price, *How to Do Things with Books*.

[100] On the history of English paper mills, see Coleman, *The British Paper Industry*, 49–54.

[101] Decorated papers were in use throughout the early modern period as wall hangings in households and taverns and as box linings, but only became common in bindings later in the seventeenth century. Marbled paper was imported from the Continent from the late seventeenth century. See Juliet Fleming, *Graffiti and the Writing Arts of Early Modern England* (London: Reaktion Books, 2001), 60; Gill Saunders, '"Paper Tapestry" and "Wooden Pictures": Printed Decoration in the Domestic Interior before 1700', in *Printed Images in Early Modern Britain: Essays in Interpretation*, ed. by Michael Hunter (Farnham: Ashgate, 2010), 317–336; and Richard J. Wolfe, *Marbled Paper: Its History, Techniques, and Patterns* (Philadelphia: Pennsylvania University Press, 1991), 48.

to a regimented or 'common style' and to conform to his study's harmonious aesthetic.[102]

Though vast quantities of unsold books would still have been purchased in bulk by grocers (particularly cheesemongers, as William Noblett has shown) waste paper was no longer pervasive in bindings across the broad social spectrum of eighteenth-century England.[103] Only used as wrappers for foodstuffs and the cheapest books, and more typically hidden beneath marbled and damasked ink or pulped for boards, the sorts of spaces in which waste was encountered shrank even as the total volume of waste in circulation continued to grow. Geoffrey Day and Amélie Junqua have outlined the illicit trade in waste paper in eighteenth-century urban environments, describing it as a 'class-marker' and as 'invisible in canonical media'—the polar opposite of its presence and pervasiveness in the sixteenth and seventeenth centuries.[104] At the other end of the social spectrum, the trade in rare and antiquarian books boomed.[105] Now, fragments of waste and sales of old libraries were more likely to stir fantasies of long-lost manuscripts and first editions of high literary and financial value, than to prompt calculations about books' worth by the pound and the uses to which each page might be put.[106]

In the eighteenth and nineteenth centuries, accounts of manuscripts and rare books being used as waste were loaded with gendered and classist assumptions: in wasting pages, servants, wives, and widows were understood to demonstrate their naivety and cultural ignorance.[107] We will go on to see that the grubby associations of waste paper preceded the Grub Street aesthetic cultivated in the works of Pope and Swift, and that the antiquarian urge to salvage old slips and scraps of paper dated back to the early Renaissance on the Continent and the decades surrounding the Reformation in England. What develops in the eighteenth century, however, is the socio-economic polarization of waste paper in both thought and practice,

[102] Howard M. Nixon, *Catalogue of the Pepys Library at Magdalene College, Cambridge, vol. 6: Bindings* (Cambridge: Brewer, 1984), xvi–xvii.

[103] Noblett, 'Cheese, Stolen Paper', 100–110; Day and Junqua, 'Á la recherche du papier perdu', 203–218. See also Werrett, 'The Sociomateriality of Waste and Scrap Paper', 46–59.

[104] Day and Junqua, 'Á la recherche du papier perdu', 203–218.

[105] For an overview of the development of the rare books trade, see Richard Landon, 'Collecting and the Antiquarian Book Trade', in *The Cambridge History of the Book in Britain*, vol. 5, ed. by Michael F. Suarez and Michael L. Turner (Cambridge: Cambridge University Press, 2009), 711–722.

[106] See, for instance, the moment in Sterne's *A Sentimental Journey* when a servant brings Mr Yorick 'a sheet of waste paper to put betwixt' his hand and a 'little print of butter upon a currant-leaf'. The 'fragment' contains parts of a story written 'in the old French of Rabelais's time', and the 'only two other sheets of it' had been 'wrapt round the stalks of a bouquet to keep it together' (97).

[107] In addition to Yorick's servant in *A Sentimental Journey*, described in the footnote above, there is a well-known anecdote in which John Stuart Mill's maid is accused of having burnt the sole manuscript of the first volume of Thomas Carlyle's *History of the French Revolution*, as well as a story in a 1921 edition of *The Strand* in which a maid sells an 'original' manuscript along with the household's waste paper to a ragman (*The Strand*, v. 61. 7 January–Jun 1921). In 1839, a Soho bookseller recorded having 'rescued' a seventeenth-century antiquarian manuscript 'from the hands of a Soho butcher who was using it as waste paper' (National Archives MON 7/43). Similar accounts also occur in the writing of Anthony Wood and John Aubrey, but the classist and sexist overtones of these anecdotes intensifies in later centuries. See *The Life and Times of Anthony Wood*, I.429.

and what disappears is the haptic sensitivity to repurposed pages that permeated the full range of domestic, commercial, and scholarly spaces. The metaphors that these waste pages prompted and the narratives that their life cycles told altered perceptibly as a result of this late seventeenth-century shift in waste practices. The following chapters recount the ways in which the imaginative life of waste paper was entangled with its material life prior to this shift.

2
Pepper and Mackerel
The Waste Paper Trope

In the 1636 printed miscellany *Fasciculus Florum*, the pseudonymous Lerimos Uthalmus offers up a dedication to 'the Great Patroness of the World, Good Acceptance'.[1] According to Uthalmus, potential patrons are 'frighted at the sight' of such prefaces, looking on them as they would a '*Lord have Mercie upon us*', the page nailed by city officials to 'doore[s] in the time of the Plague' (A3v). In these bibliographic entryways, writers would often beg for payment, sending the addressee into 'a cold Sweat', as if they had been invited to step inside a plague-ridden house (A4r). Uthalmus instead claims to seek the patronage of the general public who have the final say because, if poorly received by the populace, a book 'dwindles and pines away, discarded and rejected like an old Fashion out of date, and neglected as a last yeares Almanack' (A4r). Uthalmus closes the preface with a poem—'Ad Librum Suum' ('To His Book'), subtitled *A Pipere & Scombris* ('Of Pepper and Mackerel')—in which he dwells further on the imagined afterlives of unwanted books:

> Go Little Book, abroad, thy self alone,
> Like *Sinon*, with thy hands behind thee bound,
> To bear the broken Jests of every one;
> [...]
> From loathsome Lotions of Face-wringing-stools
> Fortune defend thee, and from Chandlers Shops,
> Pepper, and Sope, and stopping Mustard-pots. (A5v)

Uthalmus' volume brims with classical references and literary tropes. Uthalmus imitates Ovid, who famously sends his 'little book' as a hopeful envoy to Rome, comparing it to Sinon, the willing Trojan captive in Virgil's *Aeneid*.[2] The 'little book' begot by '*Paper-Parents*' who send 'forth the Infants of their Brain into the open Ayr' (A3r) participates in the long-standing trope of literary creation as childbirth (for instance, the speaker in the first sonnet of Philip Sidney's *Astrophil and*

[1] Levimos Uthalmus, *Fasciculus Florum*, STC 24,559 (London: A[ugustine] M[athewes], 1636), A3r. Further references to be given in the text in parentheses.

[2] See, for instance, John Geyssen 'Sending a book to the Palatine: Martial 1.70 and Ovid', *Mnemosyne* 52, no. 6 (1999): 718–738.

Waste Paper in Early Modern England. Anna Reynolds, Oxford University Press. © Anna Reynolds (2024).
DOI: 10.1093/oso/9780198882701.003.0003

Stella 'great with child to speak, and helpless in my throes').[3] And the title, *Fasciculus Florum*, is a commonplace about commonplacing, recalling Seneca's influential apiary model of reading: readers should approach books in a bee-like fashion, with the most nourishing portions of the textual flowers extracted and digested for later use.[4] But one other influential trope is in operation here, and its history is yet to be written. This chapter will trace the emergence of the waste paper trope, beginning with its earliest usage among poets in classical Rome and progressing to its revival and subsequent transformations in early modern England.

Uthalmus' miscellany consists of classical and contemporary epigrams and short verses, containing translations and adaptations of poets including Horace, Martial, and Statius. These authors serve, along with Catullus and Persius, as the sources for the early modern commonplace that is explored here: '*Pipere & Scombris*' is, we will go on to see, a shorthand for the waste paper fate, with reference to hot spices and greasy fish signalling a book's transformation from reading material into repurposed wrappers. Uthalmus does not quote a classical version of the trope *ad verbum* because by the early seventeenth century, '*Pipere & Scombris*' had developed into a familiar aphorism. This pithy phrase condenses the varied descriptions of repurposed pages in classical texts into a concise motto and, alongside the more everyday term 'waste paper', situates early modern mustard-stoppers and chandlers' paper scraps within a rich rhetorical heritage. Beginning with the earliest instances of the waste paper trope in the poetry of Catullus (ca. 84–54 BCE) and Horace (65–8 BCE), this chapter considers the later classical references in the works of Persius (34–62 CE) and Statius (ca. 45–96), and the frequent and various uses of the trope in Martial (ca. 40–104 CE). It then traces the re-emergence of this rhetorical figure in the sixteenth century, revealing how increased exposure to these classical texts in the wake of Renaissance humanism coincided with an increased sensitivity to the life cycle of books. The term 'waste paper' was coined during the period in which scholars on the Continent, and later in England, edited and printed classical texts, making centuries-old descriptions of repurposing pages readily available for readers across Europe.

The texts that form the focus of this chapter reveal that the waste paper commonplace is far from a straightforward commentary on an everyday material practice. Instead, the trope provided a framework through which writers negotiated an array of literary concerns. In most instances, authors used the waste paper commonplace in one of three ways: as either an insult; as a frame through which to

[3] *Sir Philip Sidney: The Major Works*, ed. by Katherine Duncan-Jones (Oxford: Oxford University Press, 2008), 153; see also *Printing and Parenting in Early Modern England*, ed. by Douglas A. Brooks (Aldershot: Ashgate, 2005) and Maria Ruvoldt, *The Italian Renaissance Imagery of Inspiration: Metaphors of Sex, Sleep, and Dream* (Cambridge: Cambridge University Press, 2004), 65–89.

[4] See, for instance, George Wood Pigman III, 'Versions of Imitation in the Renaissance', *Renaissance Quarterly* 33, no. 1 (1980): 1–32 and Randall L. Anderson, 'Metaphors of the Book as Garden in the English Renaissance', *The Yearbook of English Studies* (2003): 248–261.

flatter an addressee; or as a device through which to profess modesty. Throughout the period, the trope was adapted to prop up an author's mediation of controversial pamphleteering, systems of patronage, and the print marketplace. The waste paper commonplace is, though, more than a rhetorical aside or throwaway insult. It is grounded in an overarching narrative structured by the metaphorics of a book's ending and fragmentation. Any consideration of waste paper is necessarily a consideration of the material and temporal dimensions of a text, and, as a result, the commonplace often serves as a counternarrative to the Horatian topos of literary immortality.

The final sections of this chapter explore how waste paper provided a means through which to ask: is a text more than a paper object, and will it endure? This line of questioning builds on what Joshua Calhoun has termed 'the ecopoetic motif' of 'corruptibility' in Renaissance England.[5] As Calhoun argues, early moderns 'had no way of knowing how long a poem on paper could last', and as a result 'decay is embedded in the conversation about survival'.[6] The motif of waste paper is a related but distinct mode of corruptibility, characterized by the persistence of remnants and fragments in the environment rather than by complete and irreversible entropy. The persistence of waste in the environment means that repurposed paper exerted a particular pressure on the early modern imagination, offering at once a *memento* of a radical and accelerated version of decay and a vehicle through which to consider alternative, non-linear temporalities of lingering and return. Concluding with a reading of John Donne's waste paper poetry, I argue that references to repurposed pages are typically structured by a narrative of senescence, though one complicated by waste's continued legibility and partial endurance. Grounded in the fragmentation and dissolution of fragile paper objects, the commonplace of wasted books readily transformed into a complex conceit through which writers negotiated anxieties regarding the trajectory of matter—both human and bibliographic—within competing temporalities of entropy and resurrection.

1 Catullus, Horace, and 'Silly Paper'

Catullus (ca. 84–54 BCE) was a late Augustan poet and an early and influential craftsman of neoteric verse. Artfully constructed, erotic, playful, and often brief, Catullus' poetry shaped that of subsequent generations of Roman poets and, after its rediscovery around 1300, later neo-Latin verse.[7] Catullus' poetry is also firmly embedded in the material contexts of its composition, reception, and survival,

[5] Calhoun, *The Nature of the Page*, 2.

[6] Ibid., 138, 130.

[7] For an overview of Catullus' life, works, and reception, see Julia Haig Gaisser, *Catullus* (Oxford: Wiley-Blackwell, 2009).

meaning that familiarity with the physical structure of a Roman book-roll is necessary to understand much of his work. Such book-rolls are made up of papyrus sheets pasted together to form the required length and usually bearing text on only one side, with a wooden rod (*umbilicus*) attached to the final sheet for ease of rolling and handling. In Carmen 1, Catullus describes giving his 'pretty new book | just now polished with dry pumice-stone' (*lepidum novum libellum | arido modo pumice expolitum*) to the author Cornelius.[8] The poet then pins his hopes of the book 'endur[ing] for year on year for more than one generation' (*plus uno maneat perenne saeclo*) on either Cornelius's protection or on his 'tutelary maiden' or muse (*patrona virgo*)—an ambiguity caused by the corruption of the ninth line of the poem.[9] Regardless, Carmen 1 is concerned with the temporality and materiality of the book of poems: it is a freshly produced, organic object (a papyrus book-roll) that has only a moment ago (*modo*) been rubbed and trimmed to prevent its plant fibres from fraying. This description of newness draws attention to the life cycle of the object: once they have considered the beginning of the book's biography, a reader is likely to consider its inevitable end. Furthermore, the reader of Carmen 1 is made aware of their own role in the life cycle of the book of poems: in writing his *Chronica*, Catullus claims, his patron Cornelius 'unfolded the whole of history in three rolls' (*omne aevum tribus explicare chartis*). Reading and writing a book-roll requires the repeated rolling and unrolling of a papyrus sheet: a poem in the middle of the book cannot be accessed without the entire book-roll being opened. The book will not remain 'pretty' (*lepidum*) and 'new' (*novum*) for long because, as Florence Dupont points out, 'the papyrus book is not made to be opened and read frequently'.[10] Papyrus is fragile and friable, becoming brittle when folded or handled too often.[11] In Catullus' Rome, therefore, the act of reading blurs with that of destruction and disintegration, particularly in the case of Carmen 1. As Denis Feeney points out, 'the emblematic first poem', sitting on the outer edge of the roll, would have 'suffered disproportionately greater damage' than the poems at the centre.[12]

These material, spatial, and temporal concerns emerge again in Carmen 95. Catullus writes,

[8] *Catullus*, ed. and trans. by G. P. Goold (London: Duckworth, 1983), 1. All references to and translations of Catullus are taken from this edition.

[9] A. S. Gratwick, '*Vale, patrona virgo*: The Text of Catullus 1.9', *Classical Quarterly* 52, no. 1 (2002): 305–320.

[10] Florence Dupont, 'The Corrupted Boy and the Crowned Poet: Or, the Material Reality and the Symbolic Status of the Literary Book at Rome', trans. by Holt N. Parker, in *Ancient Literacies: The Culture of Reading in Greece and Rome*, ed. by William A. Johnson and Holt N. Parker (Oxford: Oxford University Press, 2009), 154.

[11] Patricia Roemer, 'The Papyrus Roll in Egypt, Greece, and Rome', in *A Companion to the History of the Book*, ed. by Simon Eliot and Jonathan Rose (Oxford: Wiley-Blackwell, 2009), 84–94.

[12] Denis Feeney, 'Representation and the Materiality of the Book in the Polymetrics', in *Catullus: Poems, Books, Readers*, ed. by Ian Du Quesnay and Tony Woodman (Cambridge: Cambridge University Press, 2012), 37.

> The *Smyrna* of my dear Cinna, finally published
> nine harvests and winters after it was begun,
> while the Hatrian half a million verses in a single
> *year has been belching forth, the disgusting fellow,*[13]
> the *Smyrna*, I say, will reach Satrachus' deep waters,
> the *Smyrna* will long be read till time grows old;
> but Volusius' *Annals* will die before crossing the Padua
> and many a time will furnish roomy coats for mackerel.
> Dear to my heart be the slender monument of my friend;
> but let the vulgar rejoice in their bloated Antimachus.
>
> *Smyrna mei Cinnae, nonam post denique messem*
> *quam coepta est nonamque edita post hiemem,*
> *milia cum interea quingenta Hatriensis in uno*
> *versiculorum anno putidus evomuit,*
> *Smyrna cavas Satrachi penitus mittetur ad undas,*
> *Smyrnam cana diu saecula pervoluent:*
> *at Volusi annales Paduam morientur ad ipsam*
> *et laxas scombris saepe dabunt tunicas.*
> *parva mei mihi sint cordi monumenta sodalis:*
> *at populus tumido gaudeat Antimacho.*

In these ten lines, Catullus contrasts one good poet with three bad ones.[14] Cinna has taken nine long years to produce a short volume, the now lost miniature epic *Smyrna*. This combination of poetic brevity and careful composition, according to Catullus, ensures wide circulation and material endurance. Hatriensis, however, has rapidly composed 'half a million' (*quingenta*) verses. Another bad book, Volusius' *Annals*, will not make it beyond the neighbourhood in which it was produced, staying instead within the bounds of Padua. Volusius's words do not follow the expected trajectory of a literary text: rather than being 'long read' (*diu* [...] *pervoluent*), the *Annals* will be 'many a time' (*saepe*) put to other uses. They die (*morientur*), or are repurposed, spanning the circumference of a fish rather than crossing an ocean. Finally, Antimachus' works are described as 'bloated' or 'windy' (*tumido*). What all three bad poets have in common, Catullus implies, is that they write grossly overlong books. The amount of papyrus required for such large and carelessly composed texts makes a mass of waste material available. The papyrus of Hatriensis, Volusius, and Antimachus is organic, ephemeral stuff. Torn into multiple wrappers, the fragments of each book-roll wrap fish and rapidly disintegrate. The sleek volumes of *Smyrna* will, by contrast, escape the effects of time, enduring

[13] The text is incomplete and this line has been interpolated by the translator.

[14] For a gloss of these densely packed lines, see Joseph B. Solodow, 'On Catullus 95', *Classical Philology* 82, no. 2 (1987): 141–145.

till 'time grows old' (*cana* [...] *saecula*) and so become *monumenta*, preserving its subject and author in future acts of reading.

Horace (65–68 BCE), writing in the decades immediately after Catullus and during the seismic shift from Roman republic to Empire, composed conversational, witty, and metrically sophisticated epistles, odes, and satires. In the Renaissance, Horace was highly valued as a staple schoolroom text because of the elegance and moralistic content of his verse, as well as its frequent meditation on the art of poetry itself. And like Catullus, Horace constructs a binary of good poetry as enduring monument and bad poetry as ephemeral waste wrapper. In his influential Ode 3.30, Horace claims to have 'created a monument more lasting than bronze' (*exegi monumentum aere perennius*), a text that will endure for longer than metal statues, plaques, and tablets, as well as the stony pyramids.[15] The brass and stone objects and structures that Horace evokes in his ode are designed to interrupt the boundary between life and death, immortalizing their subjects and projecting the memory of past events and the dead into the future. But as Horace makes clear, these stone and metal memorials are still subject to the destructive forces of time and the elements. This is not the case for his own verse, Horace claims, because it transcends its material form. The Ode does not go into detail regarding the mechanics of this transcendence, but we might assume that it is achieved through its iterability: as good poetry, it will be copied and recopied and will not perish along with the papyrus it is written on, and so will 'evade death' (*vitabit Libitinam*).[16]

The same cannot be said, according to Horace, for bad poetry. Elsewhere, in Epistle 2.1, Horace describes the state of contemporary literature. He makes a case for the equal value of poetry and the better-favoured plastic arts: verses can display the character and mind of illustrious men just as clearly as a sculptor can show their features in bronze (*nec magis expressi voltus per aenea signa, | quam per vatis opus mores animique virorum | clarorum apparent*).[17] Furthermore, bad poetry shares the negative capacity of sculpture to be mishandled and to decay. The final lines of Epistle 2.1 read:

> Not for me attentions that are burdensome, and I want neither to be displayed anywhere in wax, with my features misshaped, nor to be praised in verses ill-wrought, lest I have to blush at the stupid gift, and then, along with my poet,

[15] *Horace, Odes and Epode*, ed. and trans. by Niall Rudd (Cambridge, MA: Harvard University Press, 2004), 3.30. For an analysis of the sorts of bronze objects meant by Horace's lines, see Karl Galinsky, *Augustan Culture: An Interpretive Introduction* (Princeton: University of Princeton Press, 1996), 352 and on the metaphorics of bronze more broadly see Katherine Hunt, *The Brazen World* (forthcoming).

[16] For a persuasive reading of Horace's claim of iterability and the entry of his poetry into an abstract plane, see Michele Lowrie, *Writing, Performance, and Authority in Augustan Rome* (Oxford: Oxford University Press, 2010), 117–122.

[17] *Horace; Satires, Epistles and Ars Poetica*, trans. by H. Rushton Fairclough (Cambridge, MA: Harvard University Press, 1942), Ep. 2.1.248–250.

outstretched in a closed chest, be carried into the street where they sell frankincense and perfumes and pepper and everything else that is wrapped in sheets of useless paper. (2.1 264–270)

> *Nil moror officium quod me gravat, ac neque ficto*
> *in peius voltu proponi cereus usquam*
> *nec prave factis decorari versibus opto,*
> *ne rubeam pingui donatus munere, et una*
> *cum scriptore meo, capsa porrectus operta,*
> *deferar in vicum vendentem tus et odores*
> *et piper et quidquid chartis amicitur ineptis.*

An inferior poet is like an unskilled sculptor who moulds misshapen wax portraits or death masks: the verses are badly made (*prave factis*). This physicality is absent from what Horace conceives as good and so enduring poetry. Whereas Horace's own poetry exceeds the ability of metal and stone to last, bad poems are as ephemeral as wax, a substance that, soft when hot and brittle when cold, often loses its shape. Here, Horace also alludes to the wax surface of a writing tablet, and so the ease with which the text on a tablet might be erased stands as an emblem for the ease with which 'misshaped' or bad poetry will be forgotten. The final lines of Epistle 2.1 suggest a funerary procession: the poet is carried through the street in a coffin-like case whilst, simultaneously, his poems are delivered in their book-box to the shopkeepers. Because the poems are silly, trifling, and inept (*ineptis*) they will be used as scrap papyrus, *ineptis* or unsuited for any other purpose. The material is a metonym for the poetry: misshapen language has no value independent from the papyrus it is written on. Moulded clumsily like a bad poet's text or a sculptor's fragile wax, the papyrus is folded around small quantities of dry spices and herbs. After this passage through the grubby streets of Rome, it is implied, the bad poet, his subject, and his poems will rapidly pass out of memory.

Repurposing papyrus, then, was a recognizable trope as early as the first century BCE. From the earliest extant references to repurposing books, wasted pages served as an emblem with which to negotiate the material and temporal dimensions of a text, as well as a vehicle through which to rhetorically differentiate between superior and inferior poetry. Catullus and Horace were particularly influential for the next generation of Latin poets. For those writing in what has been termed the 'Silver Age' of Roman literature, then, waste papyrus wrappers already formed a readily extractable and so adaptable motif for textual failure and disintegration.

2 Togas for Fish in the 'Silver Age'

Writing in the first century CE, the acerbic and stoic satirist Persius (34–62 CE) riffs on the now well-established waste papyrus trope in his scathing attack on contemporary literature and literary taste. Persius' Satire 1 is an imagined dialogue between the author and an interlocutor, in which Persius' companion empathizes with the idea of a poet desiring enduring fame:

> Is there anyone who would disown the desire to earn the praise of the people?—or when he's produced compositions good enough for cedar oil, to leave behind him poetry which has nothing to fear from mackerels or incense?[18]

> *an erit qui velle recuset*
> *os populi meruisse et cedro digna locutus*
> *linquere nec scombros metuentia carmina nec tus?*

For Persius' interlocutor, literary reputation and longevity are emblematized by the physical treatment of the text. Longevity is achieved through the application of cedar oil, a preservative and insect repellent applied in the period to papyrus, wood, and embalmed corpses.[19] Cedar oil and fish frequently appear alongside one another in classical texts, evoking the twin ideas of textual survival and destruction. Their shared oiliness provides an evocative diptych of the contrasting fates of literary works, one rubbed in preservative oil and the other rubbing up against greasy fish. Unlike the interlocutor, however, Persius has no patience with poets who seek fame, flattery, and textual permanence. Avoidance of the waste fate, in Persius' satire, is simply one of many pointless, worldly desires. What is the material effect, he asks, of reciting a dead poet's work?

> Are your poet's ashes not blissful now? Does the tombstone not rest more lightly on his bones now? The guests applaud: will violets not spring from those remains, from that tomb and from that blessed ash now?
>
> (I. 36–40)

> *nunc non cinis ille poetae*
> *felix? non levior cippus nunc inprimit ossa?*
> *laudant convivae: nunc non e minibus illis,*
> *nunc non e tumulo fortunataque favilla nascentur violae?*

According to Persius, rereading and copying an old book of poems does as much good for the deceased author as using it as waste papyrus: none at all. In his satire, then, Persius turns Catullus' and Horace's waste papyrus trope on its head, bitterly deflating the rhetoric of 'divine poesy' (*dia poemata*, I. 31) by highlighting

[18] *Juvenal and Persius*, trans. by Susanna Morton Braund (Cambridge, MA: Harvard University Press, 2004), I. 41–43. All future references and translations will be taken from this edition.

[19] Dioscorides (ca. 40–90 CE), for instance, describes how *cedria* is used in the preservation of the dead and as an insecticide. See Lorelei H. Corcoran and Marie Svoboda, *Herakleides* (Los Angeles: Getty Publications, 2010), 49.

the entropic trajectory shared by both books and bodies, and the impotence of literary judgement in the face of inevitable decay.[20]

Persius' contemporary, Statius (ca. 45–96), wrote poetry in an entirely different mode. Often composing verses that praised the emperor Domitian and luxuriated in the aristocratic lifestyle of the time, Statius also drew attention to the life cycle of verse and the repurposing of papyrus, though in a decidedly less stoical fashion than Persius. In his 'Jesting Hendecasyllabics' (*Hendecasyllabi iocosi*), Statius describes how he has given a handsome new book of fresh papyrus (*novusque charta*) to his friend Plotius Grypus. In return, Grypus has given a book that is 'moth eaten and moldering, like the sheets that soak up Libyan olives or keep Nile incense or pepper or cook Byzantine tunny' (*tu rosum tineis situque putrem, | quales aut Libycis madent olivis | aut tus Niliacum piperve servant | aut Byzantiacos cocunt lacertos*).[21] This playful portrayal of a lop-sided gift exchange draws attention to the organic trajectory at the centre of the waste papyrus trope: that which is good is 'fresh' (*novus*), like Catullus' *lepidum novum libellum*, and that which is bad is 'mouldering' (*putrem*). Whereas good poetry escapes the decaying effects of time, handled carefully and made perpetually new in a series of freshly copied papyrus sheets, bad poetry is inescapably material, crumbling and oozing with grease rather than cedar oil. Many of Statius' readers, however, might have paused to consider how long Statius' own book would remain fresh and unworn. In gesturing towards the life cycle of a book, Statius, perhaps unintentionally, draws attention to the very likely decay of all papyrus objects, and the uncertainty of his own text's careful preservation and repeated reproduction.

But the Roman poet who most often revels in the grossness of waste papyrus and the fragility of the textual object is the polyphonic epigrammatist Martial (ca. 40–104 CE). In fact, no poet dwells so persistently on repurposed texts again until the sixteenth century, and these later writers were likely inspired in part by their encounters as students with Martial's poetry. In many instances, Martial uses the waste papyrus trope as a cutting insult. Epigram 3.50, for instance, is a jibe at a bad poet. It describes how Ligurinus invites friends to his house for dinner just so he can read them his atrocious verses (*scelerata poemata*).[22] If he doesn't hand over his books to the mackerel-seller (*quod si non scombris scelerata poemata donas*), according to Martial, he will soon dine alone (*cenabis solus iam*). Echoing the sentiment of Horace's Ode 3.30 and Catullus 95, waste papyrus also appears in this manner in 6.61: Martial claims that merely 'clever' (*ingeniosa*), or erudite, poetry will fail to secure a poet long-lasting fame. This sort of verse will become food for

[20] For a reading of Persius' rejection of traditional metaphors, see K. J. Reckford, 'Studies in Persius', in *Persius and Juvenal*, ed. by Maria Plaza (Oxford: Oxford University Press), 51–56.

[21] *Statius Silvae*, ed. and trans. D. R. Shackleton Bailey, with corrections by Christopher A. Parrot (Cambridge, MA: Harvard University Press, 2015), 4.9 1–13.

[22] *Martial: Epigrams*, ed. and trans. by D. R. Shackleton Bailey, vol. I–III (Cambridge, MA: Harvard University Press, 1993). All future references and translations will be taken from this edition.

moths and bookworms (*tineas pascunt blattasque*) and be bought by cooks (*et redimunt soli carmina docta coci*), implicitly for wrapping fish and spices. That which 'gives centuries to paper' (*donat saecula chartis*) is 'genius' (*genium*), the tutelary spirit allotted at birth that provides inspiration.[23] These two epigrams maintain the distinction established in the work of Catullus and Horace between underwhelming poetry as ephemeral stuff and literary masterpieces that transcend their material context—a framework precariously poised on the taste buds of moths and bookworms, as well on the oily and preservative acts of their readers.

In several other epigrams, however, the source of literary survival lies outside of the poet's genius. In 4.86 Martial bestows advice on his 'little book' (*libelle*): the best thing it can do is 'please' (*placeas*) Apollinaris, its potential patron. With his support, all 'Attic ears' (*auribus Atticis*) will approve. None will 'snort' or 'sneer' (*rhonchos*) at it, nor will it 'supply mackerel with "tiresome tunics"' (*scombris tunicas dabis molestas*), a reworking of Catullus' *tunicas* for *scombris* in Carmen 95. If it doesn't please Apollinaris it will end up, like the bad poetry of Horace's second epistle, in the saltfishmongers' shops (*salariorum* [...] *scrinia*), 'fit for schoolboys to plough your backside' (*inversa pueris arande charta*). The now traditional reference to mackerel is embellished: an unprotected book will wrap saltfish, and schoolboys will violate the reverse of the papyrus with their (implicitly phallic) pens. This is a different sentiment to 6.61, one in which a book's survival is determined externally, not by the quality of its verse but by the protective power of its patron. 3.2 also conforms to this patronage model of textual survival: the poet asks his little book to choose its own protector, portraying it as an animate body that might 'fly to Faustinus' bosom' (*Faustini fugis in sinum?*). If it fails to gain a patron, it is at risk of wrapping sprats in the kitchen or serving as a 'cowl' for incense or pepper (*ne nigram cito raptus in culinam | cordylas madida tegas papyro | vel turis piperisve sis cucullus*). With Faustinus' patronage, however, it will be preserved and beautified: it will be granted painted bosses and a purple cover (*pictis luxurieris umbilicis | et te purpura delicata velet*) and be anointed with cedar oil (*cedro* [...] *perunctus*) rather than the grease of sprats. This anthropomorphized object will strut about (*ambules*) fearlessly: it will become an elegantly clothed body rather than a set of fragments that clothe other stuff, so securely whole that not even the aptly named critic Probus will be able to penetrate it and tear it apart (*illo vindice nec Probum timeto*).

In these two epigrams (4.86 and 3.2), the patron acts like a binding, cover, or book-box (*scrinium*), protecting the volume from dispersal and disintegration. The cultural and economic influence of the patron is literalized in the form of expensive ornamentation and preservative practices, with the agency and skill of the poet and their effect on the circulation and endurance of the poem obscured.

[23] 'Genius', in *The Classical Tradition*, ed. by Anthony Grafton, Glenn W. Most, and Salvatore Settis (Cambridge: The Belknap Press, 2010).

This is similar to the literary culture described in Persius' satire, in which Roman society is portrayed as giving more credence to worldly fame and luxury objects than literary *genium*. In first-century Rome, both poets suggest, it is more important who you know and who supports your work than the quality of your writing. Persius describes the inevitable death and decay of books and bodies in order to satirize the superficiality of this system. Martial, however, makes explicit his participation within it. He highlights the physical vulnerability of a text in order to flatter his patrons, claiming, obsequiously, that they hold supreme influence over the fate of his books.

This is not, though, the limit of Martial's waste papyrus play. Elsewhere, such as in Epigram 13.1, he seems uninterested in the survival of his poetry. Addressing the Muses, the poet instructs them to 'waste some papyrus from the Nile' (*perdite Niliacas* [...] *papyrus*). This is a modest gesture on Martial's part, albeit a playful one: the agency of composition is surrendered to the Muses, but his poems still turn out badly. Worth only as much as the papyrus they are written on, Martial will be worse off for it (*mea damna*). This isn't a radical claim: as we have seen, authors had been rhetorically reducing bad verse to the value of scrap pages for over a century, but Martial takes the unusual step of applying this insult to his own work. Rather than praising fresh, pretty books or seeking to write a *monumentum* worthy of cedar oil, Martial portrays himself as being content with the application of olives and fish fry. His verses are as ephemeral as the stuff on which he writes, and, in this epigram at least, Martial does not seem to mind. Instead, he dwells at length on the life cycle of papyrus, drawing attention to its origin in a river plant from the Nile and its potential afterlife as a waste wrapper: the poem is written so 'That tunny-fry may not lack a gown and olives an overcoat, nor the uncleanly bookworm fear penurious hunger' (*Ne toga cordylis et paenula desit olivis | aut inopem metuat sordida blatta famem*). In a reversal of 3.2, the poems are bequeathed to anthropomorphized edibles that might otherwise catch a chill, as well as to starving bookworms—textual consumers in the most literal sense.

This is the first epigram of Martial's *Xenia*, a volume of two-line verses designed, like his *Apophoreta*, as light-hearted reading material for the carnivalesque festival of the Saturnalia. The titles of both volumes can be translated as 'gifts', and the little books pose as objects to be exchanged during the festival. In its opening verses, *Xenia* does not describe itself, like Martial 3.2, Catullus 1, and Statius 4.9, as a luxurious object or as a quasi-transcendental text as in Horace's Ode 3.30. Tied to the 'tipsy' (*inebria*) time of year that demands 'new jests' (*novos* [...] *sales*), the book of epigrams is a papyrus plaything that stands in for the nuts or dice of traditional Saturnalian pastime (*haec mihi charta nuces, haec est mihi charta fritillus*). Appropriately ephemeral, it is to be discarded when the holiday is over, or perhaps even torn up in the midst of it: Martial, in 13.3 tells his reader 'you can send these couplets to your guests instead of a gift' (*hac licet hospitibus pro munere disticha mittas*). Each two-line poem can be substituted for that which it describes. Rather than buying and giving 'incense' (*Tus*, 13.4), 'pepper' (*Piper*, 13.5), 'a small box of olives' (*Cistella Olivarum*, 13.36), or a range of fish (13.79–91), the reader might offer a poem or poems instead, and so substitute word for thing. However unlikely this course of action actually was, Martial has us imagine a reader who tears the book into fragments, making 'gift-tags' or little snippets that the recipient will, like a hungry bookworm, consume.[24] Whether or not the book remains whole, the volume consummates the waste fate anticipated in its opening verses: it is a concrete joke in which the first epigram on the outer edge of the book-roll becomes a wrapper for food, encompassing the incense, pepper, olives, and fish contained textually within.

It is not within the scope of this study to account fully for the frequent use of the waste papyrus trope in Martial's poetry or in the Roman Silver Age. Nonetheless, it is worth delving a little deeper in light of the importance of these poets' works in early modern England. Luke Roman argues that Martial's epigrams are characterized by a 'metaliterary poetics', and that this is part of his response to Horace's characterization of textuality in the literary culture of the first century.[25] Whereas Horace imagines an 'integral' and 'immortal' work that exists as an abstracted or 'general entity', Martial offers a 'sub-literary' and 'post-classical aesthetics' grounded in the 'finite, monetary value' of books and the gritty ins and outs of the patronage system.[26] Martial's portrayal of books as individual, 'use-directed' objects is, Roman argues, a self-conscious parody of the 'monumental' understanding of literature that had recently become so influential.

We should add to Roman's 'metaliterary poetics' an emphasis on the materials and mechanics of writing. In addition to reflecting on the contemporary status and understanding of literature, Martial's poetry is metamaterial. It is concerned with the technology of the book and its life cycle, particularly as it nears its end, decays, and is dismantled. This is an instance of what Lindsay Ann Reid terms a 'bibliofiction', or a 'metaphorics of literary transmission'.[27] Reid, describing the reception of Ovid in the Renaissance, argues that writers conceived of literature as both fragile and mutable, existing in individual objects and in 'multiple and materially indeterminate forms'.[28] The model for this metaphorics of textual production and circulation is, for Reid, an Ovidian 'metamorphosis', with literature kept alive by its 'reiterat[ion] in successive forms'.[29] Reid goes on to argue that recent scholarship has become 'over-materializ[ed]', fixated too narrowly on the material life

[24] T.J. Leary, *Martial Book XIII, The Xenia* (London: Bloomsbury, 2001), 17.

[25] Luke Roman, 'The Representation of Literary Materiality in Martial's Epigrams', *Journal of Roman Studies* 91 (2001): 113–145.

[26] Ibid., 123, 128, 119.

[27] Lindsay Ann Reid, *Ovidian Bibliofictions and the Tudor Book: Metamorphosing Classical Heroines in Late Medieval and Renaissance England* (Farnham: Ashgate, 2014), 3, 7.

[28] Ibid., 6.

[29] Ibid., 2.

cycle of the book and that, instead, we should attend more to the literary representations of the imagined fates of books.[30] But the bibliofiction in Martial, and in the waste paper trope as it was received and transmitted in early modern England, was already heavily materialized: the metaphorics of transmission and circulation is, in these instances, grounded in a highly specific form of metamorphosis, as texts fragment and transform into waste. 'Textual sensibilit[ies]', the waste paper trope demonstrates, go beyond what Reid describes as the literary fiction of 'bibliogenesis' and the reiteration of mutable works, revealing instead what we might term a 'biblio-entropy', or a metaphorics concerned with the dissolution and decay of textual objects.[31]

It is worth noting that Martial's metaphorics of literary transmission and disintegration was developed within the context of an emerging textual technology: the codex. Although the codex did not supersede the book-roll for several centuries, Martial 1.2 is often cited as the earliest surviving description of a parchment codex.[32] It describes how the buyer might carry the volume wherever they go because of its size and shape—a text 'that parchment compresses in small pages' (*quos artat brevibus membrana tabellis*), and that, rather than requiring bulky bookcases, can be grasped by one hand (*me manus una capit*).[33] In addition to their material features, Martial is sensitive to the different modes of reading made available by the different formats of book-roll and codex: a book-roll encourages continuous, linear reading because it must be unrolled in its entirety to reach the latter parts of the text. In 4.89, the final epigram of the volume, Martial cries out to his book 'Whoa, there's enough, whoa now, little book! We have got to the bosses' (*Ohe, iam satis est, ohe, libelle, | iam pervenimus usque ad umbilicos*), as if the poems have rolled uncontrollably to the *umbilicos*, or rod, fastened to the end of papyrus around which the book-roll is wound.[34] Elsewhere, Martial characterizes the experience of reading a codex: 10.1 refers to the shape of the page, describing how 'My small pages quite often end with the end of a poem' (*terque quaterque mihi finitur carmine parva | pagina*). More finite than a book-roll, the codex encourages the reader to dip in and out, randomly accessing passages at will and making the book as little as they like (*fac tibi me quam cupis ipse brevem*).

Waste papyrus wrappers are, therefore, an unexpected analogue of both the codex page and a book of epigrams: all three are defined by their fragmentary nature. Martial highlights the shared characteristics of a book of epigrams and a torn-up papyrus sheet in his *Xenia* and *Apophoreta*: the poems can be read lineally

[30] Ibid., 6.

[31] Ibid., 2–3.

[32] David Diringer, *The Book before Printing: Ancient, Medieval and Oriental* (New York: Dover, 1982), 165.

[33] Leila Avrin, *Scribes, Scripts, and Books: The Book Arts from Antiquity to the Renaissance* (London: The British Library, 1991), 173; Rex Winsbury, *The Roman Book: Books, Publishing and Performance in Classical Rome* (London: Bloomsbury, 2009), 23–25.

[34] Diringer, *The Book before Printing*, 139–40.

and as a whole ('to the bosses'), or as couplets detachable from their wider context. 14.2 tells the reader 'You can finish this book at any place you choose. Every performance is completed in two lines' (*Quo vis cumque loco potes hunc finire labellum: | versibus explicitum est omne duobus opus*). This atomistic mode of consumption corresponds to the motion of tearing fragments from a book-roll, an analogy which, as we have seen, Martial invites the reader to materialize. 10.1 draws a similar analogy, but in relation to a codex page rather than a book-roll: instead of the do-it-yourself approach of the *Xenia*, the poems are already available as fragments, with their endings often coinciding with the edges of the 'small pages' (*parva pagina*). This licence to make the book as 'short' (*brevem*) as the reader pleases carries more than a hint of a possible waste fate.

Endlessly polyphonic, it is difficult to find coherence in Martial's diverse play with the waste papyrus trope: it is a fate he predicts for bad poetry and one he seeks to avoid by courting patrons, as well as a fate he celebrates in the context of the Saturnalia. Waste papyrus, then, provides Martial with a surprisingly malleable framework within which to consider textual production and consumption more broadly. Throughout the epigrams, Martial uses the trope to negotiate the diverse literary culture of the first century CE and the formal novelty of the codex. Papyrus fragments become concrete metaphors for reading, turning, and unwinding a page of short verses. This attention to the physical dimensions of poetry and papyrus provides an alternative to Horace's monumental archetype: literature, Martial suggests, is a provisional thing. Though inconsistent as to whether this is to be feared or to be celebrated, Martial makes clear that poetry is bound to the materials, technologies, and kinetics of the 'little book[s]' in which it is written. It is this malleability, polyphony, and haptic sensitivity that made Martial, in close competition with Horace and Persius, the most often cited source of the waste paper commonplace in the sixteenth and seventeenth centuries.

3 Englishing Waste

As noted in Chapter 1, the practice of wasting books continued throughout the intervening centuries, but the waste paper *trope* re-emerged towards the end of the sixteenth century. This was because of the coincidence of several factors: as we will see in more detail in Chapter 3, the decades following the dissolution of the monastic libraries made available unprecedented amounts of waste parchment, at much the same time as ever-increasing numbers of paper texts—printed and manuscript—circulated in private homes and administrative spaces, as well as in printers', booksellers', and grocers' shops. This growing presence of waste paper and parchment was accompanied by a growing paper literacy, or sensitivity to its extended material biography. It is not, however, a coincidence that an early use of the term 'waste paper' can be found in Thomas Drant's 1567 translation of selections from Horace.[35] Post-dating the Privy Council's 1566 *Ordinaunce* by only a year, these two texts suggest that the phrase had become widespread by the 1560s.[36] Drant's usage also demonstrates that the term 'waste paper' did not emerge simply because of its increasing material presence in sixteenth-century England. The mid-sixteenth century saw the coincidence of the increased circulation of material scraps of waste paper and the circulation of the works of classical poets including Catullus, Horace, Persius, and Martial. Thanks to the Humanist drive to edit and print classical texts, the waste paper trope became embedded in the literary imagination. In addition to Drant's 1566 and 1567 translations of Horace, Martial was translated in 1629 and Persius in 1616.[37] Martial was first printed in England in 1615 and Persius (alongside Horace and Juvenal) in 1574; both were used as school-texts, but Continental editions of all of these poets were commonly owned by students and scholars in England throughout the sixteenth century, in addition to the manuscript translations and versions that circulated privately.[38] Furthermore, the educational culture of Renaissance Humanism shaped the ways in which these poets were interpreted: the waste paper trope was particularly ripe for commonplacing. Selected, gathered, and digested from their original contexts, the metamaterial *sententiae* of Catullus, Horace, Martial, and Persius proved particularly appropriate for a number of increasingly pressing early modern bibliographical concerns.

The continuity between the material practices of repurposing texts in the Roman and early modern periods meant that the waste paper commonplace, though developed in the classical period, still made sense in later centuries. Repurposed pages were already potentially meaningful objects, with their folds, stitchmarks, and ink drawing attention to the life cycle of texts and their tendency to disintegrate and decay. The haptic and experiential meaning of a waste wrapper was, in the sixteenth century, *enriched* by the language of Catullus, Horace, Martial, and Persius. Verse snippets from these poets became embedded in and adapted to early modern thought to such an extent that reference to materials

[35] *Horace his Arte of Poetrie, Pistles, and Satyrs*, trans. by Thomas Drant, STC 13,797 (London: Thomas Marshe, 1567), H1r.

[36] *Ordinaunces decreed for reformation*, broadside. See pp. 33–34 above.

[37] *A Medicinable Morall, that is, the Two Bookes of Horace his Satyres*, trans. by Thomas Drant, STC 13,805 (London: Thomas Marshe, [1566]); *Selected Epigrams of Martial*, trans. by Thomas May, STC 17,494 (London: [Humphrey Lownes] for Thomas Walkley, 1629); *Aulus Persius Flaccus his Satires*, trans. by Barten Holyday, STC 1977.5 (Oxford: Ioseph Barnes, 1616).

[38] *M. Val. Martialis epigrammaton libri*, ed. by Thomas Farnaby, STC 17,492 (London: Felix Kingston for William Welby, 1615); *Q. Horatii Flacci Venusini*, STC 13784 ([London]: [J. Kingston] for [Will]. Norton, 1574). On the use of editions of Martial in late sixteenth-century grammar schools, see *The Poems of Sir John Davies*, ed. by Robert Krueger and Ruby Nemser (Oxford: Clarendon Press, 1975), xxvi. John Tatham, an Oxford scholar, owned a copy of both Martial's *Epigrammata* and Persius' *Works*, each valued at 3 d. at his death in 1586. See the Folger Library's digitized 'Private Libraries in Renaissance England', last accessed August 19, 2020, https://plre.folger.edu/books.php and E. S. Leedham-Green, *Books in Cambridge Inventories: Book-lists from Vice-Chancellor's Court Probate Inventories in the Tudor and Stuart Periods*, vol. 2 (Cambridge: Cambridge University Press, 1986), 194, 436–438, 527–528, 610, 717.

such as *Pipere & Scombris* or 'waste paper', however brief or apparently marginal, became tightly coded vehicles for negotiating and representing the literary culture of the period. We will now progress to a consideration of these culturally and materially specific negotiations. After considering a number of early modern translations and editions of the classical trope, we will see how it was deployed in the context of controversial literature, in modesty and prefatory rhetoric, and in moments of textual self-reflexivity, when the author paused to consider the temporal and physical nature of the text. These relationships and concerns might, in the early modern period, be understood and experienced through the outline of a waste object that was at once everyday and replete with a rich literary history.

Classical references to repurposing books were reproduced in a range of sixteenth- and seventeenth-century texts. These included European (and later English) editions of the Latin works; printed and manuscript commonplace books containing the poems in Latin or in translation; and pedagogical works. Drant's 1567 translation of Horace renders the verses *deferar in vicum vendentem tus et odores | et piper et quidquid chartis amicitur ineptis* as 'wrytings which I thought | eternal' carried 'unto the mercers | shoppes, wher[e] francke incence is soulde, | And what soever spycerie | in waste paper is roulde'.[39] Holyday's 1616 translation of Persius asks if 'there breath[e] a man that can reiect | A gen'rall praise? and his owne lines neglect? | Lines worth immortall Cedars recompence, | Nere fearing new-sold Fish or Frankincense?'[40] Michel de Montaigne's *Essays*, first published in French in 1580, repeats Martial's *Ne toga cordyllis, ne penula desit olivis* (13.1) and Catullus' *Et laxas scombris saepe dabo tunicas* (95). In Florio's 1603 rendering of Montaigne's French translation we read of a text that might 'peradventure hinder the melting of some piece of butter in the market, or a Grocer from selling an ounce of pepper', with Martial rendered as 'least Fish-fry should a fit gowne want, | Least cloakes should be for Olives scant', and 'To long-tail'd Mackrels often I, | Will side-wide (paper) cotes apply'.[41]

Manuscript translations of selected poems from Martial include a small seventeenth-century quarto (British Library Add. MS 27,343) identified as the work Sir John Heath (1614–1691).[42] The rendering of 3.2 reveals the translator at work, weighing up certain synonyms (case/cover, dainty/fine) and syntactical choices:

> To whom wouldst bee presented? Booke?
> Quickly out for a Patron looke:

[39] *Horace*, trans. Drant, H1r.

[40] *Persius*, ed. Holyday, B4v.

[41] *The Essayes or Morall, Politike and Millitarie Discourses of Lo: Michaell de Montaigne*, trans. by John Florio, STC 18,041 (London: Val. Sims for Edward Blount, 1603), 385.

[42] On Sir John Heath, see *Same-Sex Desire in Early Modern England: An Anthology of Literary Texts and Contexts*, ed. by Marie H. Loughlin (Manchester: Manchester University Press, 2014), 244.

Else to th'black kitchin damn'd; or oyld
Thou'lt cover bee to mackrell broyld,
Or Peppers case ~~cover~~ or some such spice.
Fly to Faustinus, oh th'art wise.
There drenchd in Cedars oyle, thy brow
Sleekd with the Pumis-stone, mayst thou,
~~And in fine~~ In dainty purple cover putt
With paynted bosses, proudly strutt.
~~Swelling~~ And swell with scarlett tytle: There
not carping Probus ~~critic censures~~ you need feare.[43]

In the manuscript's rendering of 4.86, the poet tells his book that with Apollinaris' patronage:

Detractors jeeres thou needst nott dread,
Nor to bee apron to the Roast.
If nott: Tis true, thou mayst bee tost
Wast-paper; or, to th'shoppes, for ware;
Or boyes, to scrabble on, and teere.[44]

These varied translations make clear the ease with which a description of everyday life in classical Rome could be transposed to an early modern setting. Horace's *vendentem* become 'mercers shoppes', and the kitchens of both periods contain food wrapped in waste pages.

Thomas Farnaby's 1615 edition of Martial, the first printed in England, reveals this cross-period continuity especially strikingly. Epigram 3.2 is framed by a densely printed apparatus (see Fig. 2.1).[45] Glossing each line of the poem, this commentary situates Martial within his literary context: it points out that the opening of 3.2 is in *imitatio* of the opening of Catullus 1 (*Quod dono lepidum nouum libellum? etc.*); it directs the reader to Persius' first satire (*nec scombros metuentia carmina nec thus*); and, in the same marginal note, gives a neo-Latin paraphrase of the trope, *Ne damneris in culinam & aromatopolarum tabernas pro chartâ ineptâ, infelicis genii & fati* ('lest you be condemned to the kitchen & spice shops like silly paper, of unfortunate genius and fate'). This is something of a hotchpotch, borrowing Horace's *charta* [...] *ineptis* but also incorporating the *culinam* of Martial 3.2, a space suggested by any reference to pages wrapping fish. The final clause has a hint of Martial 6.61, proclaiming that a book needs *genium* to survive. The marginal commentary also relocates the poem to an early modern context with ease: one note highlights the role of Faustinus as *Assertorem, patronum*, hinting

43 British Library, Add. MS 27,343, 21r.
44 Ibid., 28r.
45 *Martialis*, ed. Farnaby, 73.

at the relevance of the verses to contemporary systems of patronage. These lines, as we will see, were of particular interest to authors wishing to seek out a new patron or address their own. *Cordyllas* is glossed as *Pisciculor, partus thunnorum* ('A little fish, the offspring of the tunny'), a fish more familiar to an early modern than a classical reader. Most strikingly, wedged between references to Persius and Horace in particularly large, black letter, we find the words 'wast paper' (see Fig. 2.1). This term leaps from the page not only because of its disproportionate size, but because it is the only vernacular term in a sea of Latin. This is unusual, as Farnaby's commentaries are usually composed entirely in the classical language; here, the editor seems to have found it useful, perhaps even irresistible, to render the process described in the poem in colloquial English. The cooking practice and the fate of books described by Martial is not just *like* an early modern practice: it is, the marginalia suggests, an ongoing one.

Farnaby isn't alone in finding continuity between the waste practices of early modern England and classical Rome. In his 1565 dictionary, Cooper references Martial 3.2 in his definition of *cucullus*, describing it in the present tense as 'a

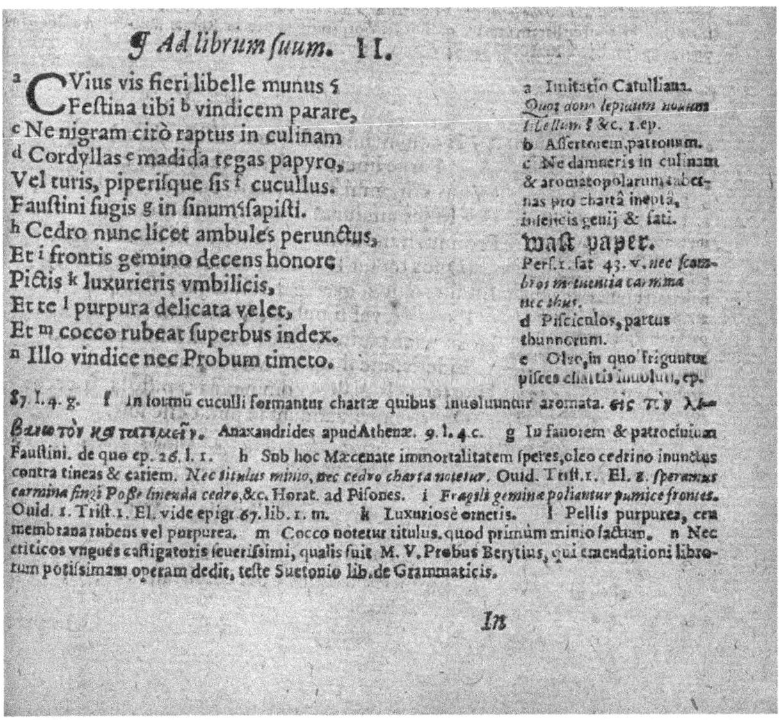

Ad librum ſuum. II.

a CVius vis fieri libelle munus ?
Feſtina tibi b vindicem parare,
c Ne nigram citò raptus in culinam
d Cordyllas e madida tegas papyro,
Vel turis, piperiſque ſis f cucullus.
Fauſtini fugis g in ſinum ? ſapiſti.
h Cedro nunc licet ambules perunctus,
Et i frontis gemino decens honore
Pictis k luxurieris vmbilicis,
Et te l purpura delicata velet,
Et m cocco rubeat ſuperbus index.
n Illo vindice nec Probum timeto.

a Imitatio Catulliana. *Quoi dono lepidum nouum libellum ?* &c. 1.ep.
b Aſſertorem, patronum.
c Ne damneris in culinam & aromatopolarum tabernas pro chartâ ineptâ, infelicis genij & fati. **waſt paper.** Perſ. 1. ſat 43. v. *nec ſcombros metuentia carmina nec thus.*
d Piſciculos, partus thunnorum.
e Oleo, in quo friguntur piſces chartis inuoluti, ep. 87. l. 4. g.
f In formā cuculli formantur chartæ quibus inuoluuntur aromata. [illegible]. Anaxandrides apud Athenæ. 9. l. 4. c.
g In fauorem & patrocinium Fauſtini. de quo ep. 26. l. 1.
h Sub hoc Mæcenate immortalitatem ſperes, oleo cedrino inunctus contra tineas & cariem. *Nec titulus minio, nec cedro charta notetur.* Ouid. Triſt. 1. El. 1. *ſperamus carmina fingi Poſſe linenda cedro*, &c. Horat. ad Piſones.
i *Fragili geminæ poliantur pumice frontes.* Ouid. 1. Triſt. 1. El. vide epigr. 67. lib. 1. m.
k Luxurioſè orneris.
l Pellis purpurea, ceu membrana rubens vel purpurea.
m Cocco notetur titulus, quod primùm minio factum.
n Nec criticos vngues caſtigatoris ſeuerisſimi, qualis fuit M. V. Probus Berytius, qui emendationi librorum potiſsimam operam dedit, teſte Suetonio lib. de Grammaticis.

In

Fig. 2.1 Detail from *M. Val. Martialis epigrammaton libri*, ed. by Thomas Farnaby, STC 17,492 (London: Felix Kingston for William Welby, 1615), fol. 73, Q*.13.12(F).

cornette of paper that Apothecaryes use'.[46] It was possible, therefore, for early modern editors and translators to render the process described by the Roman poets in a concise and current phrase: for Drant, Horace's *chartis ineptis* is best understood as 'waste paper', rather than any other synonyms or descriptors, such as personified 'silly' or 'foolish' sheets.[47] Similarly, Sir John Heath, in his manuscript translations, renders the *inversa* of the *charta* in Martial 4.86 as, simply, 'wast-paper'.[48] This does not, however, empty the trope of the meanings developed by these earlier poets: instead, it condenses them. The eye-catching phrase that sits in Farnaby's commentary, perhaps absorbed by the reader before any of the Latin verses, makes clear that this is a poem about 'wast paper'. It marks it out as relevant to anyone interested in this particular trope, directing them toward other classical examples and making it straightforward to extract and commonplace. The ambiguities and complexities of the trope are still available, but they are tightly coiled within this apparently mundane phrase.

4 Waste Paper Insults

How, then, was the waste paper commonplace deployed after its extraction? Typically, it was used in one of three ways: as an insult, to flatter a patron, or, closely related to this, as a modesty trope. It was particularly suited to use as a slur in controversial literature and so appeared with frequency in religious and political disputes. Crucially, the trope allowed an author to show off classical learning while simultaneously deploying a scurrilous and often scatological joke. Edmund Porter, for instance, took the time to translate Martial 3.2 at some length in his attack on the anonymous 1646 translation of John Crellius' *The Expiation of a Sinner*. He imagines that 'the height' of *The Expiation of a Sinner*'s 'preferment will be as *Martial* merrily writes of his own Poems':

> Make haste, and get a Patron, pretty-book,
> Before the Black guard, or the Master-Cook
> Snatch thee as waste-paper for his Kitchin,
> To put Spice, Sprats, Frankincense, or Pitch in.[49]

Rather than being read by learned divines, Porter suggests, *The Expiation* will be violently handled by either the 'Master-Cook' or the 'Black guard', a kitchen servant. These book users are only interested in the pages' capacity to fold and wrap,

[46] 'Cucullus' in Thomas Cooper, *Thesaurus linguae Romanae & Britannicae*, STC 5686 (London: Henry Wykes, 1565).
[47] *Horace*, trans. Drant, H1r.
[48] British Library, Add. MS 27,343, 28r.
[49] Edmund Porter, *Theos Anthropophoros*, Wing P2985 (London: for Humphrey Moseley, 1655), 12.

and how they might serve as a cornet or protect food from the heat of the oven. Porter adds 'Pitch' to the list of substances Martial's book might wrap, transferring the blackness of Martial's kitchen (*nigram* [...] *culinam*) onto both the kitchen servant and the paper in a process akin to what Miles Grier has termed 'the ink-face complex'.[50] This complex is a shorthand for the process by which the mark of blackness—whether tattooed or inked—denoted servility across a wide range of social and cultural groups. Porter's opponent (Crellius' translator) is embodied by his dismembered book and enters Grier's 'elastic racial category of illiterate, legible blacks'.[51] Rhetorically stained by the blackness of the kitchen environment, the translator is relegated to the status of the lowly blackguard or scullion, just as his book is transformed into fragments of waste with its ink overwritten and obscured by pitch and grease.

Most writers did not go to such scholarly and imaginative lengths to suggest that their enemies' books would end up as waste paper; most were satisfied with a marginal note or brief quotation suggesting that their target had written something so worthless that it would quickly be cast aside. Daniel Featley, for instance, issued a printed challenge to the Jesuit John Percy, claiming that Percy's most recent 'Replyer' in their lengthy dispute had not 'reply[ed] one word to the defence of my proceeding', and so 'for want to better imployment, *Ne toga cordylis, & penula desit oliuis*'.[52] Percy's pages, the quotation from Martial 13.1 implies, have failed in their intended 'imployment': they do not fulfil the criteria of successful disputation. They will, therefore, find 'imployment' elsewhere, providing gowns for fish and overcoats for olives. Several decades earlier, the anonymous author of the 1599 *Master Broughtons Letters* accused the titular Hugh Broughton, a Hebrew scholar, of 'spoyl[ing] cleane paper with base imployments'.[53] Like Percy's pages, Broughton's book will find alternative 'imployment', a prediction expressed via Horace. Broughton's *Explication of the Article peri tou katelthein eis haidou* (1599), on the descent of Christ into Hell, 'might be sent from the Printers presse to the Apothecaries shop, there to make cases for spices at the best, *Et piper & quicquid chartis amicitur ineptis*'.[54] Horace's vague description of 'everything else that is wrapped in sheets of useless paper' is well suited to use in an early modern context: it is flexible enough to encompass a range of contemporary uses not specified in earlier examples of the trope, like those listed by Robert Burton in the lengthy preface to *The Anatomy of Melancholy* (1621). In Burton's text, 'Democritus Junior' describes the symptoms of a 'scribling age', with 'putid [*putidus*,

[50] Miles P. Grier, 'Inkface: The Slave Stigma in the Early Imperial Imagination', in *Scripturalizing the Human: The Written as the Political*, ed. by Vincent L. Wimbush (Abingdon: Routledge, 2015), 193–220, esp. 207.

[51] Ibid., 195.

[52] Daniel Featley, *The Grand Sacrilege of the Church of Rome*, STC 10,733 (London: Felix Kyngston for Robert Milbourne, 1630), 282.

[53] Anon., *Master Broughtons Letters*, STC 3864 (London: [F. Kingston for] John Wolfe, 1599), 11.

[54] Ibid., 11.

stinking, ill-savoured] papers' found in 'every close-stoole and jakes [...] under pies, to lappe in spice, and [to] keepe rostemeat from burning'.[55] This catalogue of waste practices is accompanied by the marginal note, *Et quicquid chartis amicitur ineptis*. Burton aligns his catalogue of mundane stuff with the classical tradition, positioning it as another instance of Horace's 'silly paper'.

These insults form part of the aesthetic of railing, reviling, and invective that Maria Teresa Micaela Prendergast has identified as characterizing late sixteenth- and early seventeenth-century literary culture. As Prendergast argues, moralizing texts have a tendency toward rhetorical indulgence and are more often sensationalist and 'seductively scurrilous' than, as one might expect, reformative or didactic.[56] Because scholarship has focused on individual controversies, Prendergast points out, the shared qualities of this 'anti-genre' have been overlooked.[57] The waste paper insult is one such shared quality, and in equating the author's body with the author's book, controversialists imagine exacting textual violence and dismemberment on their opponents.[58] At the same time, the ferocity and linguistic excess of the waste insult highlights the vulnerability of an author's own book, alongside an opponent's. Condemning the work of a rival as badly written, illogical, or blasphemous, meant, in this 'scribling age', drawing attention to the broader life cycle of texts and their progression along a trajectory that seemed increasingly likely to include transformation into waste.[59] Waste paper insults, like Democritus Junior's catalogue of uses for waste, are part of a chorus voicing a wider sense of crisis and cataclysmic change in the textual culture of the period. In the late sixteenth century and throughout the seventeenth, writers often noted that the number of books and paper sheets in circulation increased year on year. Books seemed to be becoming increasingly brief, topical, and flimsy. Sold unbound and often remaining so, or stitched within ephemeral (often waste) wrappers, these texts seemed particularly vulnerable to disintegration and transformation into waste.[60] James Aske, for example, in his 1588 quarto, attacks 'the damned practizes' of 'the divelish popes of Rome', claiming that 'the Booke-binders shops, and every Printers presse are so cloyed and clogged with Bookes of these and such-like matters'.[61]

[55] Robert Burton, *The Anatomy of Melancholy*, STC 4159 (Oxford: John Lichfield and James Short for Henry Cripps, 1621), 8.

[56] Maria Teresa Micaela Prendergast, *Railing, Reviling, and Invective in English Literary Culture, 1588–1617: The Anti-poetics of Theater and Print* (New York: Routledge, 2012), 2–5.

[57] Ibid., 10.

[58] On this embodied, textual violence, see Jason Scott-Warren, 'Nashe's Stuff', in *The Oxford Handbook of English Prose, 1500–1640*, ed. by Andrew Hadfield (Oxford: Oxford University Press, 2013), 205 and Joad Raymond, *Pamphlets and Pamphleteering in Early Modern Britain* (Cambridge: Cambridge University Press, 2003), 97.

[59] Burton, *The Anatomy of Melancholy*, 8.

[60] On temporary binding practices, see Knight, *Bound to Read*, 3–4.

[61] James Aske, *Elizabetha Triumphans*, STC 847 (London: Thomas Orwin for Thomas Gubbins and Thomas Newman, 1588), A3r–v.

Gesturing toward a now familiar trope, he suggests that 'such-like' topical books are 'good for nothing (as they say) but to make wast-paper'.[62]

In addition to abusing an opponent, then, this trope was a vehicle through which to encapsulate an epoch in which thousands of pages, rapidly written and printed and turning out of date, seemed destined to become waste. It was, therefore, both a characteristic of controversial prose and a characterization of it. A use of the waste paper trope by John Bramhall, for instance, is framed as an insult, but also offers an insight into the early modern understanding of what constituted a book and the likelihood of its survival. Bramhall, a Royalist bishop, accused the Parliamentarian writer Henry Parker of failing to treat *His Majesties Answer to the Nineteen Propositions of Both Houses of Parliament* (1642) with sufficient respect: according to Bramhall, Parker terms this royal work 'Papers (that's his phrase) as if they were old Almanacks out of date, fit for nothing but to cover Mustard pots, *metuentia carmina scombros aut thus*'.[63] But the work that Bramhall attacks—Parker's *Observations Upon Some of His Majesties Late Answers and Expresses*—does not in fact claim that the King's publication will or should be used to stop mustard pots, or that it will only be relevant for as long as an ephemeral almanac. The reference to Persius and contemporary waste practices is Bramhall's own addition. Parker does, however, call *His Majesties Answer* 'papers', a word which seems innocuous enough to a modern reader but that evidently incensed Bramhall.[64] Bramhall uses a curtailed version of Persius' 'poems that fear mackerels and incense' (in Persius, the poems do *not* fear such stuff) to negotiate different ways of understanding the nature of a text: the description of the King's *Answer* as 'Papers' is conceived as disrespectful because it suggests that they are temporary, or as Burton puts it, 'putid', tending toward grossness and decay.[65] Elizabeth Yale has explored the expansive textual category of papers, outlining that, in the seventeenth century, they

> typically included loose sheets, notes from experiments and observations, commonplace books, correspondence, and drafts of treatises. [...] They could be classed or stored with pamphlets, unbound books, and other loose printed material. 'Papers' could also refer to bound (though not printed) books.[66]

This taxonomy is unstable, but to describe a printed book as 'papers' is a coded insult: manuscript notes, letters, and drafts were particularly vulnerable to loss, dispersal, and destruction, as Yale's study of seventeenth-century scientific archives has made clear. Although 'papers' might be bound, the term suggests a textual

[62] Ibid., A3r–v.

[63] John Bramhall, *The Serpent Salve*, Wing B4236 ([S.L: S.N], 1643), 105.

[64] Henry Parker, *Observations Upon Some of his Majestie Late Ansvvers and Expresses*, Wing P413 ([London: s.n., 1642]), passim.

[65] Burton, *The Anatomy of Melancholy*, 8.

[66] Elizabeth Yale, 'With Slips and Scraps: How Early Modern Naturalists Invented the Archive', *Book History* 12 (2009): 4.

object that sits at the opposite end of the scale to a weighty volume that is expected to endure, bound and shelved for future reading. Classifying an object as 'papers' means that it will be treated accordingly, and 'papers' are, Bramhall suggests, as short-lived as almanacs, liable to be used as mustard-pot-stoppers and wrappers for fish and incense once out of date.

The Jesuit William Lacey also employs the waste paper trope as a means of exploring this textual taxonomy. His *Judgment of an University-Man Concerning M. William Chillingworth* (1639) is a response to Chillingworth's scandalously tolerant *The Religion of Protestants* (1638). In it, Lacey describes the tendency of 'Protestant writers' to lazily term books written by Catholics as 'Popish Pamphlet[s]', and with this 'phrase of degradation', to consider the argument as good as won.[67] '[H]ow many rhemes of paper', Lacey asks, 'may vindicate a Booke from the contempt of a Pamphlet?'.[68] Lacey highlights the superficiality of valuing a book by its length. In his opinion, 'a large volume *in folio*' can contain as little of worth as the shortest work, being 'a Pamphlet in substance', and the shortest works 'may contain the waight & worth of an ample volume'.[69] This sceptical consideration of the distinction between 'Pamphlet[s]' and 'volume[s]' concludes with Lacey's use of the waste paper trope: Chillingworth's text is one such intellectually insubstantial but physically ample folio. It will only be considered 'usefull' by 'Glovers and Grossers', 'where they shall be valued according to there bulke, *Ad vicum vendentem thus & odores | Et piper, & quicquid chartis amicitur ineptis*'.[70] Echoing the sentiment of Catullus' Carmen 95, Lacey uses Horace to query the accepted nexus of value in early modern England: pamphlets are not worthless because they are small and of passing relevance, and volumes are not valuable because they are large and physically enduring. The assumption that pamphlets tend quickly to waste paper and so are valued in terms of their raw materials is turned on its head: Chillingworth's volume is especially appealing to 'Glovers and Grossers' because of its 'ample' number of pages.[71]

As extant binding waste and Henry Fitzgeffrey's plea to his binder not to be bound 'in the *Folio*: or the *Quarto cut*' makes clear, books of 'the *Smallest size*' were indeed less likely to be 'eaten vnder *Pippin-pyes* | Or in th' *Apothicaryes* shop be seene | To wrapp Drugg's: or to drye *Tobacco* in'.[72] Folio pages were, because of their larger size, more useful as wrappers. Waste paper practices, therefore, undermine the assumption that intellectual and physical insubstantiality went hand in hand. The idea that readers would waste their little pamphlet because they

[67] William Lacey, *The Judgment of an University-Man Concerning M. William Chillingvvorth his Late Pamphlet*, STC 15117 ([Saint-Omer]: The English College Press, [1639]), 11.

[68] Ibid., 11.

[69] Ibid., 11.

[70] Ibid., 11.

[71] Ibid., 11.

[72] Henry Fitzgeffrey, *Satyres: and Satyricall Epigrams*, STC 10,945 (London: Edward Allde for Miles Patrich, 1617), G4r.

were unhappy with its textual contents is revealed, at least in part, as a rhetorical construction: practical considerations, such as the size and quantity of the paper in question, were much more important to early modern grocers, glovers, and cooks than literary or theological quality. A brief glance at the insults in several controversial texts, then, makes clear that claiming a book is destined to end up as waste could signify a surprising range of things. In addition to embellishing texts with intellectual insults, the classical trope offered a frame through which to consider the classification and the intellectual and material value of books, as well as to express anxiety regarding their growing number. What is more, usage of the waste paper trope was not limited to these attacks on opponents. Instead, early modern references to repurposed pages make full use of the rhetorical range present in the classical tradition, extending, as we will see, to the sorts of modesty tropes and exuberant playfulness that we find in Martial's verse.

5 Waste, Modesty, and Patronage

In addition to the widespread use of the waste paper commonplace as an insult, many early modern writers employed the trope as a means to either profess modesty or to proclaim the power a potential patron might wield over their work. A wide range of early modern authorial gestures fall within the framework of 'modesty topoi': these include courtly 'sprezzatura' and the artful disavowal of literary skill, as well as the various negotiations of authorship and the 'stigma of print' by writers in the literary marketplace.[73] Patricia Pender and Wendy Wall have explored the relationship of these gestures to women's writing in the early modern period, highlighting the range of meanings available in these often overlooked tropes. Pender outlines the lengthy literary history of the rhetoric of modesty, arguing that although these tropes were 'already and obdurately conventional' and 'exhausted' in the early modern period, they 'do not all say the same thing'.[74] Rather than taking them at face value, Pender suggests, we should appreciate their 'literary' value.[75] Wendy Wall, for instance, has argued that the 'stigma of print' was a rhetorical framework for 'articulating' emerging conceptions of authorship and the commodification of the book: it was a means of negotiating the 'collision between manuscript and print practices' and 'aristocratic amateurism and the marketplace'.[76] Early modern writers, Wall has shown, frequently performed reluctance to enter into print, particularly in the prefatory epistles of their books.

[73] See, for instance, 'Topoi', in *The Spenser Encyclopedia*, ed. by A. C. Hamilton (Toronto: University of Toronto Press, 2003).

[74] Patricia Pender, *Early Modern Women's Writing and the Rhetoric of Modesty* (Basingstoke: Palgrave Macmillan, 2012), 3, 11–12.

[75] Ibid., 11–12.

[76] Wendy Wall, *The Imprint of Gender: Authorship and Publication in the English Renaissance* (Ithaca: Cornell University Press, 1993), 3–4.

For Michael Saenger, such 'front matter' is best understood as an early form of advertising, a clever marketing ploy that promotes the purchase of a book through 'inventive' and 'artistic methods' including, among others, the humility topos.[77] Descriptions of a waste paper fate, intriguingly, are often found amongst these modesty topoi, suggesting a rhetorical, and perhaps even financial, motive for their inclusion.

Up-and-coming clergyman Lancelot Dawes, for instance, employed the waste paper trope in the prefatory material to his 1609 printed sermon, *Gods Mercies and Ierusalems Miseries*. Newly appointed as vicar of Barton Kirk, Dawes dedicated the work to Henry Robinson, Bishop of Carlisle. The epistle is brief and pithy, written in delicately balanced, almost euphuistic prose and professing a conventional resistance to print. 'This Sermon', Dawes writes, 'was made for the Crosse, not intended for the Presse. I was by authoritie commanded the former, and by importunitie of many have at length consented to the latter'.[78] By 1699, a dictionary of 'canting' terms labelled the 'Importunity of Friends' as 'the stale Excuse for coming out in Print, when Friends know nothing of the Matter', revealing the frequency with which this transparent claim was deployed.[79] Apparently, though not necessarily convincingly, unwilling to publish his sermon, Dawes is modest about its potential for success in the literary marketplace and nonchalant about the fate of the paper objects that will emerge from the press. The book is, in the classical style, anthropomorphized:

> It may perchance heereafter complaine with the *Satyrist.*
> *Deferat in vitem vendentem thus, at odores,*
> *Et piper, et quicquid chartis amicitur ineptis.*
> But that is no great disgrace in these daies, it shal have Store of company.[80]

Dawes professes not to care if his sermons end up wrapping frankincense, perfume, and pepper in the shops of London; like Robert Burton writing a decade later, Dawes suggests that the city is already full of waste pages. Reference to 'the *Satyrist*' Horace, therefore, does a number of things in Dawes' short dedicatory epistle: with it, he displays appropriate decorum and modesty when publishing a religious work, as well as an ability to conform to the 'firmly established' conventions of entering into print.[81] Furthermore, the classical trope provides a frame

[77] Michael Saenger, *The Commodification of Textual Engagements in the English Renaissance* (Ashgate: Aldershot, 2006), 1–3, 55–57.

[78] Lancelot Dawes, *Gods Mercies and Ierusalems Miseries*, STC 6388 (London: [John Windet] for Cle. Knight, 1609), A2r–v.

[79] 'Importunity of Friends' in B. E., *New Dictionary of the Terms Ancient and Modern of the Canting Crew*, Wing E5 (London: for W. Hawes, P. Gilbourne and W. Davis, 1699).

[80] Dawes, *Gods Mercies*, A2r–v.

[81] Brayman Hackel, *Reading Material*, 103.

through which Dawes can showcase his wit and literary nous to an influential potential patron, the Bishop of Carlisle.

Joad Raymond has argued that by the seventeenth century, the 'stigma of print' had mellowed into a 'different kind of stigma'.[82] Modesty topoi became a means to 'deny' that an author's 'move to print was a sign of pride', because for the average writer print was no longer considered a 'socially inferior' alternative to manuscript circulation.[83] Raymond's chronology is supported by the fact that, at the turn of the seventeenth century, modest uses of the waste paper trope were ripe for parody and creative play. In his 1615 *Defence of Trade*, Dudley Digges offers his book for use as waste paper, but with a knowing wink to the reader who is already over-familiar with this rhetoric of modesty. Digges' work is a response to *The Trades Increase*, a critique of trade in the East Indies printed earlier that year, and it is presented as a letter from Digges to Sir Thomas Smythe, governor of the East India Company. It styles itself, as Dawes does his sermon, as not having been intended for print. In a 'Post-script to the Reader', Digges describes how he was 'farre from the ambition of [the readers'] acquaintance' and that both his own work and the work it responds to are 'a couple of Inke-wasting toies', to be used about '(if you list) *piperi & scombris*, that *Trades Increase* to packe up fish, and this *Defence of Trade* to wrappe up spice'.[84] Despite its brevity, Digges' use of the trope is multi-layered: the *Trades Increase* is largely concerned with England's herring trade, and Digges' *Defence of Trade* is concerned with the East India Company and the commerce in pepper and spices. This is, then, like Martial's *Xenia*, a concrete joke. In this instance, the modesty-cum-waste paper trope at once parodies an overused rhetorical device and frames its pages as an unusually appropriate wrapper for pepper and mackerel. The location of Digges' densely meaningful reference to waste paper is not unusual, though it provides a useful illustration of the placement of most instances of the trope: found primarily in prefatory epistles, postscripts, and printed marginalia, early modern writers most often referred to waste within the paratextual spaces of their texts. As well as being the sites in which a writer most commonly considers the materials and practices of producing and consuming texts, paratexts are the most physically vulnerable parts of a book. Like the beginning of a book-roll, the first and last pages of the volume were most likely to be damaged, hence the widespread use of (often waste) pastedowns and flyleaves in early modern bindings. Similarly, the margins that often contained references to waste were regularly trimmed when a book was rebound. What is more, the flyleaves and margins are also the spaces of the book in which a reader is most likely to annotate or doodle, and so are already waste spaces of a sort. Digges' postscript is a rather precariously attached, unpaginated folio at the end of *The Defence of*

82 Raymond, *Pamphlets and Pamphleteering*, 57.
83 Ibid., 57.
84 Dudley Digges, *The Defence of Trade*, STC 6845 (London: William Stansby for Iohn Barnes, 1615), I1r–v.

Trade. The material make-up of the book, therefore, performs the vulnerability described within its paratexts.

Heidi Brayman Hackel has highlighted the importance of the tropes found in these paratextual spaces in 'signal[ling] concerns about interpretation and [...] explicitly direct[ing] readers' experiences of a text'.[85] She points out that references to 'base and scatological uses for printed paper' are part of this conventional attempt to prescribe a reader's interaction with the book: the preliminary material seeks to prescribe a 'gentle', rather than violent, reading and handling illustrated, best of all, by the trope of a book's transformation into waste.[86] As Brayman Hackel also notes, by the turn of the seventeenth-century these paratexts often came as a pair, with one addressed to a general readership and another to an individual dedicatee. We can, therefore, add to this desire to prescribe the actions of a general readership the desire to prescribe the actions of a very specific reader—the potential patron of the text. The clergyman Thomas Phillips' 1639 *Booke of Lamentations* richly illustrates this desire. The volume contains two sermons, one with a dedicatory epistle addressed to Thomas Coventry, and another to Sir Miles Sandys. Martial, intriguingly, is a significant presence in both dedicatory epistles. The first describes how 'books have a bounded; flying from the presse into the world', and how this particular 'little manual or *Enchiridon*' hopes to 'obtaine' Coventry's 'approbation', and to 'shroud himselfe under [his] protection and patronage'.[87] The text is frequently interrupted by superscript letters that correspond to quotations, predominantly from Martial, in the margins: these include reference to epigrams 1.2, 1.3, 4.86, and, in the second dedication, to 7.85, 3.2, 1.2 (again), and 2.1.[88]

Despite the range of epigrams he draws upon, Phillips does not replicate Martial's playfulness and polyphony: the portrayal of patronage in *The Booke of Lamentations* conforms to the framework laid out in Martial 3.2 (to Faustinus) and 4.86 (to Apollinaris) exclusively. In these epigrams, the 'little book' is instructed to fly to the *sinum*, or bosom, of a protector. If it fails to find such shelter—*sinum* literally means the fold of a toga, or pocket—it will end up serving as wrappers in kitchens and merchants' shops (3.2). Phillips' allusion to these epigrams is at once an earnest bid for patronage and a symptom of the contemporary perception of a world brimming with loose papers and ephemeral little books. As Phillips himself notes, his duodecimo volume is one of many little books that 'have a bounded; flying from the presse into the world'.[89] 'A bounded' is suggestive here, indicating

85 Brayman Hackel, *Reading Material*, 88.

86 Ibid., 80–81.

87 Thomas Phillips, *Booke of Lamentations*, STC 19878.5 (Cambridge: J. D[awson] for P[eter] Cole, 1639), A3r–4v. On the structures of patronage in the early modern period see *Patronage in the Renaissance*, ed. by Guy Fitch Lytle and Stephen Orgel (Princeton: Princeton University Press, 1981), especially Werner L. Gundersheimer, 'Patronage in the Renaissance: An Exploratory Approach', 3–24.

88 Phillips, *Booke of Lamentations*, A3r–4v, G4r–5r.

89 Ibid., A3v–4r.

that, in a world of small, flimsy, or temporarily wrapped books, a patron might be fruitfully conceived of as akin to a bookbinding. Phillips hopes that Coventry will act as a 'shroud', a cover or wrapper that, like a *sinum* or pocket, might protect the book from destruction, dispersal, and repurposing.[90] One copy of *The Booke of Lamentations*, now held at Cambridge University Library, neatly encapsulates the desires, anxieties, and material practices at work in Phillips' dedicatory epistle: preserved in a library and with its seventeenth-century binding intact, Phillips' text has been successfully shrouded, though Coventry's role in this achievement is most likely minimal. The binding does, however, contain a material reminder of the potential fate that Phillips' volume both invokes and, in this instance, avoids: its flyleaves are pages removed from another printed text of the same decade, a 1631 school-text copy of Ovid's *Metamorphoses*.[91]

It is difficult to determine if there is any actual correlation between the lack of a patron and the likelihood of a text being transformed into waste. Claiming that an individual has the power to prevent such a fate is most likely, above all else, a form of flattery, and one that allows an author to demonstrate their classical education and rhetorical sophistication. At the same time, waste paper practices were a very real threat to the integrity and survival of early modern books. Anxieties regarding survival were widespread in an environment of frequent waste paper use, and might be expressed through learned carelessness, as in the case of the clergyman Dawes, or through the scholastic obsequiousness of Phillips. The prefaces and postscripts explored in this section constitute only a small fraction of the texts that employ the waste paper trope in the sixteenth and seventeenth centuries. In addition to revealing the frequent commonplacing of the classical trope and the way in which the waste paper fate haunted so many early modern writers, these instances also provide some insight into how these writers negotiated the demands of the print marketplace. The pseudonymous Uthalmus, as we have already seen, satirized the habit of addressing unknown patrons in the preface of *Fasciculus Florum*. According to Uthalmus, it is only 'Good Acceptance' in the literary marketplace that can prevent the fate of *Piperi & scombris*.[92] Both printed in the late 1630s, Thomas Phillips' and Uthalmus' distinct approaches to prefatory epistles demonstrate the various models of patronage available in these decades.[93] Although it might seem 'exhausted' through frequent use, waste paper was part of what Michael Saenger terms the 'system of rhetorical energies' that

[90] Ibid., A4v.

[91] *Ovidii Nasonis Metamorphoseon Libri XV*, STC 18,954 (Cambridge: *Ex Academiae celeberimmae typographeo*, 1631), fols. 121–122 and 143–144 have been used as flyleaves in a 1639 edition of Phillips' *Book of Lamentations*, STC 19878.5 (Cambridge: J. D[awson] for P[eter] Cole, 1639). Cambridge University Library, Syn 8.63.181.

[92] Uthalmus, *Fasciculus Florum*, A3r.

[93] Recent scholarship has worked to complicate the narrative of traditional systems of patronage as having been replaced by the print marketplace in the late sixteenth and seventeenth centuries. Instead, it has been demonstrated, the understanding of patronage 'expanded' as the period progressed. See

allowed authors to negotiate the pressures of traditional systems of patronage and the print marketplace, as well as a vehicle through which to articulate anxieties over the vulnerability of books.[94] It now remains to broaden out to a consideration of the imaginative framework within which these waste paper insults and modesty tropes so frequently occurred.

6 Waste Monuments

Whether limited to a marginal note or pervading the main body of a literary work, waste paper marks a writer's consideration of textual survival. Insults and bids for patronage claim that survival is dependent on something controllable, be it the literary, theological, or polemical quality of the text; the protection of a patron; or the text's success in the broader literary marketplace. John Webster, for instance, in the preface to his 1612 edition of *The White Devil*, hints that he has ensured textual survival through his literary skill. After the play's notorious first performance in 'so dull a time of Winter' and 'so open and blacke a Theater', which lacked 'a full and understanding Auditory', Webster submits it to 'the generall view with this confidence. *Nec Rhoncos metues, maligniorum, | Nec Scombris tunicas, dabis molestas*' ('you will not fear the sneers of the ill-disposed nor supply mackerel with "tiresome tunics"', Martial 4.86).[95] After praising contemporaries including Ben Jonson and William Shakespeare, Webster affixes an epigraph from Martial that refers ambiguously to either Webster's own play or those of his aforementioned peers: *non norunt, Haec monumenta mori* ('These are the only memorials that cannot die', Martial 10.2).[96] Here, Webster creates a binary of waste and monument, of unworthy texts that will be repurposed and grand texts that will endure.

Webster's reference to memorials, or *monumenta*, that cannot die, also recalls Horace's widely repeated Ode 3.30, in which the poet claims to have created 'a monument more lasting than bronze' (*exegi monumentum aere perennius*). According to Andrew Hui, Horace's 'immortality topos' provided the foundation for a 'poetics of ruins' in Renaissance literature.[97] Confronted by the architectural detritus of the Roman and monastic worlds, sixteenth- and seventeenth-century writers sought to craft 'more enduring artifact[s]'.[98] We might supplement

David M. Bergeron, *Textual Patronage in English Drama, 1570–1640* (Ashgate: Aldershot, 2006), 5, 12–13 and Helen Smith, *Grossly Material Things: Women and Book Production in Early Modern England* (Oxford: Oxford University Press, 2012), esp. 69–86.

[94] Saenger, *The Commodification of Textual Engagements*, 3.

[95] John Webster, *The White Diuel*, STC 25,178 (London: N[icholas] O[kes] for Thomas Archer, 1612), A2r.

[96] Webster, *The White Diuel*, A2v.

[97] Andrew Hui, *The Poetics of Ruins in Renaissance Literature* (New York: Fordham University Press, 2016), 31.

[98] Ibid., 2.

Hui's sense of a cultural 'yearning for timelessness', articulated through Horace's aphorism, with Lindsay Ann Reid's understanding of early modern 'bibliofictions', or literary representations of the life cycles and circulation of books. As Reid argues, these fictions depict a literature that 'exists simultaneously in the physical world and the supraliminal realm of imagination', and both Reid and Hui argue for an early modern understanding of textual immortality grounded in 'mutability', or the capacity of books to be reread, transformed, translated, and adapted.[99] This is, perhaps, what Webster has in mind in citing Martial's enduring *monumenta*. It is, Hui argues, what underpins Shakespeare's description, in his *Sonnets*, of textual rather than sexual reproduction. Shakespeare's 'solution' to the passage of time is no longer 'that this nameless beloved must have children', but that his poetry will endure as 'a living record of your memory'.[100] Reworking the Horatian aphorism that neither 'marble, nor the gilded monuments | Of princes shall outlive this pow'rful rhyme', Shakespeare develops an understanding of poetry as elastic and enduring, constantly evolving, in Hui's words, 'in the lived hermeneutic experience of the reader'.[101]

The bibliofiction of mutable endurance was not, however, the only bibliographic narrative current in the period. As we have seen, Reid characterizes this idea of transformative textual transmission as a 'bibliogenesis', and the bibliofiction of waste paper might be usefully understood as its opposite—as a narrative concerning the life cycle of texts grounded in their endings and entropy, rather than their generation, and in alternative, more destructive, modes of circulation. Gillian Russell's scholarship on eighteenth-century ephemera is especially useful here as it demonstrates that the binary of textual ephemera and the literary 'realm of "durable" transcendent value' was not yet as fixed and stable as it would become in the Augustan age.[102] Whereas the waste paper fate could, in the eighteenth century, be relatively comfortably consigned to the output of, for instance, Grub Street hacks, acts of repurposing haunted all manner of sixteenth- and seventeenth-century writers and texts. We might, therefore, conceive of the waste paper commonplace as a readily available counternarrative to bibliogenesis and the immortality topos: waste is at once a physical fragment or memorial of a text that has not survived, and a haunting idea that brings into doubt the capacity of any text to survive indefinitely. As often as writers invoked ideas of immortality, repeated rereading, and posterity, writers deflated such hopes by referencing the seemingly indiscriminate fate of fragmentation and an altogether different form of circulation and reception as waste. Both of these divergent fates are present,

[99] Ibid., 3, 31, 227; Reid, *Ovidian Bibliofictions*, 2–6.
[100] Hui, *Poetics of Ruins*, 9–11; Shakespeare, 'Sonnet 55', quoted by Hui.
[101] Hui, *Poetics of Ruins*, 11.
[102] Gillian Russell, *The Ephemeral Eighteenth Century: Print, Sociability, and the Cultures of Collecting* (Cambridge: Cambridge University Press, 2020), 40.

counterintuitively, in the classical quotations in Webster's preface to *The White Devil*.

In addition to early modern literature, anxieties regarding this indiscriminate vulnerability are frequently found within religious texts. This is because of the not unfounded perception that such religious texts were more likely to be used as waste than any other sort of book. Many sermons and devotional works refer to 'the *common Destinie* that hangs over the *Most writings of this Kind*', and this sense is, in part, borne out by the material record:[103] religious texts are present as binding waste in extant books far more often than any other genre or category of text. This was not, though, because of irreligiousness or a lack of devotional appetite as religious writers often claimed, but rather the opposite: religious texts dominated the early modern book market, and so the large quantities of sermons, devotional poetry, theological treatises, books of Psalms, and prayer books that left the printing presses and sat in booksellers' shops and warehouses provided a significant proportion of the period's stock of waste paper. Strikingly, sheets from the Bible are found as binding waste far more frequently than those from any other individual work, and so the word of God was often encountered dismembered and pasted onto boards, peeking out of the spine of other, far less sacred, books (see Fig. 3.1). Again, this was because of the practicalities of the book trade: because so many Bibles were printed in the sixteenth and seventeenth centuries, there were a large number of misprinted and surplus Bible sheets available for use as waste, as well as unsold copies made redundant by newer editions. Misidentifying the reason behind this high volume of religious waste—wilfully or otherwise—many religious writers proclaimed that their contemporaries no longer purchased and read, and instead threw away, good and holy books.

Furthermore, these references to holy books turned waste situate the life cycle of the book within a very particular framework. In these descriptions, or bibliofictions, the book pursues a narrative trajectory that was foundational to early modern thought and one propagated by the religious texts themselves. Waste suggests the eschatological narrative of the decline and decay of the mundane world, and, according to many early modern thinkers, its imminent end.[104] In 1626, William Vaughan, promoter of the colonization of Newfoundland and a passionately anti-Catholic polemicist, composed a peculiar text that maps out this eschatological bibliofiction. In *The Golden Fleece*, Vaughan articulates his colonial and anti-Catholic preoccupations through an unorthodox romance narrative.

[103] John Andrewes, *The Brazen Serpent*, STC 591 (London: G[eorge] P[urslowe] for Thomas Thorpe, 1621), A4r.

[104] On early modern perceptions of the end of the world, see Bernard Capp, 'The Political Dimension of Apocalyptic Thought', in *The Apocalypse in English Renaissance Thought and Literature: Patterns, Antecedents, and Repercussions*, ed. by C. A. Patrides and Joseph Wittreich (Manchester: Manchester University Press, 1984), 93–124, and Warren Johnston, *Revelation Restored: The Apocalypse in Later Seventeenth-Century England* (Woodbridge: The Boydell Press, 2011), 1–16.

Early on, the character Sir William Alexander, a writer, asks his companion, Master Elveston:

> Doe we not find by experience that the Bookes of many rare Divines lye on the Stationers hands, as it were moth-eaten, or inverted to base Offices, and solde for wast leaves to Apothecaries, to Glovers, Cookes, and Bakers [...] what then shall become of my Bookes, which I have alreadie published to the World?[105]

Surveying a topsy-turvy urban environment in which books are 'inverted' from use to misuse, Alexander fears for the future of his own works. His companion Master Elveston is, though, more optimistic. He is certain that Alexander's '*Bookes* shall never bee put to such vile and servile uses; nor any lively monument, which issues from a well tempred braine'.[106] The phrase 'lively monument' recalls the poetics of endurance identified by Hui and alludes to the Horatian immortality topos. The monuments are 'lively', Elveston's phrasing suggests, because they will be rejuvenated by the 'artistic process of continuous transmission, translation, and transformation' that Hui describes.[107] Alexander, however, immediately deflates this bibliofiction and replaces it with one of disintegration and decay. For Alexander, there is no 'mysticall *Receit*', or remedy, for ephemerality.[108] Instead, articulating a peculiar analogy, he argues that '*Bookes*' have 'their Destinies as well as Common-wealths'.[109] His books have a life cycle, he claims, but unlike the life cycle of the godly, this will not end in 'aeternall life'.[110] The same goes, according to Alexander, for nations, and for all material things: 'Must not all things under the Sunne wax old, fraile, and faile at last? *Senescente mundo consenescunt omnia*', he asks, misquoting Roger Bacon's *senescente mundo consenescunt homines*, 'the world growing old, men grow old as well'.[111] In fact, Alexander continues, good books are more likely to be turned to waste paper than ever before because 'the neerer we are to the end of the world, the more childish and doting is the judgement of the wisest man'.[112] Alexander, in Vaughan's romance, is resigned—like the stoic Persius before him—to the fate of textual disintegration. This disintegration is, after all, only part of a larger pattern of earthly entropy.

Not all early modern poets, therefore, share the 'metaphysical desire for immortality', which is, according to Hui, 'a nobler version of our basic survival instinct'.[113] Instead, we should keep in mind the pervasive effect of the early modern aesthetic

105 Vaughan, *The Golden Fleece*, 8.
106 Ibid., 8.
107 Hui, *The Poetics of Ruin*, 227.
108 Vaughan, *The Golden Fleece*, 8.
109 Ibid., 8.
110 Ibid., 8.
111 Ibid., 8; see *The Works of Francis Bacon*, vol. 2, *Philosophical Works*, ed. by James Spedding, Robert Leslie Ellis, and Douglas Denon Heath (Cambridge: Cambridge University Press, 2011), 148.
112 Vaughan, *Golden Fleece*, 9.
113 Hui, *Poetics of Ruin*, 51.

of repeated rumination over death and dissolution. This aesthetic is perhaps more difficult for us, as modern readers, to understand, but it was central to an early modern understanding of the world and an individual's place within it. Far from a despairing nihilism, contemplating the fragmentation and dissipation of matter was a legitimate and laudable devotional pastime. The thriving trade in mourning rings, burial shrouds, and the endless texts and trinkets that invited viewers to dwell on their potentially imminent death demonstrate that, in this period, narratives were as, if not more likely, to end in disintegration as regeneration or continuance.[114] As Joshua Calhoun has argued, early moderns were highly sensitive to the fact that 'books are composed of once-living organisms and that those once-living organisms are now actively, if very slowly, decomposing'; as a result, paper objects readily participate in the pervasive culture of *memento mori*.[115] This gradual decay, or 'bio-deterioration', as Calhoun terms it, is rapidly and violently accelerated by the transformation of a text into waste.[116] The 'lively monument' of textual transmission, within this context, becomes haunting fragments that rapidly disintegrate into only partially legible snippets of text.[117] For writers such as Vaughan, waste paper is both a symptom of the eschatological progression of the world, and a consequence of it. Sensitivity to the bibliofiction of waste becomes, therefore, even more pressing in the early modern imagination when, as it so often did, it becomes entangled with the question of scriptural survival and the theological chronology of endings and decay. No monument can endure, whether paper, bronze, or stone, beyond the end of times, and the repurposed pages of Bibles and sermons peeking from the edges of bindings makes visible, for many early moderns, the imminence of this ending. Although the majority of the early modern authors who deploy the waste paper trope cited so far are relatively obscure, there is one particularly canonical author whose works persistently dwell on the disintegration of paper alongside the disintegration of the world and everything within it: the writings, both religious and secular, of the poet and divine John Donne.

7 John Donne's Paper Carcasses

As John Carey notes, Donne was 'obsessively worried about being scattered and fragmentary' and, at the same time, 'wanted, and invented, a universe [...] in which all things were continually on the edge of nothingness'.[118] Similarly, Achsah Guibbory argues that 'the idea of universal, inevitable decay [...] profoundly affected

[114] On *memento mori* culture, see Nigel Llewellyn, *The Art of Death: Visual Culture in the English Death Ritual, c.1500–c.1800* (London: The Victoria and Albert Museum, 1997), esp. 95–96.
[115] Calhoun, *The Nature of The Page*, 130.
[116] Ibid., 126.
[117] Vaughan, *The Golden Fleece*, 8.
[118] John Carey, *John Donne: Life, Mind and Art* (London: Faber and Faber, 1981; rep. 1983), 220, 172.

John Donne', structuring an 'essential part' of his 'imagination'.[119] The final section of this chapter situates Donne's meditation on corporeal fragmentation in two poems, published in the same year, alongside one another. The first is the playful 'Panegyricke Verse' he contributed to Thomas Coryate's 1611 *Coryats Crudities* and the second is the funeral elegy he composed upon the death of Elizabeth Drury in *An Anatomy of the World, Wherein, by Occasion of the untimely death of Mistris Elizabeth Drury the frailty and decay of this whole world is represented* (1611). Although written for different audiences and with distinct tones, when read together, these two poems reveal the complex imaginative and intellectual work waste paper prompted within the early modern eschatological framework.

Donne's 'Panegyricke Verse' is one of a 'copious rhapsodie of poems' filling over a hundred pages that preface Coryate's 655 (numbered) page account of his travels throughout Europe.[120] These panegyrics, Andrew Hadfield notes, are 'often extremely rude and disrespectful to Coryate', but also give 'the impression that there has been something of a pact between [the] author and his circle'.[121] It is no surprise, then, that waste paper makes an appearance within this rhapsody of homosocial jibes: Hugo Holland, for instance, contributes a Latin poem he claims had been intended for John Harington's scatological *Metamorphosis of Ajax*, but was perfectly well suited for Coryate's work instead. Holland adapts Persius' *nec scombros metuentia nec tus*, suggesting that Coryate's pages might be used as wrappers for tobacco, pitch, and pepper that, when burnt, would mask the smell emerging from beneath Ajax's shield, or toilet cover.[122]

Donne's poem dwells on the subject of waste paper at greater length than Holland's. The central premise is, like many of the dedicatory poems, that *Crudities* is excessively long, an 'Infinite worke, which doth so farre extend, | That none can study it to any end'.[123] Both impossible to finish and pointless to study, its seemingly endless supply of paper leaves will be repurposed rather than read. In a reworking of the familiar Catullan theme, Donne maps the geographical trajectory of Coryate's pages: like Volusius' *Annals*, they do not circulate far, covering stuff rather than covering distance, and dispersed as wrappers rather than as reading material.[124] Apostrophizing Coryate's text, Donne claims that 'thy leav's must embrace what comes from' both the East and West 'Indies', '[t]he Myrrhe, the Pepper, and the Frankinsence' (D3v). These exotic spices and perfumes 'magnifie' the leaves, ennobling them, but also have a microscopic effect (D3v). 'Magnifie[d]', the page

[119] Achsah Guibbory, *Returning to John Donne* (Farnham: Ashgate, 2015), 19, 20.
[120] Coryate, *Coryats Crudities*, C2r.
[121] Andrew Hadfield, *Literature, Travel, and Colonial Writing in the English Renaissance, 1545–1625* (Oxford: Oxford University Press, 1998), 60.
[122] Hugo Holland, '*Ad Janum Harringtonum Badensem*', in *Coryats Crudities*, E1v–2r. John Harington, most likely in on the joke, also contributes to the 'Panegyricke Verses'.
[123] '*Incipit Joannes Donne*', in *Coryats Crudities*, D3r. Further references to this poem will be included parenthetically in the text.
[124] See Catullus, Carmen 95.

is framed as if through a lens: a fragment with its textual detail cropped and highlighted. Distributed across London, Donne suggests, it might give the illusion of the text being enlarged.

Alternatively, the leaves might 'stoope | To neighbor wares', wrapping less exotic 'Tomes of Curran[t]s, and of Figs, | Of Med[i]cinall, and Aromatique twigs', 'Convay[ing] these wares in parcels unto men' (D3v). Or they might

> [...] stoope lower yet, and vent our wares,
> Home *manufactures*, to thicke popular faires,
> If *omnipregnant* their, upon warm stals
> They hatch all wares for which the buyer cals. (D3v)

Describing stages of a journey that move ever closer to home, Donne suggests that the book's pages will come to wrap unspecified common produce, 'thicke', 'warm', and gross stuff, perhaps hot and steaming pies, and so 'thus thy leav's we justly may commend, | That they all kind of matter comprehend' (D3v). Donne puns throughout on the capacity of a page to transport both textual and physical matter. It is a 'tome' 'convay[ing]' both meaning and 'wares', 'curran[t]s' and words; full of all sorts of stuff and topics (*omnipregnant*), it is a 'Universall Booke' that 'comprehend[s]' all 'matter', both through its digressive length and as a wrapper of goods from all over the world (D3v). Furthermore, the 'Scattred [...] Booke in peeces' takes on a playfully talismanic power, like the fragments of Bible verses carried about or 'applied' to the body as medicinal spells throughout the period: 'Some shall wrap pils, and save a friends life so, | Some shall stop muskets, and so kill a foe' (D3v).[125] Here, the everyday use of waste pages in the household, alongside their use as cartridge paper, transforms into an exaggerated efficacy over life and death.

As a result of this scatteredness, *Coryats Crudities* will not be a victim of the voracious 'Critiques of [the] age' because they won't be able to 'find' it (D4r). Rather than sitting 'All in one bottome, in one Librarie, | Some leav's may paste strings there in other books' (D4r). Spread throughout multiple libraries as binding waste and lining the 'bottom' of boxes, Donne claims that Coryate's pages avoid piercing critique because, already dispersed, they can only be read in a fragmentary fashion. Donne describes what must have been common in the period: the almost accidental absorption of words and phrases when glancing over a leaf of waste. How, as Donne puts it, 'one may, which on another looks, | Pilfer, alas, a little wit from you, | But hardly much' (D4r). The microscopic effect of tearing up and wasting a book has no effect on the reading experience because, Donne concludes, 'As *Sybils* was, your booke is mysticall, | For every peece is as much

[125] See Tanya Pollard, 'Spelling the Body', in *Environment and Embodiment in Early Modern England*, ed. by Mary Floyd-Wilson and Garret A., Sullivan Jr. (Basingstoke: Palgrave Macmillan, 2007), 171–186.

worth as all' (D4r). Each fragment is a synecdoche of the whole: both a single page and the entire book are worthless, with every unit of the text and the text as a whole containing very 'little wit'. As Sarah Wall-Randell has noted, the Sibyl was associated in both antiquity and the early modern period with textual impermanence and destruction, and, in Donne's panegyric, the joke is that each leaf of the Sibylline Oracle's prophecy does *not* contain the whole text.[126] Left to be scattered by the wind outside of the Oracle's cave, the prophecy is ephemeral, rapidly disintegrating into disjointed and nonsensical fragments.

In her reading of this prefatory poem, Wendy Wall takes Donne seriously, arguing that the 'proliferation and ennobling of matter that Donne imagines [...] does not seem ironic'.[127] Instead, she argues that the poem represents an alternative form of 'circulation'—'a significant mode of production' in which 'the housewives who sell pastries and confections at fairs' amplify the book 'in its reuse'.[128] In addition to suggesting that the poem records a gendered mode of textual transmission, lost to posterity in ovens and dunghills, Wall suggests that, in reading this light-hearted poem, 'we might be guided by Donne's other evocative poetic meditations on substantiality and meaning'.[129] Similarly, David Hedrich Hirsch has connected Donne's 'satiric analysis of Coryate's "corpus"' with his 'concern elsewhere with the scattered remains of the body'.[130] Donne's 'Panegyricke Verse' certainly shares the concerns of another poem that dwells on scattered pages: the tonally distant funeral elegy in *An Anatomy of the World* that seems to have been composed at a similar time as Coryate's dedicatory poem. In this poem, Elizabeth Drury, who had died late in 1610, is at first described as being 'confine[d]' in a 'Tombe' or 'Marble chest'.[131] The mention of these stones leads Donne into an unfavourable comparison of 'Marble, Jeat, or Porphiry' with 'the Chrysolite of eyther eye, | Or with those Pearles, and Rubies which shee was' (B7r). Donne imagines all of the precious stones of 'the two Indies'—similar in origin to the wares encompassed by the waste pages of *Coryats Crudities*—accumulating in her tomb; but, he sighs, ''tis glas' anyway, cheap and fragile like all things in comparison with 'her materials' (B7r). This conventional Petrarchan conceit of a woman's superiority to jewels is shattered, however, when Donne dwells, even for a moment, on the fate of Elizabeth's 'materials': his rhythm falters with the curtailed clause—'Yet shee's demolish'd' (B7r).

[126] Sarah Wall-Randell, 'Marlowe's Lucan: Winding-Sheets and Scattered Leaves', in *Christopher Marlowe, Theatrical Commerce, and the Book Trade*, ed. by Kirk Melnikoff and Roslyn L. Knutson (Cambridge: Cambridge University Press, 2018), 20–25. The Oracle is described in Book 6 of Virgil's *Aeneid*.

[127] Wall, *Recipes for Thought*, 156. See also Katrin Ettenhuber's reading of the poem in *Donne's Augustine: Renaissance Cultures of Interpretation* (Oxford: Oxford University Press, 2011), 48–49.

[128] Wall, *Recipes for Thought*, 156.

[129] Ibid., 156.

[130] David A. Hedrich Hirsch, 'Donne's Atomies and Anatomies: Deconstructed Bodies and the Resurrection of Atomic Theory', *Studies in English Literature, 1500–1900* 31, no. 1 (1991): 69–94, 71.

[131] John Donne, *An Anatomy of the World*, STC 7022 (London: William Stansby for Samuel Macham, 1611), B7r. Further references to this poem will be included parenthetically in the text.

Here, the matter of the body blurs with the stony matter of the tomb, not because both are precious or enduring, but because, like a stone structure, the body can be 'demolish'd', reduced to ruin.

These lines prompt a pessimistic reworking of the Horatian immortality topos. Donne turns from Elizabeth's ruined 'materials' to a consideration of whether literature might outlast the lives of its author and its subject, as 'tombs'—synonymous with 'monuments of the dead'—are conventionally assumed to do:[132]

> Can we keepe her then
> In works of hands, or of the wits of men?
> Can these memorials, ragges of paper, give
> Life to that name, by which name they must live?
> Sickly, alas, short-liv'd, aborted bee
> Those Carkas verses, whose soule is not shee.
> And can shee, who no longer would be shee,
> Being such a Tabernacle, stoope to bee
> In paper wrap't: Or, when she would not lie
> In such a house, dwell in an Elegie?
> But 'tis no matter; we may well allow
> Verse to live so long as the world will now
> For her death wounded it. (B7r–B7v)

Querying whether textual 'memorials' might grant perpetual life, Donne's elegy does not reach a confident conclusion. 'Elegie[s]', 'memorials', and 'verses' are revealed not to be 'living record[s]', as Shakespeare puts it in Sonnet 55, or more lasting than bronze, as in Horace's Ode 3.30, but only 'ragges of paper'. Blurring the present state of paper with its previous life as worn-out rags, Donne also looks forward to its material end. The poem will become, like its subject, a 'Carkas'. Here, Donne gestures towards the similarities between winding-sheets and paper sheets that have been explored in depth by Sarah Wall-Randell.[133] Wall-Randell notes that early modern readers were sensitive to the intertwined life cycles of both sorts of sheets: the fact that both are, or once were, linen, and that both will, or already have, decayed and fermented in the form of rags, either in the papermaker's vat or alongside a corpse in the grave. In her reading of the publisher's preface to Christopher Marlowe's posthumous translation of Lucan, Wall-Randell argues that the image of pages used as winding-sheets provides an alternative understanding of textual immortality. Rather than paper providing 'the medium for the maker's immaterial escape from the ravages of time', the paper winding-sheet embodies

[132] In its earliest usage, 'monument' referred to a tomb or sepulchre. They continued to be synonymous throughout the early modern period. See, for instance, 'Monument' in Robert Cawdry, *A Table Alphabeticall*, STC 4884 (London: J[ames] R[oberts] for Edmund Weaver, 1604).

[133] Wall-Randell, 'Marlowe's Lucan', 11–25.

'a model of books as subject to fragmentation, decay, and loss, and [...] reminds us that, as with paper itself, mortal corruption may reside in their very material'.[134] Donne gestures toward this encroaching and total decay in the pun, ''tis no matter'. Dismissing the series of questions that he himself has proposed, Donne, in the same breath, suggests that soon there will be no matter, 'allow[ing]' that poetry might endure, but only so long as 'the world will'. This, we assume, might not be very long at all, because the decay of the world has been accelerated by its grief, or 'wound[ing]', at the loss of the young Elizabeth Drury.

The funeral elegy does not, however, conclude with this image of disintegrated sheets and paper carcasses. As is typical of Donne's poetry, we do not settle comfortably on one conclusion—here, on the understanding of both books and bodies as irredeemably ephemeral. As Joshua Calhoun has noted, much early modern poetry is 'enliven[ed]' by the 'the nature of the book and its tendency towards decay', with the corruptibility of matter simultaneously fuelling and threatening poetic expression.[135] Calhoun argues that early moderns including Donne often meditated on what 'the conditions' were 'that allow corruptible organic objects to carry messages into the distant future'.[136] This question is tied, in *An Anatomy of the World*, to the question that Donne asks persistently throughout his poetic and spiritual career: what happens to the corruptible, organic matter of the human body at the Resurrection? John Carey has written of Donne's obsessive meditations on the material intricacies of this divine moment, arguing that 'his sole intent is to emphasize the power of the reintegrative act, and the fantastic state of dispersal that precedes it', spurred on by his fear of 'extinction' and 'the thought of his body helplessly mouldering through its long years of dissolution'.[137] Fixated on the corporeal nature of the resurrection, Donne was deeply invested in the belief that, even after its dissolution into dust, each body will be made whole at the end of days.[138] This corporeal reconstitution shapes Donne's description of Elizabeth Drury's grave, which, we read 'shall restore | her greater, purer, firmer, then before' (B7v). At the Resurrection, Elizabeth again resembles, as she did in the opening lines of the elegy, the firmness of stone rather than the softness and flexibility of paper; the corporeal matter that was initially 'demolish'd' is now, like an architectural structure, 'restored' (B7r, B7v).

The winding-sheet and the paper 'ragges' of the elegy itself are, we imagine, left behind at the Resurrection, rotting along with the rest of the world's matter. The poem does, though, offer another bookish analogue to the body of Elizabeth Drury in its final lines. An imagined reader 'come[s] to reade the booke of destiny' and is taken aback by the premature end of Elizabeth's story, and so assumes 'that some

134 Ibid., 17, 20.
135 Calhoun, *The Nature of the Page*, 127.
136 Ibid., 136.
137 Carey, *John Donne*, 221, 226.
138 Carey, *John Donne*, 219–230; Hirsch, 'Donne's Atomies and Anatomies', 82.

leaves were torne out of the booke' (B8v). This is, though, a misconception, and the divine text is far from incomplete or fragmentary; rather than resulting from the violent destruction of God's book, Elizabeth's death in fact *adds* to it, 'filling up' the 'blanks' in the pages of those who will follow her modest example (B8v). 'They shall make up that booke' (B8v), Donne writes, employing a metaphor he will use again in his *Devotions Upon Emergent Occasions*, when he promises that 'All mankind is of one Author, and is one volume; when one Man dies, one Chapter is not torne out of the booke, but [... God's] hand shall binde up all our scattered leaves againe'.[139] It is telling that the divine book that rounds off Donne's funeral elegy is not explicitly papery, and that the manner in which the book is 'ma[d]e up', or materially constituted, is unclear. It remains distinct from the rags of paper that serve as its decaying counterpart at the poem's opening, and that make up the pages of *The Anatomy of the World* itself.

The imperviousness of the Book of Destiny to fragmentation does not give Donne or the reader similar hope for the survival of the pages that they hold in their hands. At the same time, the disassembled and fragmented pages of books serve Donne as a useful model through which to conceive of what Carey terms 'the fantastic state of dispersal' and reconstitution of the body at the Resurrection.[140] As is established early on in the poem, paper carcasses and human corpses are materially similar, and so the capacity of loose sheets and fragments to be gathered up and rebound after their transformation into waste provides Donne with a substantial pattern for the fate of the human body after its death. The dust to which the body decays is, for Donne, a form of waste. Waste, as William Viney has theorized, persists in an 'anachronic tense, one that commingles pasts and presents and leaves things lingering, loitering and persisting'.[141] This peculiar 'tense' means that waste objects readily move in and out the temporality of disuse, 'crisscrossing in and out of the projective time of use and the slack indeterminate and disoriented time of waste'.[142] Waste, as Viney argues, does not always remain waste: it has the capacity to be resurrected and restored to its prior state. Fragments of repurposed paper, then, readily figure for the scattered atoms of dust left behind after death, with the scattering of a book's pages serving as a typically paradoxical conceit in Donne's imagination: variously corporeal and transcendent, fragmented and whole, paper pages stand in for both the mundane dissolution and the miraculous reconstitution of the human body.

But where does this leave the pages, made from rags, slowly decaying, in the reader's hands? The 'scattered leaves' of the dissolved body prior to the Resurrection, described in the *Anatomy* and in *Devotions Upon Emergent Occasions*, recall

[139] John Donne, *Devotions Upon Emergent Occasions*, ed. by Anthony Raspa (Oxford: Oxford University Press, 1987), 86.
[140] Carey, *John Donne*, 221.
[141] Viney, *Waste*, 178.
[142] Ibid., 60.

Coryate's 'Book in peeces' (D3v), and so suggest a similar, although much more modest, potential for bibliographic reconstitution.[143] This hope does not extend to a Horatian faith in literary endurance and monumentality, but instead to a resurrected legibility more closely resembling the alternative, domestic circulation identified by Wall, and grounded in the 'pilfer[ing]' of a 'little wit' from dispersed fragments of waste (D4r).[144] Because waste lingers, it has the capacity to be gathered up and restored to some degree of textual wholeness. This is not, though, an especially grandiose hope, and it is further undercut by the emphasis, in Donne's funeral elegy, on the fact that earthly texts—whether whole or fragmentary—will only endure as long as the world itself. After this, Donne suggests, there will be 'no matter' other than heavenly matter.[145]

*

This chapter has traced the rich rhetorical history of waste paper as it was transported from classical Rome to early modern England. The material practice of repurposing pages coincided, in the mid-sixteenth century, with the reception of the poetry of Catullus, Martial, and Persius, and the enduring influence of Horace. Towards the end of the sixteenth century, English authors lifted these waste paper tropes, either from recent translations or Latin editions often studied at school and university, and played with them, using them to mediate religious and political controversy, the patronage system, and the dynamics of the literary marketplace. Far from being exhausted, stale, or straightforward, waste paper tropes were flexible and capacious. Waste was an object and an emblem with which to articulate textual desires and anxieties, and to think about the relationship of corporeal and spiritual matter to time. Early modern writers were imaginative in their reuse of this classical heritage: the handling of millennia old narratives of textual mishandling became increasingly urgent, lively, and provocative in the face of multitudinous flimsy printed books and manuscript papers, the crumbling ruins of the nation's past *monumenta*, pressing concerns about the imminent end of the world and, in the case of John Donne, the reconstitution of the scattered body at the Resurrection. 'Comprehend[ing]' or encompassing everything imaginable (frankincense, figs, medicines, pies, bullets, and other books), the life cycle of waste paper enabled its users to 'comprehend' the life cycle of 'all kinde[s] of matter'.[146] Paper fragments, particularly when looked upon through the lens of the waste paper trope, told a narrative of dissolution and disintegration, typically complete but, in the case of John Donne, only partial, with the legibility and modest endurance of scattered fragments suggesting the potential for the miraculous reconstitution of the body. In the next chapter, we will continue to consider the

143 Donne, *Devotions Upon Emergent Occasions*, 86.
144 Wall, *Recipes for Thought*, 156.
145 Donne, *Anatomy of the World*, B7r.
146 Donne in *Coryats Crudities*, D3v–4r.

capacity of waste to structure and articulate particular understandings of the trajectory of the world and everything within it. We will turn, however, to another form of waste that circulated widely in sixteenth- and seventeenth-century England and the conceptions of the world and of time that it prompted: the monastic manuscripts that were dispersed in the wake of the Reformation, and the ways in which this waste is recorded and represented in the writings of two early modern antiquarians, John Bale and John Aubrey.

3

Butterflies and Binders' Shops

Reading Monastic Waste

In his 2009 *Stuff*, the anthropologist Daniel Miller offers an influential account of the relationship between humans and things. Things, he writes, 'direct our footsteps, and are the landscapes of our imagination'.[1] Material culture is a frame for our thoughts and our behaviours: in sum, 'objects make people'.[2] Miller's theorization of the relationship between the human and the stuff that surrounds them is a useful frame through which to consider the material and imaginative landscapes of post-dissolution England, and how repurposed monastic manuscripts, like ruined and reclaimed monastic buildings, shaped thought.[3] But where Miller argues that objects are formative precisely because they are 'invisible and unremarked upon', or 'familiar and taken for granted', the fragments explored in this chapter were vividly present for the early moderns who encountered, remembered, and imagined them.[4] Although manuscript waste may often have been unremarkable—going largely unremarked—it was nonetheless a category of object upon which significant thought might be expended. The experience of monastic waste might move from a scarcely perceived prop at one extreme, to a thoughtfully apprehended object at the other.

This chapter is concerned with post-dissolution waste practices and, in particular, how that waste was received and interpreted in the sixteenth and seventeenth centuries. It offers case studies of two early modern antiquarians and their encounters with repurposed manuscripts. The first is a reading of *The Laboryouse Journey & Serche of John Leylande, for Englandes Antiquities*, heavily edited by John Bale and printed in 1549. Bale's experience of manuscript fragments in stationers' and binders' shops, it is argued, granted him a sensitivity to the material history of waste and its subsequent figurative value. Then, shifting attention to the continued afterlives of monastic waste in the seventeenth century, we turn to John

[1] Daniel Miller, *Stuff* (Cambridge: Polity Press, 2009), 53.

[2] Ibid., 53.

[3] Throughout, the episode typically labelled the 'Dissolution of the Monasteries' is referred to without capitalization. This is to highlight that, as Harriet Lyon writes, the 'long process of dissolution' was only understood to be 'a major Reformation event' retrospectively, and as part of a 'particular polemical vision' (*Memory and the Dissolution of the Monasteries in Early Modern England* (Cambridge: Cambridge University Press, 2021), 23).

[4] Ibid., 50.

Waste Paper in Early Modern England. Anna Reynolds, Oxford University Press. © Anna Reynolds (2024).
DOI: 10.1093/oso/9780198882701.003.0004

Aubrey's account of his lifelong encounters with old manuscripts. Within his *Naturall History of Wiltshire* sits a uniquely bio-bibliographical 'Digression'. In it, Aubrey builds a series of bibliofictions, or 'biblio-biographies', from extant fragments.[5] Mingling his experience of repurposed manuscripts with autobiography, local events, and a national history, Aubrey makes visible the imaginative work that might be done with waste. Bale and Aubrey, it will be shown, were sensitive and responsive readers of waste: their writings demonstrate how waste fragments might structure an antiquarian and an archaeological understanding of the past and anticipation of the future. Whereas, for Bale, the tendency of fragments toward loss and oblivion fitted within the eschatological narrative we have already encountered in Chapter 2, it was their capacity to endure that afforded Aubrey food for thought. He understood waste as manifesting multiple times and narrating, for those who took the time to apprehend them, biographies of dissolution, ephemerality, and occasional survival.

1 The 'Remnaunt' of the Reformation

Old books were repurposed prior to the dissolution of the monastic libraries: scribes scraped clean the parchment surfaces of manuscripts, particularly administrative and legal texts, throughout the medieval period, readying them for the addition of new ink. Worn out and redundant books were cut up to bind new ones, and the bindings of extant incunabula reveal that particularly large numbers were wasted in the late fifteenth century, as old manuscripts were gradually replaced by printed texts.[6] The 'great cataclysm' that took place between 1536 and 1540 intensified these palimpsestic processes, and, a decade later, Edward VI's 1550 Act against Superstitious Books and Images led to a second wave of large-scale book destruction.[7] The seventeenth-century antiquarian Anthony Wood describes how Edward VI's Visitors ravaged Oxford's libraries, removing

[5] On this characteristically antiquarian genre, see John Leland, *De uiris illustribus/On Famous Men*, ed. James P. Carley with Caroline Brett (Oxford: Bodleian Library, 2010), xxvii.

[6] Ryley, *Re-Using Manuscripts*; Pickwoad, 'The Use of Fragments of Medieval Manuscripts', 1–2.

[7] John Leland, *De uiris illustribus*, civ. The foundational studies of the dissolution of the monasteries are David Knowles, *Bare Ruined Choirs: The Dissolution of the English Monasteries* (Cambridge: Cambridge University Press, 1976) and Joyce Youings, *The Dissolution of the Monasteries* (Fakenham: Allen and Unwin, 1971). On the dissolution of the monastic libraries see James P. Carley, 'Monastic collections and their dispersal', in *The Cambridge History of the Book in Britain*, vol. 4, ed. by John Barnard and D. F. McKenzie (Cambridge: Cambridge University Press, 2002), 339–347; Ronald Harold Fritze, '"Truth Hath Lacked Witnesse, Tyme Wanted Light": The Dispersal of the English Monastic Libraries and Protestant Efforts at Preservation, ca. 1535–1625', *The Journal of Library History* 18, no. 3 (1983): 274–291; Nigel Ramsay, '"The Manuscripts flew about like Butterflies": The Break-Up of English Libraries in the Sixteenth Century', in *Lost Libraries: The Destruction of Great Book Collections since Antiquity*, ed. by James Raven (Basingstoke: Palgrave Macmillan, 2004), 125–144; and C. E. Wright, 'The Dispersal of the Monastic Libraries and the Beginnings of Anglo-Saxon Studies. Matthew Parker and his Circle: A Preliminary Study', *Transactions of the Cambridge Bibliographical Society* 1, no. 3 (1951): 208–237.

any printed books or manuscripts 'guilty' of 'red letters', 'controversial or scholastic Divinity', and 'Angles, or Mathematical Diagrams' condemning them as 'superstitious' and 'Popish' and so turning them to 'base' and 'servile uses'.[8] The dramatic influx of 'Popish' texts into the waste paper trade made the violence of the dissolution and the Reformation palpable, capturing the imaginations of readers and historians and altering the parameters within which the processes of wasting and repurposing might be understood.

By the mid-sixteenth century, all types of medieval books were vulnerable to becoming waste. These included service-books, theological treatises, and historical chronicles, whether recent and relatively plain or luxuriously illuminated and dating from the twelfth century—texts soon to be sought out by antiquarian collectors.[9] Additionally, an array of administrative documents and financial records, stored haphazardly around the monastery and rarely catalogued, entered the waste market, though this material has been largely overlooked by book historians.[10] Whatever their origins, these objects were now valued according to their bulk rather than their contents, and any attribution of historical or aesthetic worth was rare: in Anthony Wood's words, 'books were dog cheap, and whole libraries could be bought for an inconsiderable nothing'.[11] His account is based largely on the writings of John Bale who, in 1549, described 'a merchaunt man [...] that bought the co[n]tentes of two noble lybraryes for .xl. [40] shyllynges pryce', which he has used 'in the stede of graye paper by the space of more than these. x. yeares, & yet he hath store enough for as many yeares to come'.[12] A decade later, in 1557, John Dee records purchasing a manuscript from the dispersed Duke Humfrey's Library at Oxford 'par le poys' (by the pound weight) and, in 1564, a Mr Seeres paid 24 shillings 'for old Parchment books weying cc [200] pounde'.[13]

It is difficult to quantify the number of manuscripts dismembered and destroyed during these decades. Few monastic catalogues survive from the period leading up to the dissolution, and Henry VIII's surveyors rarely included books within their

[8] Anthony Wood, *The History and Antiquities of the University of Oxford*, vol. 2, ed. John Gutch (Oxford: Printed for the Editor, 1796), 106–110.

[9] For the chronological trends of sixteenth- and seventeenth-century manuscript destruction see N. R. Ker, *Fragments of Medieval Manuscripts*, ix–x.

[10] See Barbara Harvey, *The Obedientiaries of Westminster Abbey and their Financial Records, c. 1275–1540* (Woodbridge: Boydell, Press, 2002).

[11] Wood, *History and Antiquities*, 108.

[12] John Bale and John Leland, *The Laboryouse Journey & serche of John Leylande, for Englandes Antiquities*, STC 15445 (London: S. Mierdman, 1549), B1v. All further references will be given parenthetically in the text.

[13] Dee's note, found within the volume itself, reads '*et a ceste heure voyre en L'an de notre seigneur 1557 a moy Jehan Dee Angloys: lequel ie achetay par le poys*'. This suggests that much of Duke Humfrey's collection, sold in bulk, found its way into binders' and stationers' shops. Fragments of manuscripts traceable to Humfrey's library have been found in other bindings: see A. C. de la Mare and Stanley Gillam, *Duke Humfrey's Library and the Divinity School, 1488–1988* (Oxford: Bodleian Library, 1988), 124; David Rundle, 'Habits of Manuscript Collecting: The Dispersals of the Library of Humfrey, Duke of Gloucester', in *Lost Libraries: The Destruction of Great Book Collections Since Antiquity*, ed. by James Raven (Basingstoke: Palgrave Macmillan, 2004), 116; Ker, *Fragments of Medieval Manuscripts*, x.

inventories of monastic goods.[14] Neil Ker's *Medieval Libraries of Great Britain* lists 5,200 library and service-books extant from the eight hundred or more religious houses of England. Of these, 1,800 belonged to the secular cathedrals that remained relatively unscathed, leaving a total of 3,400 surviving books from the dissolved monasteries.[15] As Nigel Ramsay writes: '[t]hat tens, even hundreds, of thousands of library books and service-books were destroyed in the course of a few years is undeniable'.[16] Although these fragments are not as conspicuous as the crumbling stones of dissolved abbeys and monastery buildings, a visit to a rare books library reveals the fate of many of these manuscripts. Ker has catalogued 2,200 Oxford bindings that contain medieval manuscript pastedowns, and old parchment leaves were also used to form flyleaves, limp or stiff covers, stab-stitched wrappers, laminated to form boards, and trimmed into guards in binderies throughout England until the early seventeenth century.[17]

Bale was right in estimating that his unnamed merchant would have 'store enough' of monastic waste for 'many yeares to come': large quantities of medieval manuscript waste was used as flyleaves, pastedowns, and wrappers in bindings until the 1580s and 1590s (see Fig. 1.2), and it was only towards the turn of the seventeenth century that binders began to use their medieval waste parchment more sparingly. To conserve durable parchment waste, binders reserved scraps of old manuscripts for the spines of new books—the area of the binding in need of most reinforcement. This eking out of a dwindling supply is visible in the binding structures of several books, printed between 1608 and 1614, held in Bishop Cosin's Library: these are striking hybrids, containing printed waste paper guards in addition to a combination of twelfth-century and early seventeenth-century waste parchment manuscript spine supports (see Fig. 3.1). By the middle of the seventeenth century, medieval manuscript waste is almost entirely absent from new bindings.

Binding waste might give the impression that waste fragments were static, stitched tightly within other books. This was, however, far from always the case: as we saw in Chapter 1, waste moved in and out of a variety of contexts and spaces, and this was true for waste parchment as well as waste paper. Richard Layton, a principle commissioner of the dissolution, delightedly informed Thomas Cromwell that, on his return to New College Oxford, he had 'fownde all the gret quadrant court full of the leiffes of Dunce [the 13th century theologian Duns

[14] Carley, 'The Dispersal of the Monastic Libraries', 256, 265. The catalogue of the house of Syon, compiled between ca. 1500 and ca. 1524, is the only extant and intact monastic library catalogue from these decades. It reveals what was incorporated and what was disposed by monastic librarians between the introduction of print and the advent of the dissolution. By 1504, the collection was over 1300 volumes strong, but only thirty books of Syon providence have since been identified. As with most monastic collections, the exact fate of the collection is unclear. In an impressive feat of bibliographical detective work, Christopher De Hamel has argued that the collection most likely sat mouldering in the abandoned library for a number of years after the monastery's dissolution in 1539. He determines this by the fact that the surviving books predominantly come 'from the middle of the alphabetical range of class-marks', and so hypothesizes that the outer edges of the bookcases fell victim to thieves, rodents, and the elements. See Vincent Gillespie, *Syon Abbey: Corpus of British Medieval Library Catalogues*, vol. 9 (London: The British Library, 2001), lxv and Christopher de Hamel, *Syon Abbey: The Library of*

Fig. 3.1 Strips of a fourteenth-century manuscript reinforcing printed pages of the Geneva Bible used as guards, in the binding of *The Works of the Very Learned J. Jewell*, STC 14580 (London: [Eliot's Court Press for J. Norton], 1611), Cosin K.2.14.
Reproduced by permission of Durham University Library and Collections.

Scotus], the wynde blowyng them into evere corner'.[18] Duns Scotus' windy and worthless words are, for Layton, both rhetorically and literally lightweight: this manuscript's material qualities marry with its contents. He goes on to describe how a student, Mr Grenefelde, was found 'getheryng up part of the saide bowke leiffes (as he saide) therwith to make hym sewelles or blawnsherres [scarecrows or scaring sheets] to kepe the dere within the woode' in his home county of Buckinghamshire.[19]

the Bridgettine Nuns and their Peregrinations after the Reformation (London: Roxburghe Club, 1991), 113.

[15] N. R. Ker, *Medieval Libraries of Great Britain: A List of Surviving Books* (London: Royal Historical Society, 1964).

[16] Ramsay, '"The Manuscripts flew about like Butterflies"', 138.

[17] David McKitterick, *A History of Cambridge University Press*, vol. 1, *Printing and the Book Trade in Cambridge, 1534–1698* (Cambridge: Cambridge University Press, 1992), 9; Rowan Watson, 'Medieval Manuscript Fragments', *Archives* 13, no. 61 (1988): 70.

[18] On the concerted wasting of the works of Duns Scotus, see Jennifer Summit, *Memory's Library: Medieval Books in Early Modern England* (Chicago: University of Chicago Press, 2008), 88–90.

[19] Quoted in Margaret Aston, *Lollards and Reformers: Images and Literacy in Late Medieval England* (London: Hambledon Press, 1984), 327.

No longer bound and chained within libraries, manuscripts moved beyond the highly literate and homosocial centres of monasteries and universities. John Bale, in an often-quoted passage, describes how

> A great nombre of the[m] whych purchased those superstycyouse mansyons, reserued of those lybraye bokes, some to serue theyre jakes [toilets], some to scoure theyr candelstyckes, & some to rubbe their bootes. Some they solde to the grossers and sope sellers, & some they sent over see to ye bokebynders, not in small nombre, but at tymes whole shyppes full, to the wonderynge of the foren nacyons
>
> (B1r)

Bale highlights the presence of a thriving waste paper and parchment trade in sixteenth-century England, and its saturation after the dissolution of the monasteries: there was simply not enough demand for the amounts of waste made available in the 1530s and 1540s, and so 'some' was 'sent over see[s]'. This is a source of shame and regret for Bale who, as we will see, had a vested interest in recovering documents of the nation's past. His description also details the nature of waste parchment practices: the predominantly parchment fragments, like paper ones, served 'base ends', but with some notable differences. Parchment is particularly strong for its weight and, as the conservator and bookbinder Nicholas Pickwoad observes, is durable and resistant: soft when wet, it can be moulded into any shape, hardening and retaining its form when dry.[20] This meant that, as well as forming wrappers and flyleaves, parchment could be twisted to form tackets, laces, sewing supports, and end bands.[21] Its capacity to withstand more friction than paper meant that, as well as serving as toilet paper and wrappers for 'grossers' and 'sope sellers', monastic manuscripts were used as wipes and scourers, and were particularly valuable for binders.

Despite retaining legible text, these sheets of manuscript were no longer experienced primarily as textual objects. Instead, the old manuscripts participated in more practical forms of knowledge, or, to borrow Matthew Hunter's phrase, provoked a 'materialized intelligence'.[22] Those involved in the production and use of books had always been sensitive to the material capacities of parchment, but the dissolution dispersed this knowledge more widely. Furthermore, this parchment literacy was no longer neutral: it had become imbued with social and religious meaning. Dismembered books prompted a series of 'generative [...] imaginings' through their 'physical manipulation', as well as their cognitive apprehension.[23]

[20] Pickwoad, 'The Use of Fragments of Medieval Manuscripts', 2.
[21] Ibid., 1, 18.
[22] Matthew C. Hunter, 'Picture, Object, Puzzle, Prompter: Devilish Cleverness in Restoration London', *Art History* 36, no. 3 (2013): 549.
[23] Ibid., 164.

The fragments were known haptically: for Mr Grenefelde, through their lightness, their capacity to be stitched together and shiver in a Buckinghamshire breeze; for other users described by Bale, through their pliability and relative durability, their capacity to wipe, rub, fold, and wrap. We have no record of the 'imaginings' of many of these users, but their material knowledge of monastic waste would have been intermingled with a knowledge of the fragments' religious pre-histories. Earlier stages of the objects' life cycles were simultaneously available alongside the present, forming palimpsests of multiple times.

Monastic waste, then, would have been apprehended in similar ways to the other remnants of the religious past in the post-Reformation landscape: the weathered ruins, repurposed altar stones, monumental brasses, plate, textiles, paintings, and statues that persisted in either presence or use.[24] Archaeologists and historians have long hypothesized the responses to these leftovers: David Stocker, in his influential 'Rubbish Recycled: a study of the re-use of stone in Lincolnshire', proposes three categories such incorporation might fall under.[25] 'Casual' reuse is when old stuff is re-used because of its appropriate size, shape, and material capacities, such as when altars were turned into paving slabs; 'functional' reuse occurs when an object continues to serve its intended purpose, such as when a monastic building continues to stand, but is secularized; the third, and the only that Stocker believes generates meaning, is 'iconic' reuse, when an object is recontextualized in such a way as to suggest significance beyond practical use.[26]

Within this schema, all waste paper practices are 'casual': old books are used in the place of plain white or brown paper, fresh parchment, or fabric. But, as Sarah Tarlow has argued, within the 'heated symbolic environment' of post-Reformation England, the majority of pre-Reformation material culture might be read iconically.[27] Sensitively describing iconoclasm as a stage in the life cycle of an object, Tarlow argues that, when trodden on as doorsteps, repurposed altars might 'articulat[e] a rejection' of a Catholic 'understanding of holiness'.[28] Monastic manuscripts, bound within Protestant volumes, might have articulated a similar renunciation of past devotional practice. These hybrid objects were relatively common. The Huntington Library, for instance, holds mid-sixteenth-century textual

[24] The performativity and partiality of iconoclasm has been explored in detail in recent scholarship, and the previously dominant theory of a Protestant 'iconophobia' has been nuanced. See Margaret Aston's *England's Iconoclasts: Laws Against Images*, vol. 1 (Oxford: Clarendon Press, 1988); Tara Hamling, *Decorating the 'Godly' Household: Religious Art in Post-Reformation Britain* (New Haven: Yale University Press, 2010); and Alexandra Walsham, '"Like Fragments of a Shipwreck": Printed Images and Religious Antiquarianism in Early Modern England', in *Printed Images in Early Modern Britain: Essays in Interpretation*, ed. by Michael Hunter (Abingdon: Routledge, 2010), 87–109.

[25] David Stocker, 'Rubbish Recycled: A Study of the Re-Use of Stone in Lincolnshire', in *Stone Quarrying and Building in England AD 43–1525*, ed. by David Parsons (Chichester: Phillimore, 1990), 83–101.

[26] Ibid., 83–101.

[27] Tarlow, 'Reformation and Transformation', 115.

[28] Ibid., 116.

hybrids that are far from unusual, including a 1550 funeral sermon preached by the Protestant martyr John Hooper bound within monastic flyleaves, and a copy of the Calvinist Theodore Bèze's *Confessio Christianae fidei* (1565) bound within fragments of a missal, with musical notation visible.[29] Such objects might have supported emerging historiographical narratives, such as the ongoing erasure of multi-sensual Catholic worship by the Protestant religion of the book.[30] The symbolic potential of Catholic breviaries, missals, and commentaries used for 'base' and 'servile' ends, as food wrappers, bookbindings, and toilet paper, is immediately apparent.

But, as Tarlow goes on to argue, fragments from the pre-Reformation past might also have provided a focal point for 'passive resistance'.[31] Iconoclastic practices rarely obliterated their targets entirely; instead, they circumvented, concealed, translated, disguised, and transformed objects. In these altered forms, objects might still communicate their 'older meanings', potentially appropriated to fit within the 'spiritual way of being' of individuals and communities.[32] Like devotional objects hidden away under floorboards, monastic waste might have provided a link between the lost past and its anticipated restoration: often, complete prayers and portions of the liturgy were legible in waste sheets (see Fig. 3.2) and so might have been read and even recited by the highly literate. Although there are no records of such resistant use, extant waste does bear witness to its thoughtful apprehension by later users: early moderns might trace the lines of illuminated text in the bindings of their books (see Fig. 1.2), and, as we will see when we come to consider Aubrey's 'Digression', the sight of broken, monastic books might provoke sadness and a sense of loss, even when the viewer could not interpret their textual contents.[33]

Stocker's 'iconic' reuse might, then, be usefully expanded using Jonathan Gil Harris' framework of responses to 'untimely matter': monastic waste might demonstrate the 'supersession' of the Catholic past, but it might also 'explode' into the present, taking precedence over its current use, either because of the user's confessional identity or because of their desire to recuperate valuable, old books from their waste use.[34] And for some, past and present might sit in conjunction, provoking a recognition of the distinct events and times manifested in waste.[35]

[29] The edition of Hooper (STC 13755) is held at Henry E. Huntington Library, 438880 and the Bèze (STC 2007) at Henry E. Huntington Library, 327725.

[30] See James Kearney, *The Incarnate Text: Imagining the Book in Reformation England* (Philadelphia: University of Pennsylvania Press, 2009), esp. 8–40; Matthew Milner, *The Senses and the English Reformation* (Farnham: Ashgate, 2011), esp. 125–205.

[31] Tarlow, 'Reformation and Transformation', 118.

[32] Ibid., 118.

[33] On remnants of the Catholic past and a sense of loss, see Margaret Aston, 'English Ruins and English History: The Dissolution and the Sense of the Past', *Journal of the Warburg and Courtauld Institutes* 36 (1973): 231–55.

[34] Harris, *Untimely Matter*, esp. 29–30, 91–94.

[35] Ibid., esp. 143–46.

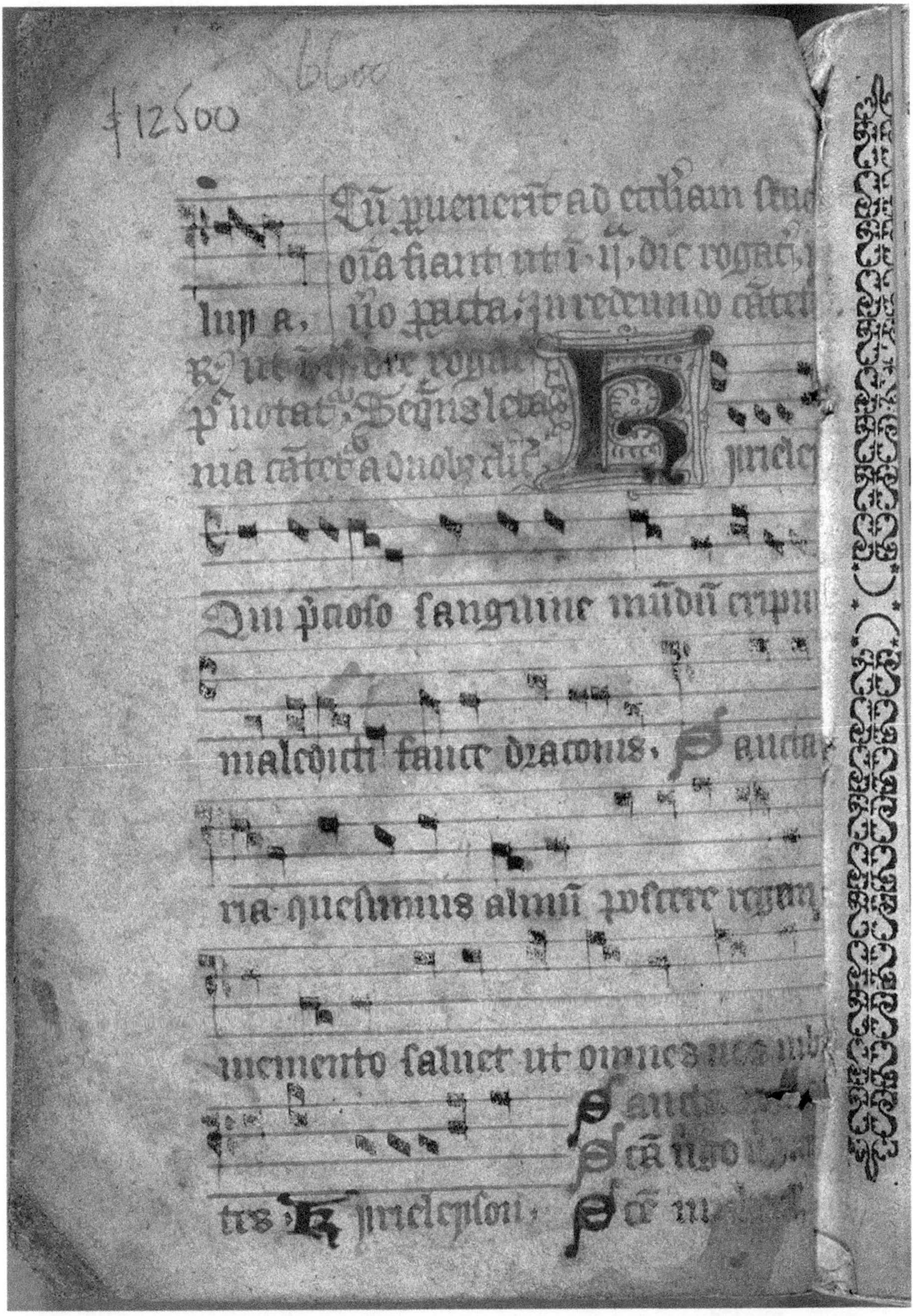

Fig. 3.2 Medieval manuscript service-book used as a wrapper for Sebastian Münster, *A Briefe Collection and Compendious Extract of Straunge and Memorable Thinges*, STC 18243 (London: Thomas Marshe, 1574), 492913. The Huntington Library, San Marino, California.

The direction in which a waste palimpsest might point depended on what type of monastic manuscript it was (was it a missal or a chronicle, for instance, pointing to church ceremony or to the national past?), and on who was using it. Furthermore, an individual user might fluctuate between each of these temporal modes. The rest of this chapter outlines two encounters, or series of encounters, with monastic waste recorded in the work of Bale, editing Leland, and in Aubrey. These writers attempted to make sense of waste fragments and to fit them within their own narratives: Bale celebrated the supersession of popish books, but lamented the supersession of chronicles and histories, seeking to selectively 'revive' their past lives as texts. Aubrey, writing over a century later, surveyed the telescopic history of waste, dwelling on the extended life cycle of the fragments. Their works demonstrate the capacious imaginings that, for more than a century, were generated by monastic waste.

2 John Bale: 'The thynges dyssypated were dyuerse'

John Bale and John Leland's *The Laboryouse Journey* is shaped by monastic fragments. It sheds light on waste practices and the complex statuses of monastic waste in the mid-sixteenth century, as well as the layers of figurative meaning that might accrue around the fragments when, because of their tendency to 'decaye' and 'dyssypat[e]' (A8r) and their presence in grocers' and stationers' shops, they frustrated attempts to collect, catalogue, and construct a Protestant narrative of the Reformation. John Leland was a humanist scholar and author of Latin verses, 'personally loyal' to Henry VIII and eager to uncover the literary triumphs of the nation's past.[36] He sought out Old English manuscripts that might provide theological support and a historical precedent for Henry VIII's divorce and break with Rome. Leland travelled throughout England between 1533 and 1536, and again in 1541 and 1544 in the immediate aftermath of the dissolution.[37] He envisaged producing an immense body of work based on the natural sites and manmade artefacts he encountered on these 'itineraries', beginning with *De uiris illustribus* (a chronology of Britain's literature), derived from the manuscripts and manuscript fragments he had uncovered, as well as a history of Britain (in fifty volumes), a history of the islands neighbouring Britain, a topography of British place names, a history of British universities, and an ecclesiastical history. Instead, several weeks after Henry VIII's death in 1547, Leland was incapacitated. According to a friend, quoted by Bale, Leland had fallen into madness because 'he was

[36] Trevor Ross, *The Making of the English Literary Canon: From the Middle Ages to the Late Eighteenth Century* (Montreal: McGill-Queen's University Press, 1998), 51–64; James Simpson, *Reform and Cultural Revolution* (Oxford: Oxford University Press, 2002, 7–33.

[37] This chronology is taken from Carley, *De uiris illustribus*, xxv–cxxxviii and John Chandler, *John Leland's Itinerary: Travels in Tudor England* (Stroud: Sutton Publishing, 1993), xii–xxiv.

vayne gloryouse, and that he had a poetycall wytt' (B4r). Leland did not recover and died in 1552. He left behind some printed Latin verse, a mass of unedited manuscript material (*De uiris illustribus*), and a letter composed subsequent to his 1543 itineraries and perhaps presented to the King as a 'New Year's Gift' on January 1st 1544. It described his bibliographic achievements and outlined plans for the expansive projects listed above.

John Bale's friendship with Leland seems to have begun in 1533 when Bale, at this time a monk, travelled between the Carmelite and Austin houses compiling *Anglorum Heliades*, a bibliographic catalogue of the houses' monastic authors.[38] Both Leland and Bale, therefore, worked independently of one another to record the nation's literary and religious history on the eve of the dissolution, halting their peregrinations as the lesser houses began to fall. This might have been to avoid suspicion of pro-Catholic sympathies, but this accusation was unlikely to have been levelled at Bale for long. It was around this time that he defected from the Carmelites, taking up the post of priest at Thorndon, Suffolk, and setting to work composing morality plays that allegorized Protestant reform. With the fall of Cromwell in 1539 and a backsliding, from Bale's perspective, to popish practices, Bale fled to the continent. He penned a range of polemical and bibliographic works, including his *Summarium* of British manuscripts. On receiving news of Leland's incapacitation in 1547, Bale hurried this catalogue into print. Returning to England at the accession of the reformist Edward VI, Bale began to edit Leland's incomplete *De uiris illustribus* whilst pursuing his own ambitious bibliographic work, assembling a list of the authors, titles and opening lines of all extant and noteworthy British books. The product was, eventually, his *Catalogus*, printed between 1557 and 1559.[39] This was his second itinerary, undertaken between 1548 and 1552 in the midst of Edward's iconoclastic reforms, and before a second period of exile during the reign of Mary I.

As with Leland's second journey, carried out after the dissolution, there was an increased sense of urgency: no longer just recording, Leland and Bale saw themselves as salvaging British history from the oblivion of mould, decay and waste. Both sought to produce what James Carley has termed 'bio-bibliographies', catalogues that outlined the history of Britain through its ancient writers. Both also saw the products and evidence of Britain's history slipping through their fingers, turned instead toward 'base ends', because neither writer had the means to gather up, transcribe, and store every manuscript that they came across.[40] In the midst of his prolonged itinerary, in 1549, Bale printed and enlarged Leland's manuscript

[38] See Leslie P. Fairfield, *John Bale: Mythmaker for the English Reformation* (West Eugene: Wipf and Stock Publishers, 1976); John N. King, *English Reformation Literature: The Tudor Origins of the Protestant Tradition* (Princeton: Princeton University Press, 1982); Carley, *De uiris illustribus*, cvii–cxii.

[39] Bale's notebooks are held at Bodleian Library, MS Selden Supra 64.

[40] Carley, *De uiris illustribus*, xxvii.

letter, a 'New Year's Gift', which had probably been sent to him a few years earlier by the humanist and reformer Sir John Cheke. Inserting a lengthy dedication, commentary, and conclusion as well as details of his own antiquarian labours, Bale retitled it *The Laboryouse Journey.*

This text intermingles two somewhat contradictory perspectives of the dissolution. Leland, like his patron Henry VIII, remained religiously conservative and, as a humanist, sought to recover old manuscripts primarily for their philological value. But Bale was a passionate reformer who celebrated the dissolution: he only wished its 'dyssypat[ion]' had been less 'diuerse' (A8r). Bale was happy for popish trash to be sold as waste, writing how

> Yf the byshop of Rome[']s lawes, decrees[,] decretals, extravagantes, cleme[n]tines and other suche dregges of the devyll [...] and frutes of the bottomlesse pytte, had leaped out of our libraries, and so become coverynges for bokes [...] we might wele haue ben therwith contented. (G3r)

Such wasting would represent the proper state of affairs, with the Catholic manuscripts' corrupt nature neatly transforming into a new usefulness. As we have seen, monastic waste binding Protestant books might offer an emblem of the Reformation, and here Bale expresses hopes for a practice that did in fact come to be commonplace. The Huntington Library holds a particularly pertinent manifestation of this supersession of the Catholic past: their 1560 edition of Bale's *The Acts of English Votaries* (first printed in Antwerp in 1546) is bound in fragments of a monastic service-book (see Fig. 3.3). A work that attacked the corrupt practices and institutions of the monasteries is 'guarded' by the remnants of those very institutions.

If only, Bale laments, the wasting hadn't been so indiscriminate and efforts had been made to safeguard the 'liuelye memoryalles of our nacyion' (A7v), the '[m]oste olde and autentyck', and largely Old English, 'Chronycles' (C2v). These might provide ample evidence for 'the usurped autoryte of the Byshopp of Rome and hys complyces' (C5r), and, by demonstrating the chronological precedence of Protestantism, legitimize the new religion in the face of Catholic attacks on its novelty. England, as Leland's (sometimes erroneous) discoveries make clear, was in the process of returning to a purer, pre-Popish past, where, for instance, 'Kynge Athelstane' had 'the scriptures [...] translated into the Saxonysh or Englyshe speche' (D2v). Although an ancient vernacular Bible was never actually recovered, Bale proclaims that Leland's labours uncovered both popish lies and English truths from the rubble of the dissolution. This narrative was not Bale's alone, but Bale was a key early proponent. Ronald Harold Fritze has outlined how a Protestant humanist approach to history led to the birth of antiquarianism and, in the work of Bale, a Protestant chronology and hagiography that would prove crucial

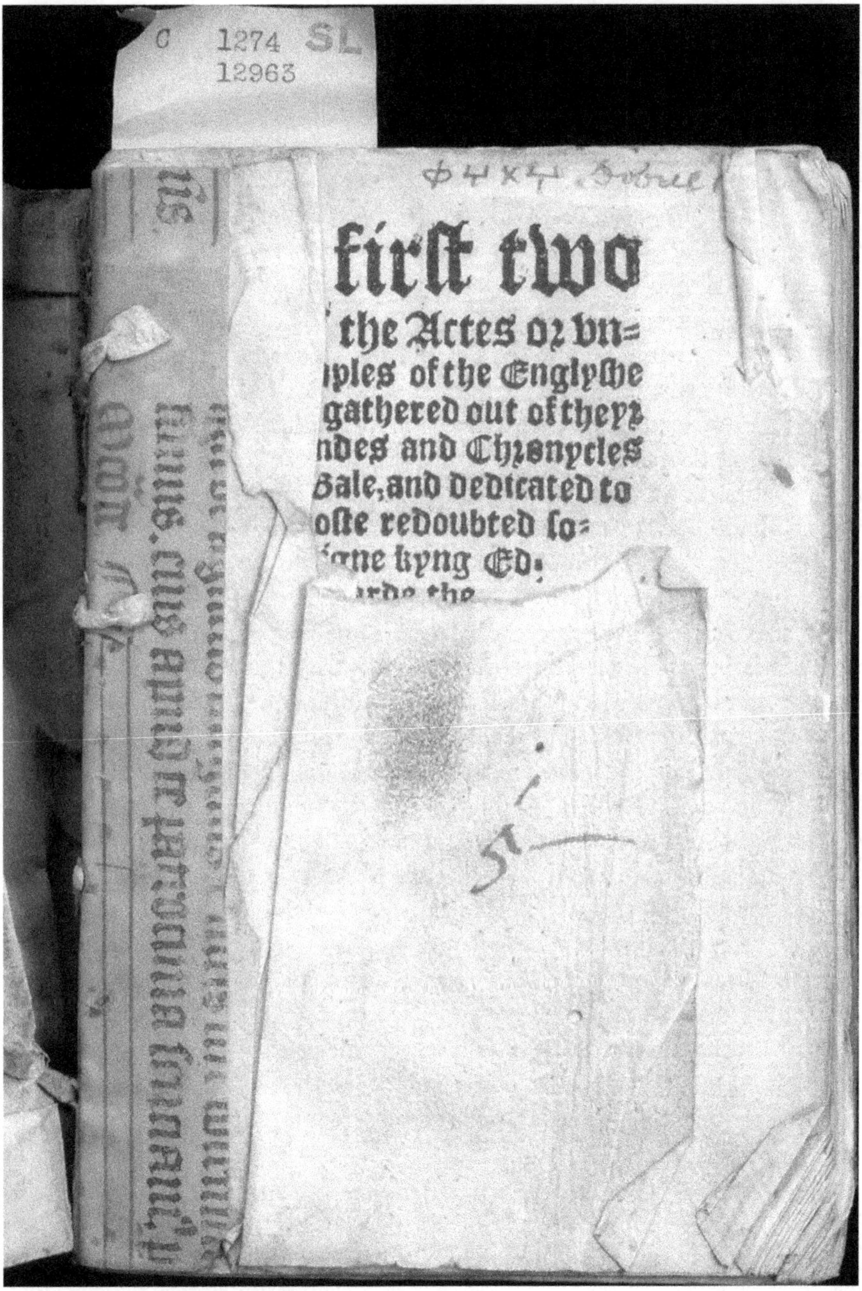

Fig. 3.3 Parchment fragments of a Catholic service-book used as guards in John Bale, *The First Two Partes of the Acts or Vnchaste Examples of the Englyshe Votaryes*, STC 1274 (London: Iohn Tysdale, 1560), 12963. The Huntington Library, San Marino, California.

to later reformers such as John Foxe.[41] Scholars such as Leland, Bale, and Matthew Parker gathered newly accessible histories and Old English texts from the dispersed libraries, selecting from these the malleable material with which to write a patriotic history of Protestant renewal.

The *Journey*, then offers a complex combination of Protestant triumph and an antiquarian sense of loss.[42] Bale situated Leland's bibliographic itineraries, and implicitly his own, within the dominant narratives of the Reformation. Leland, in his 'New Year's Gift', described how he had brought 'the monumentes of auncyent wryters [...] out of deadly darkenesse to lyvelye lyght' (B8r), and Bale elaborated further: these books had been previously 'tyed up in cheanes' (C3r), concealed, like the scriptures and the true faith, in 'uncertayne shadowes' and popish deceit (C6v). Leland's labours were, according to Bale, Christ-like, as he harrowed the 'deadly' dark spaces, dragging their contents into the Protestant light. Leland 'wold clerely redeme them from dust and byrdfylynges' (C2v), granting the manuscripts salvation from a uniquely bibliographic kind of damnation. This narrative is, though, undermined by the lived experience of Bale in the post-Reformation landscape: the dispersal of scripture and knowledge for the benefit of the nation was accompanied by a more literal 'scatter[ing], or spread[ing] abroade' of the pages of old manuscripts.[43] The *Journey*, rather than demonstrating Protestant supersession, makes visible the bifurcating value of waste in this historical moment. Most book-reusers in sixteenth-century England did not adhere to Bale's taxonomy of monastic manuscripts, valuing both historical chronicles and Catholic service-books according to their material capacities and their weight and not their textual contents. This was, according to Bale, a symptom of 'Auaryce', with the waste practitioners resembling the sinful monks. Those who participated in the trade in monastic manuscripts are 'bellygoddes' or gluttons (A8r), a slur more usually employed in anti-Catholic polemic.[44] Like the Catholic authorities who were guilty of avarice in their sale of indulgences and exploitation of the doctrine of Purgatory, those who repurposed the wrong sorts of old books, according to Bale, sought only financial gain.

But the distinction between quasi-popish 'bellygoddes' who transformed old books into waste and the philologically and theologically driven antiquarians who sought to preserve a highly valued sub-section of manuscripts was not so clear-cut.

[41] Fritze, '"Truth hath lacked witnesse"', 274–291. See also Phillip Schwyzer, *Literature, Nationalism and Memory in Early Modern England and Wales* (Cambridge: Cambridge University Press, 2004), 49–75 and Summit, *Memory's Library*, 101–35.

[42] On the relationship between the dissolution and the development of antiquarian thought, see Angus Vine, *In Defiance of Time: Antiquarian Writing in Early Modern England* (Oxford: Oxford University Press, 2010), 7–9; Aston, 'English Ruins and English History', 231–255; Yale, 'With Slips and Scraps', 1–36; and Graham Parry, *The Trophies of Time: English Antiquarians of the Seventeenth-Century* (Oxford: Oxford University Press, 2007), 2–19.

[43] 'Disperse' in Cawdry, *A Table Alphabeticall*, D2r.

[44] See Elena Levy-Navarro, *The Culture of Obesity in Early and Late Modernity: Body Image in Shakespeare, Jonson, Middleton, and Skelton* (Basingstoke: Palgrave Macmillan, 2008), 116.

In the mid-sixteenth century, antiquarians like Leland, Bale, and Parker did not collect old manuscripts to keep them whole.[45] The volumes in Parker's library are notoriously composite: he glued and stitched leaves from Old English and medieval manuscripts into his own volumes according to his political and theological needs.[46] We can only assume that the copies from which these fragments were extracted were otherwise wasted and discarded. Similarly, Bale, in the *Journey*, describes a fantasy in which the 'notable Antiquyte[es]' should 'be stayed in time, and by the art of pryntynge be brought into a no[m]bre of coppyes' (B2r). Old manuscripts, then, were valued as vehicles for iterable text, rather than as objects: once the textual content had been extracted, the parchment and ink became an empty shell, worth only as much as a 'bellygodde' was willing to pay for it. This process of de-accession and replacement was a continuation of scribal practice and is visible in the manuscript 'proof' sheets, surviving in bindings, from the shops of early printers such as Aldus Manutius and William Caxton.[47] It was only several decades later that a sense of old manuscripts as historical artefacts, aesthetically as well as philologically valuable, emerged.[48] The library of Sir Robert Cotton, for instance, indicates the complexity of antiquarian interactions with monastic manuscripts: he dismembered duplicate or unwanted works, inserting them as 'stuffing' or binding waste in other, partially disassembled books.[49] But this '"cut and paste" approach' was often aesthetically driven, with fragments of highly illuminated works used as decorative borders, frontispieces and end-leaves in other manuscripts and printed books.[50] We begin to see a more consistent valuation of old manuscripts as unique objects, collected as both textual and physical records of the past, later in the seventeenth century in the practices of antiquarians such as Anthony Wood, John Bagford, and, as we will see, John Aubrey.[51]

Bale, writing in the years immediately following on from the dissolution, evidently hoped that the valuable, old texts might be extracted and abstracted, and

[45] Kearney, *The Incarnate Text*, 136; Simpson, *Reform and Cultural Revolution*, 14–17; Summit, *Memory's Library*, 109–110.

[46] Knight, *Bound to Read*, 41–50; Timothy Graham, 'Matthew Parker and His Manuscripts: A Study of an Elizabethan Library and Its Use', in *The Cambridge History of Libraries in Britain and Ireland*, vol. 1, eds. Elisabeth Leedham-Green and Teresa Weber (Cambridge: Cambridge University Press, 2006), 328.

[47] See Lotte Hellinga, *Texts in Transit: Manuscript to Proof and Print in the Fifteenth Century* (Leiden: Brill, 2014), 108–111, 146–47, 420.

[48] Daniel Woolf, *The Social Circulation of the Past: English Historical Culture, 1500–1730* (Oxford: Oxford University Press, 2003), esp. 141–44. See also Vine, *In Defiance of Time*, 16–17.

[49] James P. Carley and Colin G. C. Tite, 'Sir Robert Cotton as Collector of Manuscripts and the Question of Dismemberment: British Library MSS Royal 13 D.I and Cotton Otho D. VIII', *The Library* 6, no. 2 (1992): 94–99.

[50] Michelle P. Brown, 'Sir Robert Cotton, Collector and Connoisseur?', in *Illuminating the Book: Makers and Interpreters, Essays in Honour of Jane Backhouse*, ed. by Michelle P. Brown and Scot McKendrick (London: The British Library, 1998), 291–98.

[51] On the malleability of early modern books more broadly see Knight, *Bound to Read* and Adam Smyth, '"Shreds of holinesse": George Herbert, Little Gidding, and Cutting Up Texts in Early Modern England', *ELR* 42, no. 3 (2012): 452–481.

therefore preserved indefinitely by print. The belief that they would then be 'stayed [...] in time' (B2r), frozen and immutable, was in part a response to the fragmentation and decay of the manuscripts that Bale encountered throughout his itineraries. His description of the 'auncient Chronicles' as 'monumentes of lernynge' (D6v) hints at an understanding of textuality that resembles Horace's enduring *monumentum*: often blurred with 'muniment', a legal document that preserved rights and privileges (from the Latin *mūnīmentum*, meaning fortification), a monument might refer to either a written text or an object.[52] The term is always laden with a particular set of temporal and material attributes, present in Horace's ode. It was, from its earliest use, synonymous with 'tombe', and later, with 'Sepulchre, Statue, Pillar, or the like'.[53] Bale's use of the term might, then, be understood as expressive of a desire for old manuscripts to take on, or even exceed, the fixity and hardiness of a stone effigy or a brass inscription, so that they might preserve and uphold his narrative of England's religious past. But the monastic waste of England's present interrupts and explodes into the *Journey*, undermining Bale's hopes for transcendent printed texts and a Protestant mythos upheld by complete and accessible Old English histories. Bale, unlike Parker, never actually edited and reproduced an Old English manuscript. His personal library, comprising largely chronicles, was lost when he fled from Catholic priests in Ireland in 1553, just as Henry VIII's Royal Library (and its manuscripts, gathered by Leland) was dispersed after the King's death.[54] The fate of these two libraries suggests the broader vagaries of monastic manuscripts after the dissolution, and demonstrates the vulnerability of paper and parchment objects in the sixteenth century even after they had been salvaged by antiquarians and entered into a collection. Bale's own textual output (the *Journey*, *Summarium*, and *Catalogus*) are, above all else, monuments to lost books and records of ghostly waste fragments.

Bale describes the experience of gathering the material for his *Catalogus*, a work that, like Leland's *De uiris illustribus* and his earlier *Summarium*, would be a list of 'the names of the[m]' who 'hath bene learned and who hath written from tyme to tyme in this realme' (C7v): in addition to seeking out works in the decaying monastic libraries, according to the *Journey*, Bale sought out fragments of manuscripts that had already been sold as waste, and set to work deciphering what remained of their textual content. He writes how,

[52] Hui, *The Poetics of Ruin*, 195.

[53] Nigel Llewellyn, *Funeral Monuments in Post-Reformation England* (Cambridge: Cambridge University Press, 2009), 218; Edward Phillips, *The New World of English Words*, Wing P2068 (London: E. Tyler or Nath. Brooke, 1658), CC4r.

[54] James P. Carley, 'The Royal Library under Henry VIII', in *The Cambridge History of the Book in Britain*, vol. 3, eds. Lotte Hellinga and J. B. Trapp (Cambridge: Cambridge University Press, 1999), 274–82; Honor McCusker, 'Books and Manuscripts Formerly in the Possession of John Bale', *The Library* 16, no. 2 (1935): 144–65.

> Among the stacyoners & boke bynders, I found many notable Antiquitees, of whom I wrote out the tytles, tymes, and begynnynges, that we myghte at the leaste shewe the names of them, though we have not as now, their whole works to shewe. (G2v–3r)

He goes on to describe his broader experience of waste in Norfolk and Suffolk in the 1540s: how sheets were used, not only as pastedowns, flyleaves, guards, and limp-bindings, but also 'turned to the vse of their grossers, candelmakers[,] sope sellers, and other worldly occupyers', and other such 'office[s] of subjeccyon & vtter conte[m]pte' (G3r–v). These fragments can only produce fragmentary texts: the 'tytles, tymes, and begynnynges', not 'their whole works' (G3r). If only, Bale laments, 'ye had their whole workes in dede, as they were in substaunce & fashyon, whyche now for the more part are peryshed, ye shoulde have seane most wonders of all' (H5v). These works, 'for the more part [...] peryshed', were perhaps the leaves of manuscripts catalogued by Bale that are no longer extant, such as Frithegod of Canterbury's *De visione beatorum*, *Contemplationes*, or *Life of Ouen*, or Gerard of Nazareth's *De conversatione*.[55] Their fate resembles that of the fragments of pre-Reformation manuscripts that have been removed from early modern bindings, either by late seventeenth-century antiquarians such as John Bagford, or the medievalists and rare book librarians of later centuries. Bagford's fragments, for instance, held at the British Library, contain a fragment of a ninth-century manuscript of Justinus' *Epitome* which appears to have been trimmed and folded to form the wrapper for a pamphlet, and later bibliographers have discovered numerous other Old and Middle English treasures, including a fragment of a ninth-century Bible used as a stiffener in a semi-limp binding, and a fragment of the Metrical Psalms in the twelfth-century Eadwine Psalter.[56] We catch glimpses of these objects in Bale's *Journey*, in and amongst his idealized 'monumentes of lernynge' (D6v), and, even in these brief descriptions, they manifest their extended life cycle and oscillating value. They make visible how such fragments moved from the status of monastic text to post-Reformation waste, but how they might also be salvaged as objects of value and restored as texts, albeit partial ones, by later collectors and antiquarians.

The monastic waste described in the *Journey*, then, tells a history of use and shifting value rather than one of 'redem[ption] [...] from dust and byrdfylynges'

[55] Michael Lapidge, 'Frithegod of Canterbury', *Anglo-Saxon England* 17 (1988): 48–49; Andrew Jotischky, *The Perfection of Solitude: Hermits and Monks in the Crusaders States* (University Park: University of Pennsylvania Press, 1995), 18.

[56] Julia Crick, 'An Anglo-Saxon fragment of Justinus' *Epitome*', *Anglo-Saxon England* 16 (1987): 181–196; Michelle P. Brown, 'A New Fragment of a Ninth-Century English Bible', *Anglo-Saxon England* 18 (1989): 33–43; Patrick P. O'Neill, 'Another Fragment of the Metrical Psalms in the Eadwine Psalter', *Notes and Queries* 233 (1988): 434–436.

(C2v). But Bale does more than describe his frustrating encounters with these fragments: he is also sensitive to the material biographies visible in waste. In addition to containing the 'tytles, tymes, and begynnynges' (G3r) of manuscripts that Bale wishes still survived, the waste fragments of the *Journey* possess figurative significance and provide Bale with a potent set of concrete metaphors: Bale considered undiscerning waste practices, as we have already seen, as symptoms of avarice. Prioritizing 'belly ba[n]kettes & table tryu[m]phes' over 'the conservacyon of [...] Antiquytees' (B1v–2r), these greedy 'bellygoddes' abandon manuscripts to 'rotte in vyle corners, or drowne them in [their] jakes' (E7r). Found 'amonge wormes and dust' (E7r) in monastic libraries, rotting in private hands or privies, Bale's language highlights the organic nature of the manuscripts, and how, exposed to the elements, they have begun to decay: they are animal skin or pressed vegetable matter turned text and then waste, and are rubbed against the leather of 'boots' and the soap, spices, and foodstuffs of the 'grosser's shop' (B1r). When turned to 'serve our jakes', they are rubbed against human skin and excrement (B1r). Bale goes on to describe how 'we abhorre & throwe fourth' our 'noble workes [...] as most vyle, noysome matter', regarding them as little as 'ye parynges of our nayles' (E7v). By comparing discarded manuscripts with excremental off-cuts, Bale expresses his disgust at those who incorrectly value ancient texts. The phrase also suggests his haptic attentiveness to the objects that he encountered in binders' and stationers' shops. Waste strips of parchment, cut up to form guards and not yet discoloured by centuries of waste use, perhaps reminded Bale of nail trimmings (see, for instance, Fig. 3.3). Parchment and fingernails, after all, share a distinctive, milky off-white colour, and both are easily ingrained with dirt. Both are stiff but flexible and might contain traces of other corporeal textures and matter.

These material histories and organic textures are, at the conclusion of the *Journey*, used to colour another narrative that Bale often returned to in his religious writings: he complains that whereas men of old laboured

> to holde thynges in remembraunce, whych otherwyse had most wretchedly peryshed. Our practyses now are [...] to destroye their frutefull fou[n]dacyons. [...] [W]e in these dayes are as prompte to plucke down (I mean the monumentes of lernynge) as though the worlde were now in hys lattre dottynge age, nygh drawynge to an end. (E6v–7r)

The 'fou[n]dacyons' of early modern England and its 'remembraunce' of the past are unstable and under threat after the dispersal of the monastic libraries. They have been corroded because people find it hard to 'hold' fast onto godly 'thynges', and instead abject and expel them as waste. Though it might at first seem peculiar that an ardent reformer such as Bale would lament the destruction of 'fou[n]dacyons' and 'pluck[ing] down' that took place during the dissolution of the monasteries, he is here making a nuanced distinction between the 'perysh[ing]' of books ('the monumentes of lernynge') and the 'fou[n]dacyons' of the monasteries themselves. This bibliographic cataclysm, though in part of Catholic books, is symptomatic of a decidedly Protestant narrative in which the 'practyses' of 'pluck[ing]' and pulling apart books takes on an eschatological significance. As Leslie Fairfield has demonstrated, Bale subscribed loosely to a Lutheran religious chronology, espousing a 'sense of senescence and decay' resulting from 'living at the "latter end of the world"'.[57] Bale's catalogues, including his editorial work on Leland's *De uiris illustribus*, were structured according to the progression of the 'seven ages of the world', laid out at length in his *Image of Both Churches* (1545). This chronology progressed through a procession of British authors, beginning in apostolic purity, moving through the dawning of an enlightened age with the writings of Wycliffe, and concluding in the corrupted present day—the sixth and penultimate age—with the seventh and last age fast approaching.[58] The waste 'remnaunt[s]' (F4v) of monastic manuscripts, familiar to Bale through his search for England's antiquities in binders' shops, as well as more everyday encounters with the wrappers of food, spices, and soap, materialize this apocalyptic historiography. It seemed reasonable to think that the post-Reformation world was teetering on the brink of destruction: it was full of corruptible things that were, like waste sheets, 'nygh drawynge to an ende' (E7r).

Bale's relationship with monastic fragments was, the *Journey* makes clear, a vexed one: the wasting of popish 'dregges' was, from the perspective of a reformer such as Bale, a necessary consequence of the Reformation, but his desire to collect and 'stay' a subsection of these 'dyssypated' manuscripts 'in time' produced contradictory representations of such objects (B2r). While he fantasized about the possibility of solid and stable textual monuments, with the old manuscripts that conformed to reformist ideals imagined as fixed in print and time, the vulnerability and fragmentation of all the monastic manuscripts is revealed to be the reality in sixteenth-century England. In addition to celebrating the triumphs of Protestant, antiquarian labour, then, Bale's edition of Leland's 'New Year's Gift' tells the material and imaginative history of monastic waste as it fluctuated radically in value and came to structure broader narratives of temporality and decay.

3 John Aubrey: Reading Wrack and Ruin

A number of Leland and Bale's near-contemporaries also gathered manuscripts from the dispersed monastic libraries, and an account of early modern encounters with binding waste might limit itself to this lineage of sixteenth-century antiquarians. One of the striking features of monastic waste is, however, its capacity to

[57] Fairfield, *John Bale*, 75–85.
[58] King, *English Reformation Literature*, 66–67; Fairfield, *John Bale*, 99–120.

endure, and so an assessment of these objects should account for their manifold afterlives that reach into the seventeenth century as well as into contemporary rare books libraries and collections. As we have already seen, Anthony Wood read Bale's account of the fates of monastic books carefully and incorporated it into his history of the University of Oxford, a project he worked on from the late 1660s until his death in 1695. Wood was not alone in seeking out evidence of the distant past in these decades: the origins of English antiquarianism are often traced back to the itineraries of Leland in the 1530s, but the nature of antiquarian thought and practice altered radically over the subsequent centuries. By the end of the seventeenth century, Daniel Woolf has argued, antiquarians no longer valued manuscripts, monuments, and inscriptions only for the philological or religious information they might provide.[59] Such objects were less often discarded after their use, as in the case of Leland, Bale, and Parker, and instead antiquarians prized antiquities as valuable items in their own right. Although collection still necessitated selection and disposal, there was a growing emphasis on material preservation, and duplicates were often transferred between networks of scholars rather than wasted or discarded. Woolf emphasizes that this was the result of a changing understanding of time: old objects were no longer everyday 'fellow-travellers' or 'casual survivors' of the past.[60] Instead, they were perceived as being 'out of time', exotic curios and 'intruders on a mental landscape' that was now concerned with the innovation and discovery that might be achieved through empirical observation and natural philosophy.[61]

Monastic manuscript waste was one such time-traveller. These objects proved particularly compelling to John Aubrey, Wood's contemporary and on and off collaborator. In fact, Aubrey, despite the chronological distance from which he wrote, was more sensitive to the multiple histories contained within waste manuscripts than any of his contemporaries or predecessors. These imaginative encounters are recorded in an unusual 'Digression' within the 1690–1 version of his *Natural History of Wiltshire*, begun in 1656 and re-copied at this later date at the request of the Royal Society.[62] The 'Digression' stands out because in it, old manuscripts are not simply textual sources, but neither are they precisely curios, artefacts taken out of time and slotted into a *Wunderkammer* of exotic things.[63] Aubrey views the fragments neither as a vehicle through which to straightforwardly access the past, nor as relics, as Woolf's argument might suggest, from a pre-modern time distinct from his own. Instead, Aubrey reads them as dynamic palimpsests that participate in his own world, but which also contain overlapping layers of history sedimented within

[59] Woolf, *The Social Circulation of the Past*, esp. 141–144. See also Vine, *In Defiance of Time*, 16–17.

[60] Woolf, *The Social Circulation of the Past*, 141, 12. Woolf's understanding of objects has much in common with Harris' *Untimely Matter*. See pp. 21–22 above.

[61] Woolf, *The Social Circulation of the Past*, 184, 141.

[62] Aubrey's 'Digression' is reproduced in Yale's 'Of Slips and Scraps', 1–2.

[63] Harris, 'The New New Historicism's Wunderkammer of Objects', 117. See pp. 16–17 above.

their surfaces and folds. From these, he shapes a series of stories that incorporate the past into the present.

'Story-telling' and 'incorporating past into present' are key terms with which archaeologists Michael Shanks and Christopher Tilley 're-construct[ed]' and theorized their discipline in the late 1980s.[64] They argued that archaeology records 'a world where experience exists as continuity and flow', and where 'meaning is established by constructing figurations out of successions of events': these narrations gather around the solid stuff of material culture, but objects are not in themselves 'locked' within a 'fixed and unchanging' past, and their meanings are not finite.[65] Shanks and Tilley's argument would have made perfect sense to Aubrey, writing as he was in the decades in which archaeological, as well as antiquarian, practice and thought, were rapidly developing.[66] His writings show a sensitivity to the temporal continuum inhabited by objects, as well as the manner in which this temporality is characterized by moments of rupture, conflict, and discontinuity. His narrative is not 'coherent' and 'consoling', with a beginning and an end.[67] Instead, Aubrey tells a story of personal encounters with waste things that pre-date him, and will, in many instances, outlast him. He narrates how, over a century and a half, waste fragments moved from the category of transient 'rubbish' to 'durable' artefacts, newly inscribed with ever-increasing aesthetic and financial value.[68] But these fragments always, in Aubrey's 'Digression', remember their previous state and suggest the precariousness of the present. Through them, Aubrey looks back across two centuries of bibliographic turmoil, but this past is neither straightforwardly linear, nor wholly distinct from his present. Recollecting growing up in the decades in which monastic waste was still in widespread use, he structures a miniature biography, a local chronology, and a broader, national history grounded in the nature and shape of fragments as they exist in time.

Aubrey begins his narrative with the often-quoted phrase: '[i]n my grandfathers dayes, the Manuscripts flew about like Butterflies'.[69] These were the days of his maternal grandfather, Isaac Lyte (1576–1660), and this makes clear that as late as the last quarter of the sixteenth century monastic waste was perceived as ubiquitous, highly mobile, and ephemeral. At this time, 'All Musick bookes, Account bookes' and 'Copie bookes &c'. were, according to Aubrey's grandfather, 'covered with old Manuscripts. [...] And the Glovers at Malmesbury made great Havock of them; and Gloves were wrapt up no doubt in many good pieces of

[64] Michael Shanks and Christopher Tilley, *Re-Constructing Archaeology: Theory and Practice* (Cambridge: Cambridge University Press, 1992), esp. 17–20.

[65] Ibid., 17–20.

[66] Vine, *In Defiance of Time*, 22–50.

[67] Shanks and Tilley, *Re-Constructing Archaeology*, 19.

[68] Thompson, *Rubbish Theory*, 26–29; Frow, 'Invidious Distinction', 35.

[69] Yale explains that, for Aubrey, 'manuscripts' are documents 'written on parchment or vellum before the advent of printing' and 'papers' are 'contemporary 'loose sheets, notes from experiments and observations', 'Of Slips and Scraps', 4.

Antiquity'. Although by the second quarter of the seventeenth century the world no longer brimmed with fluttering fragments, monastic waste was a memorable participant in Aubrey's childhood. He recounts that in 1633, aged seven and a pupil of 'the Latin-Schoole at Yatton-Keynel', it was the 'fashion then [...] to save the Forules of their Bookes with a false cover of Parchment sc{ilicet} [that is to say] old Manuscript'. Suggesting that the monastic waste was both a functional practice and a 'fashion' or custom, he recalls paying detailed attention to these objects and credits his childhood self with a deep appreciation of them. Although he 'was too young to understand' the textual content of the sheets, he 'was pleased with the Elegancy of the Writing and the coloured initiall Letters', perhaps tracing, as we know other readers did, the illuminated letters on the surfaces of their bindings. Aubrey recalls a sensual interaction with the manuscripts which would prove to be a prominent antecedent for the antiquarian engagements of his adult life.[70]

The school that Aubrey next attended in the nearby parish of 'Leigh-Delamer' undertook 'the like use of covering Bookes', but 'Blandford-Schoole' in Dorset, which he attended from the age of 12, did not. Although the schoolboys covered their books with 'old Parchments' such as 'Leases &c'., Aubrey 'never saw any thing of a Manuscript there'. This was because 'Here about were no Abbeys or Convents for Men'. The 'Digression', therefore, records both the longevity and the geographical distribution of monastic waste. In parishes containing and bordering dissolved monasteries, ecclesiastical manuscripts were available for wasting long into the seventeenth century. Elsewhere, waste practices were founded on a more diffuse range of disposable, parchment texts, such as expired legal documents. Aubrey's local environment brimmed with ecclesiastical estates, 'for within half a dozen miles of this place' were Malmesbury, Stanley, Monkton Farleigh, Bath, and Cirencester Abbeys, along with Bradenstoke Priory. Of these, only Malmesbury and Bath had been adapted for parish use. The others had been 'digged up', plundered for building materials and left to ruin. Stone ruins and malleable manuscript sheets were experientially interconnected, both manifesting, in the words of Margaret Aston, 'the gashes' and 'scars of earlier destruction' and shaping a nostalgic and historically driven 'sense of the past'.[71] Aubrey imaginatively reconstructs these scars, repopulating the library shelves and reading the absent wholes of potentially precious library books into the waste fragments encountered in his schooldays: 'it may be presumed the Library' of Malmesbury Abbey 'was as well furnished with choice Copies, as most Libraries of England', he writes, teasing himself with what-might-have-beens, conjecturing that 'perhaps in this Library we might have found a correct Plinys Naturall History'.

[70] Kate Bennett, for instance, records how Aubrey removed a parchment 'Cover' (a ninth or tenth century text of Ecclesiasticus) from a contemporary astrological work, inserting it into his manuscript *Monumenta Britannica*. He sought to analyse the manuscript as part of his larger palaeographical project ('John Aubrey and the Printed Book', *Huntington Library Quarterly* 76, no. 3 (2013): 94).

[71] Aston, 'English Ruins and English History', 243.

The 'Digression' moves on to delineate local personalities through their encounters with loose parchment pages. He describes the rector of Malmesbury, William Stump, great-grandson of a wealthy clothier who had purchased the site of the Abbey and its neighbouring lands after the dissolution. '[S]everall Manuscripts of the abbey' had passed down the generations of Stumps, surviving for more than a century, stowed away for special use. '[W]hen He brewed a barrel of speciall Ale', Aubrey recalls, 'his use was to stop the bung-hole (under the Clay) with a sheet of Manuscript: He sayd nothing did it so well which me though it did grieve me then to see'. Whereas the young Aubrey relished the luxurious (if faded) shapes and surfaces of waste sheets, Stump appreciated their capacity to mould and fit within a bung-hole whilst remaining relatively water-tight. Although this 'grieve[s]' the grown-up Aubrey, both engagements, ostensibly dissimilar, are rooted in the hapticity of waste. There is also a strong whiff of iconoclasm in Stump's saving of monastic manuscript waste for 'speciall' occasions, recalling Sarah Tarlow's reading of the 'heated symbolic environment' of post-Reformation England.[72] It seems likely that the rector drew deliberate attention to the rejection and supersession of the Catholic past on religious holidays, relishing the sight—and perhaps even the imagined taste—of medieval manuscripts subordinated to his special beer.

Stump's sons, we learn, took after their father. In 1647, when Aubrey was 21 years old and Civil War had ravaged the landscape for several years, he returned to the rector's house 'out of curiosity to see his Manuscripts, whereof I had seen some in my Childhood'. They were, however, 'lost, and disperst: His sonns were Gunners, & Soldiers, and scoured their Gunnes with them'. This anecdote is indicative of Aubrey's fluctuation between different scales of storytelling. He intermittently expands from the autobiographical and local to the wider cultural histories told by waste sheets. Aubrey's description of a secondary 'los[s] and dispers[al]' of manuscripts in the 1640s blurs the religious and political turmoil of the sixteenth and seventeenth centuries. 'Before the late warres', Aubrey writes, 'a World of rare Manuscripts perished here about'. At first glance, he seems to be referring to the relatively recent events of the Civil War, when ecclesiastical buildings and objects underwent a new wave of iconoclastic violence.[73] It is, after all, a critical commonplace to credit the Civil War with shaping Aubrey's antiquarian mindset, instilling in him an 'acute sense of [...] impermanence'.[74] But for Aubrey, the chronology of destruction was not so clear-cut. Civil War iconoclasm was primarily directed at the early seventeenth-century 'innovations', though in practice it was largely indiscriminate, with late medieval survivals destroyed alongside Laudian introductions. It is unlikely, however, that large numbers of monastic manuscripts survived

[72] Tarlow, 'Reformation and Transformation', 115.
[73] Aston, *England's Iconoclasts*, 84–95.
[74] Parry, *The Trophies of Time*, 276.

outside of private hands in the 1640s.[75] Instead, it is probable that 'Before the late warres' refers to a much longer, more vaguely defined stretch of time, reaching backward across a century of dissolutions and reformations from the period of Civil War to the dissolution of the monasteries. Throughout these decades, the old parchment books comprised an inhabitable but gradually shrinking 'World', undergoing repeated acts of dispersal and dismemberment, until, following the violence of the Civil War, the stockpiles of available monastic waste finally dwindled.

The 'Digression' is a multitemporal text. Moving between descriptions of the days of his grandfathers, his great-grandfathers, and his own experiences, Aubrey dips in and out of traditionally discrete events and periods of time. This multitemporality is grounded in the waste objects themselves. As Harris argues, we cannot 'separate time into a linear series of units'.[76] Instead, 'objects collate many different moments'.[77] The manuscript fragments progress through and partially record the decades, at once accumulating traces and wearing away as they pass through diverse hands and contexts. The sheets' surface layers condense common historiography as they offer up a palimpsest of several centuries, from 'medieval' composition, binding and storage; to dissolution and dispersal; and to an indefinite period of fragmentation and wasting. This might include Reformation violence, Elizabethan butterfly-like fluttering and circulation, Carolean school-book bindings, Civil War gun-swabs, and Restoration antiquarian practice, as well as anticipated future preservation or further reuse and disintegration. Far from providing a linear narrative that transitions from dispersal to collection, each fragment tells its own story: at times, this is of loss and ephemerality, at others, of unexpected endurance and survival.

Waste is most likely to endure when it is bound into other books. This category of reuse is particularly future-oriented, suggesting to the sensitive viewer further narratives of collection and archiving. Aubrey is attentive to the same stories that are visible in rare books rooms today, describing how '[o]ne may also perceive by the binding of old Bookes, how the old Manuscripts went to wrack in those dayes'. Shifting to the present tense, Aubrey puns on the violence endured by the manuscripts, eliding the frame on which parchment was stretched with its namesake, the instrument of torture. This observation also echoes a motto, borrowed from Francis Bacon, who in turn had borrowed it from the fifteenth-century Italian

[75] Julie Spraggon records several instances of the destruction of 'old books' during the Civil War, for instance, in Peterborough, Lichfield and Winchester in 1642 and 1646 when 'divers larg[e] p[ar]chm[e]nts' were used to make 'Kytes w[i]thall to flie in the Ayre'. These books, however, seem to have been administrative documents, such as charters and parish registers, *Puritan Iconoclasm during the English Civil War* (Woodbridge; Boydell Press, 2003), 209, 235.

[76] Harris, *Untimely Matter*, 2–4.

[77] Ibid., 2–4.

antiquary Flavio Biondo, that Aubrey employs frequently throughout his writing: *Tanquam tabula naufragii*, 'like planks from a shipwreck'.[78] According to Bacon,

> ANTIQVITIES, or Remnants of History, are, as was said, *tanquam Tabula Naufragii*; when industrious persons by an exact and scrupulous diligence and observation out of Monume[n]ts, Names, Wordes, Proverbes, Traditions, Private Recordes, and Evidences, Fragments of stories, Passages of Bookes [...] doe save and recover somewhat from the deluge of time.[79]

In this formulation of historical practice, remnants of the past are salvaged by heroic antiquarians who record textual and anecdotal fragments. The fragments are both the salvaged antiquity or object, and the vehicle through which Bacon's object, absent in the final clause, is 'save[d] and recover[ed]' from oblivion.

For Aubrey, binding waste is a type of fragment that can represent all other fragments, as well as a productive emblem of Bacon's process of historical salvage: even his own life can be conceived of as a fragment of waste. An instruction in his *Brief Lives*, written on the sheets that contain his short autobiography, describes how they are 'to be interponed [interposed] as a sheet of waste paper only in the binding of a book'.[80] Employing the waste paper version of the common modesty topos, Aubrey subordinates his own biography to the multitude of lives he has recorded. But, as we have discovered, far from playing an insignificant role, waste was an active participant in Aubrey's life. This self-effacing gesture is a concrete metaphor because binding waste is preservative stuff, protecting a book's vulnerable pages from friction and oblivion. This role can be striking, for instance, when waste paper or parchment binds a chronicle or history book: the Huntington's copy of Polydore Vergil's *De Inventoribus Rerum*, a history of inventions, beginnings, and creations, is bound with an illuminated manuscript, used as guards, that might manifest, to an attentive user, a series of religious and historical endings (see Fig. 3.4). What is even more striking is the text's colophon: it describes how the book that is now bound in monastic waste, was 'Imprinted [...] VVithin the precinct of the late dissolued house of the Grey Friars' by Richard Grafton, eight years after its dissolution. This object, then, carries traces of multiple times, and manifests the ways in which past events have shaped its production and its material parts. But both the manuscript waste and the Greyfriars' house partially endure alongside the 'present' that they encompass, as binding or as secularized building. The wasted manuscript, therefore, provides a vehicle and a guard for its

[78] Kelsey Jackson Williams, *The Antiquary: John Aubrey's Historical Scholarship* (Oxford: Oxford University Press, 2016), 103.

[79] *The tvvoo bookes of Francis Bacon*, STC 1164 (London: Thomas Purfoot and Thomas Creed, 1605), CC3r.

[80] John Aubrey, *Brief Lives with An Apparatus for the Lives of our English Mathematical Writers*, 2 vols, ed. by Kate Bennett (Oxford: Oxford University Press, 2015), 1:429.

continued preservation in the future. Aubrey's own biography, his rhetorical gesture suggests, follows the pattern of such a protective wrapper: though seemingly humble and peripheral, waste ensures that other things, names, and stories survive. This is exactly what Aubrey sought to do. Characteristically modest, Aubrey described himself as a 'wheatstone', only useful for sharpening the wit of others, and his works as 'only Umbrages' and 'ruines'.[81] But it was Aubrey and his scattered papers that, like fragments and waste sheets turned wrappers, shored up the past against the 'deluge of time'.[82] Binding waste might, in the future, transform once again and become more than a vehicle for survival: it is often the object that, eventually, is prized for having survived.

In his 1658 *Hydriotaphia, or Urne-Buriall* Aubrey's contemporary and correspondent Thomas Browne describes how 'Time, which antiquates Antiquities, hath an art to make dust of all things'.[83] Meditating on a group of Roman burial urns recently uncovered in Norfolk and mistakenly identified as Anglo-Saxon in origin, Browne elegiacally describes how 'Time [...] hath yet spared these minor Monuments'.[84] These containers, full of dust and 'humane fragments', survive whereas funeral monuments, made of stone and brass, do not.[85] He continues: 'In vain we hope to be known by open and visible conservatories, when to be unknown was the means of [the burial urn's] continuation and obscurity their protection'.[86] In another reworking of the Horatian topos of monumentality, Browne deflates the 'perpetuity' offered by 'bare Inscriptions' and 'Aenigmaticall Epithetes', mostly rubbed away by time and only 'studied by Antiquaries'.[87] Both texts and material monuments are fleeting, and 'duration [...] maketh Pyramids pillars of snow, and all that's past a moment'.[88] The study of binding waste demonstrates that the contents of books, like the contents of Browne's urns, often survive *because* they are hidden and obscure. Apparently trivial and unimportant, monastic waste might unexpectedly endure because it has been pasted or stitched into a binding: they are, in Aubrey's imagination, like the flotsam of a shipwreck. Borrowing again from Bacon, Aubrey describes such survivals 'as planes and lighter things': they 'swimme, and are preserved, where the more weighty since are lost. [...] In like manner is it with matters of Antiquitie'.[89] It is in this way that waste fragments might become the only, and therefore most important, remnants of a lost past.

Furthermore, the fragments are not as silent as Browne's account of the burial urns might suggest. Browne's 'sad and sepulchral Pitchers [...] have no joyful

81 Aubrey, *Brief Lives*, 1:433–434.
82 *The tvvoo bookes of Francis Bacon*, CC3r.
83 Thomas Browne, *Hydriotaphia, urne-buriall*, Wing B5154 (London: for Hen. Brome, 1658), 24.
84 Ibid., 24.
85 Ibid., 3.
86 Ibid., 24.
87 Ibid., 74, 75.
88 Ibid., 74.
89 Aubrey, *Brief Lives*, 2:761.

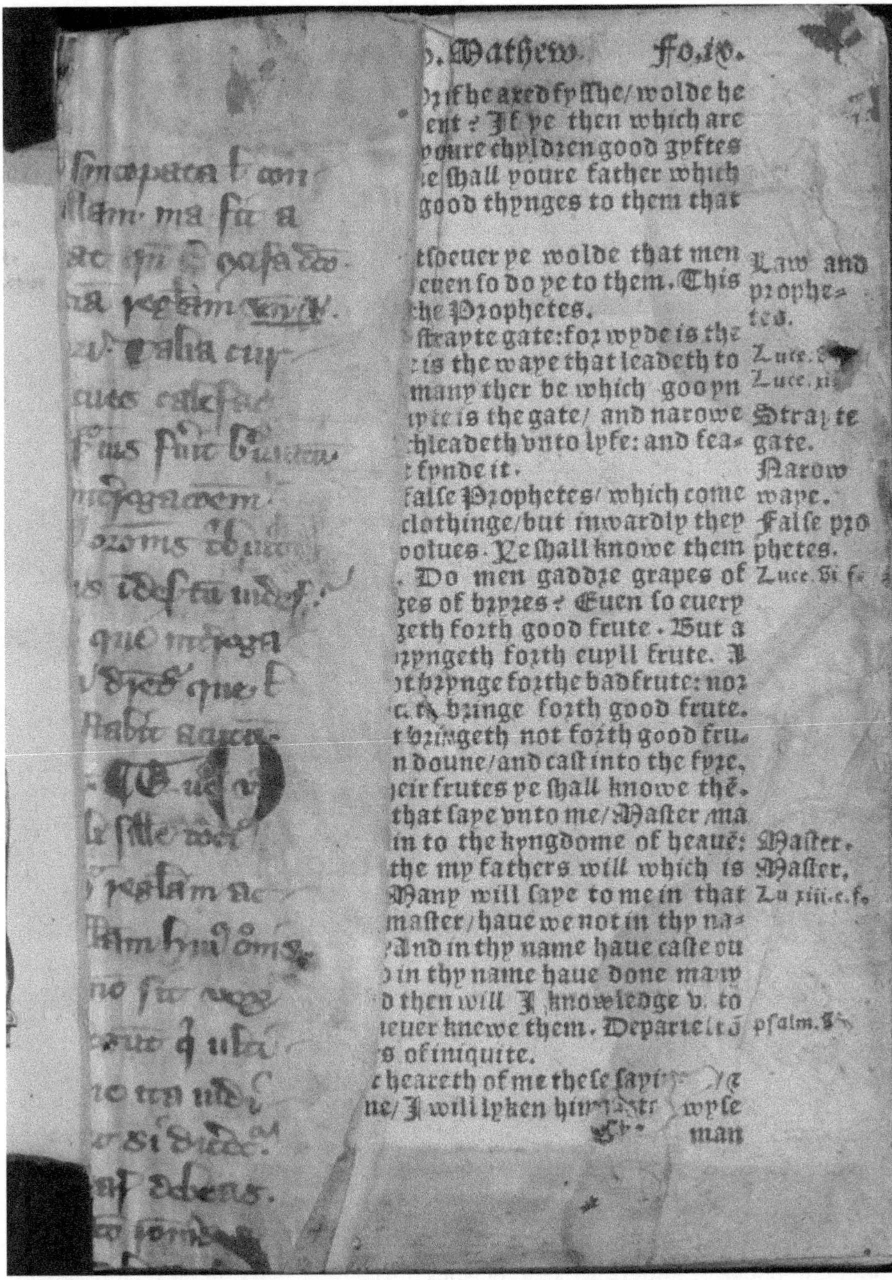

Fig. 3.4 Fragments of an illuminated manuscript used as guards and leaves of Matthew's Gospel used as flyleaves in the binding of Polydore Vergil, *De inventoribus rerum*, STC 24644 (London: VVithin the precincte of the late dissolued house of the Grey Friars, by Richard Grafton [1546]), 69747. The Huntington Library, San Marino, California.

voices', and instead sit 'silently expressing old mortality, the ruines of forgotten times', 'only' telling 'how long in this corruptible frame, some parts may be uncorrupted'.[90] But Browne mistakes muteness and a lack of legible text for an absence of narrative and meaningfulness. In fact, mortality, ruin, and corruption are material stories frequently read and retold in the early modern period, as we have seen in the work of Bale and of Donne. Furthermore, Aubrey's 'minor Monuments' are not, like Browne's, from a 'forgotten time'.[91] Instead, they are things with which Aubrey shapes a palimpsestic narrative of the last century and a half, partly from the objects themselves, partly from personal and recorded memories of them, and with which he fashions his own antiquarian identity. Waste fragments are, for Aubrey, more than a source of historical knowledge or an exotic curio to collect and display. Instead, they structure an understanding of how objects might manifest multiple moments, and provide a model through which to conceive of one's own relationship to the passage of time and the remnants left in its wake.

*

Situating the descriptions of dispersed and repurposed manuscripts in *The Laboryouse Journey* and Aubrey's 'Digression' side by side does more than demonstrate the longevity of monastic waste. It also makes clear that the widespread handling of parchment waste provoked a series of generative imaginings across the period. These were grounded in the ways in which waste fragments recorded historical moments and processes in their surfaces and layers, and how these material records of time can be used to prop up certain narratives. These narratives are always, we should remember, accompanied by their concrete counterparts. Where Bale's narrative of a triumphant recovery of knowledge and nationhood falters in the face of the fragments, an eschatological account of corporeal corruption and entropy emerges in its place. Aubrey, separated from the dissolution by more than a century, is not as deeply invested in the period's religious and political agendas as Bale, and so did not work to overwrite the fragmentary nature of the dispersed manuscripts. Instead, he meditates on them, in a metamaterial turn, as representatives of the relationship between objects and time more generally. He dwells on how they might serve as emblems for antiquarian practice, or the identity of an antiquarian, and how they might themselves salvage objects that were, in certain moments, either ephemeral or enduring. However dissimilar their interpretive frameworks, Bale's *Journey* and Aubrey's 'Digression' demonstrate that waste fragments were both commonplace and meaningful: whether bound in books or fluttering like butterflies through the early modern landscape,

[90] Browne, *Hydriotaphia*, A2r–v.
[91] Ibid., 69, A2v.

monastic waste provoked serious imaginative work during and after the dissolution. We will now turn to a late sixteenth-century writer who found in waste a not wholly dissimilar set of meanings from Bale and Aubrey. In the works of Thomas Nashe, we more regularly encounter waste paper than waste parchment, but as with Bale's manuscript remnants, these serve as a concrete metaphor for a religious worldview grounded in disintegration and decay. At the same time, as with Aubrey's fragments, Nashe's waste structures a particularly multitemporal understanding of how fragile objects begin, end, and, on occasion, endure.

4
Pickled Paper
Thomas Nashe's Poetics of Waste

No early modern writer's works are so populated with texts transforming into waste as those of Thomas Nashe. Nashe is known to have produced nine prose works between 1589 and 1599, as well as contributing several prefaces to other authors' works, collaborating on an unknown number of plays, and composing manuscript verse of which only 'The Choice of Valentines' has been identified. His career was dominated by his pamphlet war with Gabriel Harvey and came to an end because of the Bishops' Ban of 1599, which took specific aim at his and Harvey's books.[1] Throughout this decade of literary activity, his controversial pamphlets, the picaresque fiction *The Vnfortunate Traueller* and his unorthodox sermon, *Christs Teares Ouer Ierusalem*, spoken radically by Christ himself, depict books as being perpetually on the brink of disassembly and repurposing.

Broadly speaking, these depictions conform to the typical rhetorical use of waste paper as either an insult or a modesty trope. In his frequent and persistent attacks on the brothers Richard and Gabriel Harvey, for instance, Nashe constructs gross fictions in which their texts are dismantled and repurposed, sometimes used to 'wrappe the excrements of huswiuerie', and at other times by 'manie cholericke Cookes' who 'thrust [them] piping hot into the ouen vnder the bottomes of dowsets', or tarts, and use the 'torne sheetes [...] for basting paper' to keep meat from 'burning'.[2] On occasion, Nashe claims responsibility for the act of wasting himself: on reading Richard Harvey's *Lamb of God*, for instance, he 'could not refraine, but bequeath it to the Priuie, leafe by leafe as [he] read it, it was so vgly, dorbellicall, and lumpish'.[3] Nashe does not hesitate, though, to suggest that his own texts may end up as waste. The first edition of *The Vnfortunate Traueller*, for instance, is dedicated to Henry Wriothsley, Earl of Southampton, who is credited with the power to determine the material fate of the text: 'Vnrepriueably perisheth that booke whatsoeuer to wast paper', Nashe tells the Earl, 'which on the diamond

[1] On Nashe's biography, see Charles Nicholl, *A Cup of News: The Life of Thomas Nashe* (London: Routledge, 1984).

[2] Thomas Nashe, *Strange Newes, of the Intercepting Certaine Letters, and a Conuoy of Verses, as they were Going Priuilie to Victuall the Low Countries*, STC 18377 ([London]: [J. Danter], 1592), G3v.; Thomas Nashe, *Haue vvith You to Saffron-VValden, or, Gabriell Haruey's Hunt is Vp*, STC 18369 (London: John Danter, 1596), C1r.

[3] Thomas Nashe, *Pierce Penilesse his Supplication to the Diuell*, STC 18373 (London: Abell Jeffes for J. B[usby], 1592), E1v. All quotations are from this edition, unless stated otherwise.

Waste Paper in Early Modern England. Anna Reynolds, Oxford University Press. © Anna Reynolds (2024).
DOI: 10.1093/oso/9780198882701.003.0005

rocke of your iudgement disasterly chanceth to be shipwrakt'.[4] In this peculiar but still conventional reworking of the widely used trope, Wriothsley's literary judgement is materialized in the form of a refined rock that threatens to shatter Nashe's textual ship into flotsam if it doesn't meet Wriothsley's approval and so gain his patronage.

Nashe's writings are, though, more than a rich repository for the waste paper tropes that we encountered in Chapter 2. An examination of the material and intellectual contexts within which Nashe's fixation with waste developed deepens our understanding of the re-emergence of the trope in the late sixteenth century. As Georgia Brown notes, Nashe 'is a voracious critic and intimate observer of late Elizabethan culture', and so 'central to any study of the 1590s'.[5] An examination of Nashe's waste, therefore, illuminates the cultural concerns at stake in early modern descriptions of repurposed paper. What is more, in tracing the trajectory of waste paper in Nashe's imagination, we uncover what is likely the most extended and thoroughly worked poetics of waste in early modern literature. Nashe's engagement with the waste paper that populated his environment constitutes a carefully crafted framework through which he formulates a morally virtuous and materially prodigal mode of being. Most succinctly expressed in his deployment of the waste paper commonplace in the prefaces to *The Vnfortunate Traueller* and *Christs Teares Ouer Ierusalem*, this poetics of repurposed paper pervades Nashe's entire corpus and expands the range of waste intertexts available to Nashe's contemporaries and later generations of writers.[6]

Building on scholarly work concerning the physicality and prodigality of Nashe's prose style and persona, this chapter revises the consensus that Nashe is best understood as a 'creature of print', and that, through his references to waste paper, he conveys the 'shallowness', 'superficial[ity]', 'degradation', 'debasement', or 'descent' of the literary text in the print marketplace.[7] Instead, it draws attention to

[4] Thomas Nashe, *The Vnfortunate Traueller, or, the Life of Iacke Wilton*, STC 18380 (London: T. Scarlet for C. Burby, 1594), A2v. All references are to this edition unless otherwise specified.

[5] Georgia Brown, *Redefining Elizabethan Literature* (Cambridge: Cambridge University Press, 2004), 57.

[6] On the influence of Nashe on early modern literary culture, particularly comic drama and prose, see Neil Rhodes, *Elizabethan Grotesque* (London: Routledge, 1980), 5.

[7] On the physicality of Nashe's prose, see Rhodes, *Elizabethan Grotesque*, 33; Jason Scott-Warren, 'Nashe's Stuff', 207; and Brown, *Redefining Elizabethan Literature*, 88. On its prodigality, see Lorna Hutson, *Nashe in Context* (Oxford: Clarendon Press, 1989), passim; and Brown, *Redefining Elizabethan Literature*, 60–99. Jennifer Richards refers to the 'shallowness' and 'superficiality' of the printed page in her reading of the performative qualities of Nashe's prose (*Voices and Books in the English Renaissance: A New History of Reading* (Oxford: Oxford University Press, 2019), 258 and 263); Alexandra Halasz reads waste paper as symptomatic of how Nashe 'embraces' the 'degradation' of the print marketplace (*The Marketplace of Print: Pamphlets and the Public Sphere in Early Modern England* (Cambridge: Cambridge University Press, 1997), 96); Georgia Brown states that Nashe's references to waste draw attention to 'the potential debasement of printed text' and their 'scatological descent' (*Redefining Elizabethan Literature*, 53–54). A significant proportion of scholarship on Nashe focuses on his place in print culture: see, for instance, Steve Mentz, 'Day Labor: Thomas Nashe and the Practice of Prose in Early Modern England' in *Early Modern Prose Fiction: The Cultural Politics of Reading*, ed. by Naomi

the life cycle of paper that underpins Nashe's material and intellectual encounters with all manner of texts—be they the poetry, like 'The Choice of Valentines', 'written in all sorts of humors priuately';[8] the works 'spred abroade in written Coppies' and circulating among friends ahead of entering into print;[9] the letters exchanged and likely rapidly repurposed, imagined as 'deliuered to Megge Curtis in Shorditch to stop mustard pots';[10] or the range of paper documents employed in the playhouse environment that we know Nashe intermittently occupied throughout the 1590s.[11] These diverse handwritten texts would have been produced and consumed alongside printed ones—perhaps circulating unbound or bound in waste—and the already loose sheets used as waste in households, privies, and grocers' shops. Waste paper metaphors are not, in Nashe's writing, secondary to his negotiations of print culture. Instead, the inky output of the printing press forms part of his broader negotiation of the material culture of textuality. The paper that underpins this material culture—or, more precisely, Nashe's haptic encounters with this paper at various points in its life cycle—structures, by the mid-point of his career, a carefully crafted poetics of waste.

Furthermore, this poetics controverts the persistent sense that, in the words of C. S. Lewis, 'Nashe "says" [...] Nothing'.[12] This influential though contested reading and many that have come after it understand Nashe's prose as luxuriating in the superficiality of surfaces and the mundanity of base matter: his writing is either 'pure' style devoid of substance and meaning or, alternatively, focused on the 'purely physical'—the 'insubstantial substances'—at the expense of the lofty and the spiritual.[13] These readings have led to persuasive articulations of what Nashe *does* say, such as Neil Rhode's demonstration that, through the grotesqueness of his prose, Nashe depicts a 'vision of the new urban society which was both horrible and fascinating'; Lorna Hutson's identification of a political and moral

Liebler (London: Routledge, 2004), 18–32; Rachel Stenner, *The Typographic Imaginary in Early Modern Literature* (London: Routledge, 2019), 169–184; Julian Yates, *Error, Misuse, Failure: Object Lessons from the English Renaissance* (Minneapolis: University of Minnesota Press, 2003), 103–137; David J. Baker, *On Demand: Writing for the Market in Early Modern England* (Stanford: Stanford University Press, 2010), 35–60.

[8] Nashe, *Strange Newes*, K3r.

[9] Thomas Nashe, 'Preface' to *Syr P.S. His Astrophel and Stella*, STC 22536 (London: [John Charlewood] for Thomas Newman, 1591), A2v–A3r.

[10] Nashe, *Strange Newes*, B4v.

[11] See Tiffany Stern, *Documents of Performance in Early Modern England* (Cambridge: Cambridge University Press, 2009). In addition to *Summers Last Will and Testament* (1592), Nashe collaborated with Ben Jonson on the lost *Isle of Dogs* (1597) and is credited on the 1594 title page of *The Tragedie of Dido Queene of Carthage*. He is often cited as a contributor to *The First Part of King Henry VI* (1592).

[12] C. S. Lewis, *English Literature in the Sixteenth Century, Excluding Drama* (Oxford: The Clarendon Press, 1954), 416.

[13] On Nashe's attention to surfaces and the 'purely physical', see Rhodes, *Elizabethan Grotesque*, 34–35 and Richards, *Voices and Books*, 258–263. In analysing how Nashe 'forces our attention' onto his stylistic 'vein', Jason Scott-Warren complicates the modern 'depth ontology' that places style and substance within a hierarchy, arguing that surfaces were not 'trifling' in early modern thought. He does, though, conclude that Nashe's waste paper is one of many 'insubstantial substances' denoting the 'empty windiness of language' (209; 217).

commentary within Nashe's anti-humanist squandering of language; and Georgia Brown's understanding of Nashe's reconfiguration of prodigality into a force for good that transforms shame and marginality into 'modes of cultural renewal'.[14] My reading is indebted to each of these accounts, but challenges an assumption on which the majority of Nashe scholarship is based. In Nashe, waste and waste paper is far from empty, worthless, meaningless, or base. Instead, waste contains depths—'metaphorical meanings', as the archaeologist Christopher Tilley writes of material culture more broadly, 'linked to the performance' of its use and residing in its 'very materiality'.[15] The meanings encoded into artefacts, Tilley argues, work 'to reproduce or transform the social contexts in which they are encountered and move'.[16] This chapter will demonstrate that Nashe sought, through the representation of waste, to instigate radical and restorative transformation rather than to reproduce the sinful and stagnant status quo of early modern society. Though often scatological and scurrilous, Nashe's poetics of waste paper performs wholly serious imaginative work.

1 Waste Discourse in the 1590s

An examination of the explosion of the waste paper trope in the final decades of the sixteenth century leads us to the works of Thomas Nashe and, notably, his nemesis Gabriel Harvey. This chapter will begin by explaining why, in the early 1590s, these two writers violently traded bibliographic insults and, in the case of Nashe, came to celebrate the dispersal of a book's pages. Nashe and Harvey need to be situated within their literary and cultural contexts—in particular, within the context of the pamphlet controversies that the two writers participated in, as well as the interconnected concerns of proper literary and economic consumption that dominated Elizabethan public discourse. The 1588–9 Martin Marprelate controversy, like the Nashe–Harvey quarrel that arose from it, is inextricable from this wider literary and economic discourse concerning textual usefulness and the appropriate use rather than the prodigal waste of one's ink, paper, time, and intellectual potential. Nashe's conceptualization of waste paper, we will see, is both heavily indebted to and shaped in response to the representation of paper and waste in the works of both Martin Marprelate and Gabriel Harvey. The waste paper insults that Nashe hurls at Martin and Harvey are in fact inspired by these two authors' own fixation with the material dimensions of books, which in turn influences Nashe's demonstration of the usefulness and wholesome qualities of textual waste,

14 Rhodes, *Elizabethan Grotesque*, 3; Hutson, *Nashe in Context*, 7–8; Brown, *Redefining Elizabethan Literature*, 55.

15 Tilley, *Metaphor and Material Culture*, 75.

16 Ibid., 76.

materials later bequeathed to 'the dapper Mounsier Pages of Court', in the preface to *The Vnfortunate Traueller* and his patron's daughter in *Christs Teares*.[17]

The Martin Marprelate tracts radically altered the terms of religious debate and the sense of what printed objects should and could be in the late 1580s.[18] The publication of seven tracts between 1588 and 1589 by a Puritan or Puritans under the pseudonym Martin Marprelate, printed on a secret, itinerant press, forced the religious authorities to reconsider how best to address and engage the broader public. Were large, scholarly tomes really the best vehicle, or might brief, chatty, salacious, and rapidly produced pamphlets better capture the minds and souls of their readers? After a brief flirtation with meeting Martin on his own terms in pamphlets, jigs, and interludes, to which Nashe contributed the belated *An Almond for a Parrat* in 1590, the religious establishment concluded that no such popular site for religious debate should exist. In November 1589, the authorities' patronage of the scurrilous response to the Marprelate tracts came to an abrupt end.

It is widely recognized that the Marprelate controversy shaped Nashe's career.[19] It set the terms of engagement for the quarrel that would rage unabated between Nashe and Gabriel Harvey for a decade. This quarrel was sparked by the 1590 *Lamb of God*, a work written by Gabriel Harvey's brother Richard, which attacked both Martin and those who wrote against him on behalf of the establishment, including Nashe.[20] The Marprelate tracts were also formative reading for Nashe, modelling as they did a lively and effective prose style grounded in a range of radical literary techniques. As Travis Summersgill noted in 1951, the 'spontaneously conversational (or, on occasion, ranting) brilliance' of Martin's prose style enlivened Nashe's own syntax and voice, and in more recent decades, Martin has been credited as the source for a range of Nashe's stylistic attributes:[21] the physicality of his voice and his violent linguistic attacks on bibliographic bodies;[22] his use of oral and popular techniques such as jokes, conversational anecdotes, idioms, ballads, and may-games;[23] and his imaginative and flexible take on the conventions of print.[24] We might add to this list of borrowings Nashe's deeply rooted concern for the material form of the text and the manner in which it is handled by the

[17] Nashe, *The Vnfortunate Traueller*, A3v.

[18] The authoritative account of the Martin Marprelate controversy is *The Martin Marprelate Tracts, A Modernized and Annotated Edition*, ed. by Joseph L. Black (Cambridge: Cambridge University Press, 2008), esp. lvi–cxii.

[19] See Rhodes, *Elizabethan Grotesque*, 4, 37–38; Halasz, *The Marketplace of Print*, 84–86; Scott-Warren, 'Nashe's Stuff' 205–206.

[20] For a detailed account of the origins and execution of the Nashe–Harvey quarrel, see Eric D. Vivier, 'Thomas Nashe's Unprofitable Satire', *Modern Philology* 117, no. 4 (2020): 423–444.

[21] Travis L. Summersgill, 'The Influence of the Marprelate Controversy Upon the Style of Thomas Nashe', *Studies in Philology* 48, no. 2 (1951): 145–160, 160.

[22] Scott-Warren, 'Nashe's Stuff', 205–206.

[23] Neil Rhodes, 'On Speech, Print, and New Media: Thomas Nashe and Marshall McLuhan', *Oral Tradition* 24, no. 2 (2009), doi:10.1353/ort.0.0036; Richards, *Voices and Books*, 230.

[24] Jesse M. Lander, 'Martin Marprelate and the Fugitive Text', *Reformation* 7 (2002): 135–185.

reader: Martin is unusually preoccupied with the physical dimensions of the text and the quantity and quality of its raw materials.

In the 1588 epistle *Oh Read Ouer D. Iohn Bridges*, Martin parodies the immense size and weight of Bridges' *Defence of the Government* (1587), the established Church's most recent attempt to defend itself against Puritan critiques. 'The Epistle & the preface' of Bridges' work is sarcastically described as being 'not aboue 8. sheets of paper and very little vnder 7.', and the book as a whole adds up to 'an hundred threescore and twelue sheets of good Demie paper'.[25] Bridges, Martin claims, has 'daube[d]' so much 'paper' that his text is 'very briefely comprehended in a portable booke if your horse be not too weake'.[26] Demy is one of the largest paper sizes, and so Martin implies that Bridges has consumed an excess of large, and therefore costly, raw material, soiling, or 'daub[ing]', them with a crudely applied and thick coat of ink.[27] Martin's own texts, however, are appropriately small, brief, and portable: hurriedly produced as the authors and printers moved between safe houses, they waste neither large sheets of paper, vast quantities of ink, nor many hours in the print shop. In fact, all are printed on poor-quality pot paper, and the quality of printing noticeably declines in the later tracts.[28] Within his works, Martin frames the material disadvantages of clandestine publication as material virtues: they contribute to his nimble and potent persona and contrast the wasteful and unwieldy output of the established Church and the mainstream press.

In a 1594 attack on Gabriel Harvey, Nashe reworks this insult of Martin's. Harvey's 1593 *Pierces Supererogation* is described as '[s]ixe and thirtie sheets of mustard-pot paper', with Nashe, like Martin, highlighting the length of his opponent's book with comic precision.[29] Nashe goes further than Martin, though, in replacing good, demy paper with a fictionalized category: pot paper, the smallest and so cheapest size, becomes 'mustard-pot paper'.[30] Rather than wasting good paper, Harvey's pages are already the waste object that Nashe predicts them becoming, with multiple points in the biography of the sheets condensed into one loaded phrase. In 1596, Nashe again echoed Martin in an assault on Harvey. In *Haue vvith You to Saffron-Walden*, a text that frames the Nashe–Harvey quarrel as an exchange of letters, Harvey's *Pierces Supererogation* becomes a 'packet of

[25] Martin Marprelate, *Oh Read Ouer D. Iohn Bridges*, STC 17453 ([Fawsley, Northants.]: [Robert Waldegrave, 1588]), B1r.

[26] Marprelate, *Oh Read Ouer*, B1r.

[27] On paper sizes, see Mark Bland, *A Guide to Early Printed Books and Manuscripts* (Oxford: Blackwell, 2010), 25.

[28] On Martin's pot paper, see Cyndia Susan Clegg, 'The Authority and Subversiveness of Print in Early-Modern Europe', in *The Cambridge Companion to the History of the Book*, ed. by Leslie Howsam (Cambridge: Cambridge University Press, 2015), 135. See, for instance, the blurred ink on the title page of *Theses Martinianae*, STC 17457 ([Wolston, Warks]: [John Hodgkins], 1589), British Library, C.36.b.21.

[29] Thomas Nashe, *Christs Teares Ouer Ierusalem*, STC 18367 (London: [James Roberts and R. Field] for Andrew Wise, 1594), **1r.

[30] On literary representations of pot paper, see Katharine Craik, 'John Taylor's Pot-Poetry', *The Seventeenth Century* 20, no. 2 (2003): 185–203.

Epistling, as bigge as a Packe of Woollen cloth, or stack of salt-fish'.[31] The recipient asks the epistle's carrier, 'didn't thou bring it by wayne [a large wagon], or on horse-backe?' and the carrier replies, 'By wayne sir, & it hath crackt me three axeltrees'.[32] Whereas Bridges' bulky volume weighs down a horse, Harvey's breaks the wheels of a horse-drawn wagon. In scaling up Martin's jest, Nashe also depicts Harvey's text as equivalent to other mundane wares such as woollen cloth and salted fish in its unwieldiness and bulk. *Pierces Supererogation* is not, though, useful or saleable like fish or cloth, and so Nashe's fictional reader goes on to consider how best to repurpose it. Outlining a grotesquely extravagant rather than an actual waste paper practice, though one that prefigures the mixing of millions of Mills and Boon novels with tarmac to line the surface of the M6 in the early 2000s, the fictional reader suggests that he might 'make mud walls with them, mend high ways, or damme vp quagmires with them'.[33]

In contrast, as Neil Rhodes has noted, Nashe's texts are a 'highly mobile, gymnastic verbal performance', and their lightness is tied to our sense of their physical form: we imagine them as weighing significantly less than Harvey's books.[34] Just as Bridges 'daube[d]' good paper, Harvey is 'a lumpish leaden heel[e]d dawber' whose lengthy texts are made heavy by an excess of paper, poorly applied ink, and a ponderous style.[35] Though Nashe claims that his own 'stile with treading in [Harvey's] clammie steps is grown [...] heauie gated', he draws attention to his own natural mobility and nimbleness: previously, his 'stile' had been so 'light' as to need 'lead' in its 'shooes for feare the winde should blowe him into another Countrie'.[36] The agility of Nashe's prose is here embodied, raising, perhaps, the ghost of Martin, who had moved undetected around the country three years prior. 'Lead' had, after all, finally grounded Martin: in August 1589, the Puritan printers dropped a case of lead type as they loaded their cart, raising the suspicions of local villagers and resulting in their capture ten days later.[37] Martin, then, supplied Nashe with the figuration of a voice and a set of insults tied to the materiality of the text. The persona of a quick-witted, nimble, and spontaneous writer, skilfully disassembling the over-long and materially wasteful work of a stuffy and conventional elder was influential for Nashe, though the similarity between his and Martin's texts was entirely rhetorical: Nashe's books, unlike Martin's, are professionally printed and on good-quality paper, and many of Nashe's works are extremely lengthy, and so weigh as much, if not more, than Harvey's daubed 'packet[s]'.[38]

[31] Nashe, *Haue vvith You*, F1r.
[32] Ibid., F1v.
[33] Ibid., F1v.
[34] Rhodes, *Elizabethan Grotesque*, 21. On the M6 toll road, see p. 50 above.
[35] Nashe, *Strange Newes*, K4r
[36] Ibid., K4r.
[37] *The Martin Marprelate Tracts*, ed. by Black, lv.
[38] Nashe, *Haue vvith You*, F1r.

Although Nashe did not find the waste paper trope itself within Martin's works, the controversy contributed to a wider set of concerns regarding what constituted a waste of time, talent, and material resources. References to waste paper in printed texts rocketed in the early 1590s, and the majority of these usages are found in texts written by either Nashe or Gabriel Harvey and are in response to the Marprelate controversy and, subsequently, part of their own quarrel.[39] In his 1590 contribution to the Marprelate controversy, Nashe describes Martin's most recent epistle as 'his last waste paper', and proceeds to turn Martin's insults back on their originator.[40] He is 'the pistle-monger' who covers 'whole reames of paper' with ink, plodding through 'ten cart loade[s]' of the stuff and causing the 'death of ten thousand pound of candels' in the process.[41] Exhibiting a similar concern for the waste of writing materials in this pamphlet controversy, Gabriel Harvey describes John Lyly's anti-Marprelate tract, *Pap with a Hatchet* (1589), as 'good for nothing but to stop mustard pottes, or rub gridirons, or feather rattes neastes, or such like homely vse'.[42] Gabriel Harvey, like his brother Richard, lumps together the mass of textual material that emerged in the wake of the Marprelate controversy, both Puritan and anti-Puritan, claiming that 'Stationers are already too-full of such Realmes [punning on "reams"], and Commonwealthes of Wast-paper'.[43] Both Nashe and Harvey were also conscious of the fact that, if Martin was wasting valuable resources, he caused them to do so as well. The origins of one controversy in the other meant that this wastage continued for many more years as Nashe and Harvey's quarrel raged on. In *Strange Newes* (1592), Nashe claims that his textual attacks on Harvey will 'make a dearth of paper in Pater-noster-rowe'.[44] Here, Nashe seems to brag about his excessive consumption of material resources, but he elsewhere sounds

[39] Nashe's, and to a lesser degree Harvey's fascination with waste paper is revealed by an *EEBO* keyword search of 'waste NEAR paper', including textual variants and omitting results containing the verb form of 'waste'. This results in twenty-four hits across eighteen texts. Five of these texts (and nine of the hits) are from *Ordinaunces* concerning the destruction of illegally printed and imported texts (see pp. 33–34 above) and from dictionaries (see p. 46 above). The remaining results demonstrate the sporadic use of the waste paper commonplace from the 1560s onwards, with a notable increase after 1589. Five of the texts (and seven of the hits) are from Nashe's works, and one is from Gabriel Harvey's. This constitutes more than a third of the rhetorical references to waste paper within *EEBO*'s database. *EEBO* keyword searches are, of course, reflective rather than definitive in demonstrating literary trends: not all printed texts are keyword searchable, and manuscripts are wholly absent. Searches will also only show up specific phrasing, as is demonstrated by Gabriel Harvey's isolated appearance in these results, despite his frequent reference to repurposed paper using different terms. For an explanation of frequency analysis that assesses the strengths and limitations of *EEBO* as a resource, see Barbara McGillivray and Gábor Mihály Tóth, *Applying Language Technology in Humanities Research: Design, Application and the Underlying Logic* (Basingstoke: Palgrave Macmillan, 2020), 35–46.

[40] Thomas Nashe, *An Almond for a Parrat, or, Cutbert Curry-knaues Almes*, STC 534 ([London?]: [Eliot's Court Press?], [1590]), E3v.

[41] Nashe, *Almond*, B1v, D2r.

[42] Gabriel Harvey, *Pierces Supererogation, or, a New prayse of the old asse*, STC 12903 (London: John Wolfe, 1593), S2v. This section of the text likely circulated in manuscript in 1589.

[43] Ibid., S2v.

[44] Nashe, *Strange Newes*, I4r.

a more repentant note: in his final contribution to the quarrel, in *Haue vvith You* (1596), he acknowledges that these pamphlets inevitably 'drink some inck' and 'prodigally dispend manie Pages, that might haue been better employd'.[45]

Harvey's attitude towards the fact that 'waste was the matter of his own writing', as Eric Vivier has noted, is more consistent than Nashe's:[46] Harvey repeatedly stresses that his own time might have been better spent. In *Foure Letters and Certaine Sonnets* (1592), Harvey asks Nashe to give 'lesse occasion [...] to interrupt better exercises', and borrows one of Nashe's earlier insults: 'So be it: Pray Penne, Incke, and Paper on their knees, that they may not bee troubled with him any more'.[47] He then gives the younger author unsolicited advice: Harvey tells Nashe to abandon the 'fantasticall mould' and to instead 'vse heavenly Eloquence [...] and with heroicall Cantoes honour right Vertue, & braue valour [...] as noble Sir Philip Sidney, and gentle Maister Spencer haue done, with immortal Fame'.[48] This is a theme that preoccupied Harvey, whose 1593 *Pierces Supererogation* centred on the importance of managing one's national usefulness and social credit. What hope does Nashe have, he asks, of achieving either 'publike vse, or priuate credite'?[49] Rather than following the model of good Protestant humanist poets such as Sidney and Spenser, Nashe instead pursues 'vaine Fantasticallity', and so produces 'paper more currently fit for the basest mechanicall vses' and is, like his pages, 'good for nothinge'.[50] 'Time is precious', Harvey concludes, and 'would not be prodigally wasted in wastepaper'.[51]

In seeking to limit Nashe's textual prodigality, Harvey participates in contemporary discourse concerned with the use and waste of national resources, both material and intellectual. In the late sixteenth century, commentators highlighted the way in which England wastefully exported its raw materials, such as leather trimmings and rags, and imported foreign wares, such as Spanish gloves and French or Italian paper, made from the very same stuff.[52] In response to this loss of labour and profit, the government sought to encourage domestic production, in part by granting monopolies and patents for projects that sought to transform the dregs and offcuts of English industries into profit. As Lorna Hutson demonstrates in her seminal 1989 *Nashe in Context*, this economic policy was reflected in the literary culture of the period.[53] Building on the work of Richard Helgerson, Hutson argues that the growing emphasis on the morality of thrift

[45] Nashe, *Haue vvith You*, I2r.
[46] Vivier, 'Unprofitable Satire', 442.
[47] Gabriel Harvey, *Foure Letters and Certaine Sonnets*, STC 12900.5 (London: John Wolfe [and R. Robinson], 1592), F3v–F4r. This line is originally from Nashe's *Pierce Penilesse*, E1v.
[48] Ibid., F4v.
[49] Harvey, *Pierces Supererogation*, C4r.
[50] Harvey, *Foure Letters*, H2r.
[51] Ibid., H2v.
[52] See Joan Thirsk, *Economic Policy and Projects: The Development of a Consumer Society in Early Modern England* (Oxford: Clarendon Press, 1978), esp. 13–18.
[53] Hutson, *Nashe in Context*, esp. 18–23.

and the immorality of over-consumption shaped the Protestant humanist faith in the capacity of literature to serve the public good: as poetry became loaded with moral signification, writers in turn began to narrativize their careers around a trajectory of prodigality and reformation, transitioning from the light and amorous poetry of their youth to serious and useful work in their later years.[54] English literature was, like wool, leather, fish, and labour, a national resource that should not be wasted. Harvey situates his and Nashe's quarrel firmly within this economic discourse when he instructs the younger writer to reform himself and turn his texts to 'publike vse, or priuate credite'.[55] His advice in *Foure Letters* concludes with a magnanimous offer: if Nashe can 'be entreated, to quallifye his distempered veine; and to reclaim his vnbrideled selfe', Harvey will wipe away all memory of Nashe's prodigal pamphlets: 'I assuredly would be the first', Harvey writes, 'that should wrapp-vp such memorials, not in a sheete of wast-paper, but in the winding-sheete of Obliuion'.[56] Here, Harvey alludes to the waste paper commonplace before performatively dismissing it as inappropriate for his gesture of forgiveness and reformation. As Joshua Calhoun has shown, absolution is more typically represented through textual technologies of complete erasure or obliteration.[57] The partial endurance of waste would, therefore, undermine Harvey's apparent act of benevolence, hence his rejection of it. Harvey instead settles on the closely related image of 'winding-sheete[s]' as an appropriate wrapper for Nashe's discarded books—a visually and materially analogous counterpart to paper, constituting its raw materials. These linen sheets, decomposing out of sight in the grave, better suit Harvey's symbolic purposes because they allow him to insult Nashe (through apophasis of the waste paper trope) while also performing his own usefulness and ability to police the Protestant humanist literary ideal.[58]

Lorna Hutson persuasively argues that Nashe situates himself wilfully without this discourse of utility, and it seems likely that Nashe did so at least partly in response to Harvey: Nashe and Harvey, as Eric Viviers has shown, often sound a lot like one another, and this is no more the case than when they are expressing concerns regarding their own and one another's usefulness and wastefulness.[59] I do not conclude, however, as Viviers does, that Nashe becomes increasingly uncomfortable regarding his own wastefulness, or that he is unable to ever fully 'separate himself from Gabriel Harvey'.[60] I instead agree with Lorna Hutson's and Georgia Brown's understanding of Nashe's celebratory prodigality, formulated in diametric

[54] Richard Helgerson, *The Elizabethan Prodigals* (Berkeley: University of California Press, 1976).
[55] Harvey, *Pierces Supererogation*, C4r.
[56] Harvey, *Foure Letters*, G2v.
[57] Calhoun, *The Nature of the Page*, 81–98.
[58] See the discussion of paper and winding-sheets on pp. 90–91 above and Wall-Randell, 'Marlowe's Lucan', 11–25.
[59] Viviers, 'Unprofitable Satire', 424.
[60] Ibid., 443, 424.

opposition to Harvey's performance of profitability. As Hutson outlines, Nashe parodies the contradictions inherent to an economic discourse that praised production and demonized consumption, and that he instead resists 'absorption to any productive end'.[61] For Hutson, this is achieved through Nashe's employment of an older system of 'popular festive' imagery, in which all human and non-human bodies exist 'in a vital, mutually generative relationship, capable of disintegration and decay'.[62] Georgia Brown is indebted to Hutson's influential reading, understanding as she does Nashe's reformulation of wastefulness and excess into a process of traditional, free-flowing gift exchange, although Hutson and Brown come to differing conclusions regarding Nashe's attitudes towards patronage and humanism.[63]

My own reading is grounded in Nashe's representation, central to both Hutson's and Brown's readings, of open, organic systems and his celebration of the paradoxical profit to be found in redundancy and excess. It departs from these readings, though, in what might at first seem a minor aspect: Nashe's pages do not, as Hutson argues, 'refuse from the outset to be serviceable' in that they 'are offered by their author as waste paper to be consumed in various idle forms of wrapping and packaging'.[64] Similarly, repurposed pages are not, as Brown understands them, wasteful in the sense of being redundant and profitless.[65] Instead, as we have seen, waste paper is far from unserviceable or idle, and has the capacity to perform, in Nashe's works, as highly positive and productive matter. Like the rags and leather scraps coveted by economic commentators, waste paper proves unexpectedly valuable: in *Pierce Penilesse* Nashe playfully suggests that John Windet, the printer of Richard Harvey's *Lamb of God*, 'get a priuiledge betimes [...] forbidding all other to sell waste paper but himselfe'.[66] This might seem a flight of fancy to a modern reader, but it was close to a reality in which discarded stuff was a sought-after commodity: as we saw in Chapter 1, early modern householders were advised to gather and sell their 'waste paper of all sorts' to binders' shops, and monopolies for the collection of rags were, in the 1590s, coveted by England's papermakers.[67]

[61] Hutson, *Nashe in Context*, 83.

[62] Ibid., 119, 83.

[63] Brown, *Redefining Elizabethan Literature*, 65. Hutson argues that Nashe came to throw off the shackles of the socially conservative patronage system at the mid-point of his career, replacing it with the burgeoning print marketplace (*Nashe in Context*, 131–151, 175–196). In contrast, Georgia Brown argues that this widespread scholarly narrative tells us more about our own bias toward the free market and our sense of patronage as archaic 'enslavement' than Nashe's own attitude (*Redefining Elizabethan Literature*, 59–60).

[64] Hutson, *Nashe in Context*, 33.

[65] Brown, *Redefining Elizabethan Literature*, 54–55.

[66] Nashe, *Pierce Peniliesse*, E1v.

[67] Gabriel Plattes, *The Profitable Intelligencer*, Wing P2414 ([London?]: for T.U., [1644]), A4r. On monopolies in the papermaking industry, see John Bidwell, 'French Paper in English Books', in *The Cambridge History of the Book in Britain, 1557-1695*, vol. 4, ed. by John Barnard and D. F. McKenzie, with the assistance of Maureen Bell (Cambridge: Cambridge University Press, 2002), 584–85.

Nashe's systems of imagery are not, then, wholly grounded in an older, festive world, but are firmly based in the contemporary and broadly urban discourse surrounding waste, thrift, and reuse. As Jason Scott-Warren has demonstrated, Nashe 'was repeatedly drawn to the dialectic of poverty and recycling', offering a literary equivalent of the contemporary 'dearth science' in which projectors, scientists, and householders grappled with the valuable potential of waste products and practices of reuse.[68] Scott-Warren offers Hugh Platt as a figure comparable to Nashe, who, as Ayeesha Mukherjee has demonstrated, sought to transform 'penury into plenty' by using leftover materials such as sawdust, parchment shreds, and 'waste soap ashes' to produce fertilizer and preservatives.[69] We might also usefully situate Nashe within the framework of imaginative scavenging and recycling that Francis Dolan outlines in her examination of Hester Pulter's 'dunghill poetics'.[70] Pointing out that 'every house and shop had its dunghill', Dolan argues that, in the seventeenth century, waste was reconceptualized 'as a form of wealth'.[71] Waste heaps were, Dolan writes, 'productive and inspiring', providing a model for both theological transformation and literary 'composting'.[72] This reconceptualization of waste was well underway in the 1590s, and nowhere is it more apparent than in the preface to Nashe's 1594 *The Vnfortunate Traueller*, addressed 'to the dapper Mounsier Pages of Court'. In it, Nashe launches into an unprecedented set of instructions regarding the licensed transformation of his text into waste. In doing so, he refigures the dominant discourse of prodigality and reformation imposed on his works by Gabriel Harvey, into a celebration of the value of waste matter within the social and material environment of early modern London.

The preface is framed as a message from Jack Wilton, the roguish hero of Nashe's picaresque fiction, to his fellow Pages: after a polite greeting, we are told that Wilton 'bequeath[s] for wast paper here amongst you certaine pages of his misfortunes', entwining, through this pun, the fate of the book's pages and Wilton's own body.[73] Wilton is in fact repeatedly threatened with dissection and destruction in the main body of the text as he travels through Europe, most explicitly when held captive by the anatomist Zachary. In this episode, his body is on the brink of transformation into valuable raw matter akin to the pages of a book that, once dismantled, take on a new sort of value.[74] Whereas Wilton avoids such transformation in the central narrative of *The Vnfortunate Traueller*, the pages of the book are recategorized from textual to waste objects in the preface, and so they implicitly

[68] Scott-Warren, 'Nashe's Stuff', 212.

[69] Ayeesha Mukherjee, *Penury into Plenty: Dearth and the Making of Knowledge in Early Modern England* (London: Routledge, 2015), esp. 71–88.

[70] Francis E. Dolan, 'Hester Pulter's Dunghill Poetics', *Journal for Early Modern Cultural Studies*, 20, no. 2 (2020): 16–42.

[71] Ibid., 18, 24.

[72] Ibid., 20, 29–32.

[73] See Brown, *Redefining Elizabethan Literature*, 91.

[74] Nashe, *The Vnfortunate Traueller*, M1v–M3v.

brim, like Platt's waste soap ashes, Pulter's dunghill, and an anatomist's cadaver, with organic and economic potential. A multitude of waste uses are prescribed and proscribed:

> In any case keep them preciously as a *Priuie* token of his good will towards you. If there be some better than other, he craues you would honor them in their death so much, as to drie and kindle *Tobacco* with them: for a need hee permits you to wrap veluet pantofles in them also, so they be not woe begone at the heeles, or weather-beaten like a blacke head with graye haires, or mangie at the toes like an ape about the mouth. But as you loue good fellowship and ames ace [a double ace, the lowest throw of the dice, meaning 'bad luck'], rather turne them to stop mustard-pots, than the Grocers should haue one patch of them to wrap mace in: a strong hot costly spice it is, which above all things hee hates. To anie vse about meate or drinke put them too and spare not, for they cannot doo their Countrey better seruice. Printers are madde whoresons, allow them some of them for napkins.[75]

Here, the reader, granted the persona of a 'dapper Mounsier Pag[e]' of 'the Court', enters into an unusual contract with the narrator: constructing a community of cooperative gallants, rather than the traditional vertical hierarchy of patron, anonymous reader, and author, the dapper pages are instructed to disassemble the text and defend it against 'dispraise' after 'perusing' it.[76] The dismembered sheets will serve willingly as toilet paper, or '*Priuie* token[s]', and the fate of drying and kindling tobacco is framed as an honourable 'death'. Moving downward along a hierarchy of waste uses, Wilton 'permits' the pages to wrap and so protect pantofles (fashionable, foreign slippers or overshoes), frugally extending their life span.[77] The stopping of mustard pots is preferred to the wrapping of mace, and their 'use about meate or drinke' is unconditionally celebrated. This is, Wilton states, the best 'service' they can 'do their Countrey', situating the preface firmly and satirically within the dominant cultural discourse of nationalistic usefulness. Finally, printers are 'allow[ed]' the waste pages as 'napkins', blurring their use in the print shop as frisket sheets and wrappers with their more carnivalesque use 'about meate or drinke'.[78]

Though this is in part, as Hutson argues, a parody of 'the page's obligatory promise to give service and profit to the commonweal', Nashe's emphasis is on the

[75] Ibid., A3v.
[76] Ibid., A3v.
[77] For evidence for the use of waste paper in the lining of clothing see pp. 45–46 above.
[78] For a compelling interpretation of Jack Wilton's napkins as a satire on the reputed textual and corporeal gluttony of printers, see Stenner, *The Typographic Imaginary*, 177. On the use of waste paper in the printshop, see p. 45 above.

usefulness of waste paper to a particular social group:[79] the book's pages participate in an urban subculture of knaves, rogues, and wastrels. In addition to wasting *The Vnfortunate Traueller*, this imagined group of readers is instructed to 'sweare' on it; to use it as a surface for 'play[ing] with false dice'; not to 'refuse it for a pawne in the times of famine and necessitie'; and not to say the name of the text 'within fortie foote of an ale-house' unless 'the tauerne is honorable'.[80] The imagined readers fight, cheat, drink, and live on the brink of indigence, with nothing stored up for times of dearth. The pages help to maintain a performance of modishness by protecting fashionable footwear ('pantofles') and wrapping and kindling tobacco, as well as granting the means with which to purchase a few pennies-worth of food and drink. Instructed to eke value out of an object often undervalued or discarded, the pages are to be 'pawne[d] in the times of famine and necessitie', perhaps sold as part of the second-hand book or waste paper trade to a binder's shop, or to a grocer in need of cheap wrapping paper. Nashe's waste pages, therefore, have a surprising amount in common with Hugh Platt's waste soap ashes. According to Platt, these ashes can be strewn over the ground to 'enrich' any pasture: devoured by 'lean' and 'hungry' ground, this waste stuff remedies famine.[81] Nashe's waste paper is, in an analogous act, scattered for the benefit of lean and hungry city-dwellers. It might enrich them by a few pennies, and so is imagined as directly or indirectly providing sustenance and succour.

Nashe is not, though, so earnest as Platt, and, though similar in many ways, the preface to *The Vnfortunate Traueller* is not a literary version of contemporary agricultural treatises that sought plenty in little and usefulness in the unwanted by-products of local industries. Thrift and frugality are, for Jack Wilton's fellow pages, part of an otherwise wasteful and prodigal way of life: the 'dapper Mounsiers' are frivolous and only just subsist, with practices of reuse and patchwork repair allowing them to sustain a prodigal cycle of expenditure and consumption. The paradox of thrift enabling waste is encapsulated in the preface's reference to mustard and mace. Prohibited from wrapping mace, because it is 'a strong hot costly spice [...] which aboue all things [Jack Wilton] hates', the pages are instead permitted to 'stop mustard-pots'.[82] Commonly glossed as a reference to the staff carried by the city's magistrates, and so as a gesture toward carnivalesque misrule, this penal pun is also part of the continued allusion to contemporary concerns regarding the morality of

[79] Hutson, *Nashe in Context*, 147.

[80] Nashe, *The Vnfortunate Traueller*, A4r.

[81] Hugh Plat[t], *Sundrie New and Artificiall Remedies Against Famine*, STC 19996 ([London]: P[eter] S[hort], 1596), C1r.; *The New and Admirable Arte of Setting of Corne*, STC 19993 (London: P. Short, 1601), D1r; and *The Second Part of the Garden of Eden*, Wing P2392 (London: for William Leak, [1659]), 108.

[82] Nashe, *The Vnfortunate Traueller*, sig. A3v.

nationalistic consumption.[83] Locally grown, mustard was available in abundance throughout England's gardens, waysides, and riverbanks, and is preferred to the 'costly' spice imported from the East Indies.[84] But a preference for local produce is not consistent throughout the preface: the 'dapper Mounsier Pages' wear foreign footwear and smoke tobacco, a controversial importation from the New World. How, then, do we reconcile the exclusion of mace with the page's 'crau[ing]' of foreign and 'honor[able]' tobacco?[85] The answer lies in the lifestyle of the Mounsier Pages: whereas the smoking of tobacco and the following of modish fashions is necessary for the performance and extravagant visibility of dapperness, the consumption of mace is not. In a preface advising methods for keeping up appearances in 'times of famine and necessitie', mace was a useless luxury contrasting starkly with the utility of repurposed pages. The 'strong hot' spice of mace shares the culinary and humoral properties of mustard, and so the latter is a frugal substitution that props up an otherwise prodigal pattern of consumption.[86] Mustard, then, has much in common with waste paper: both are prosaic and often overlooked, readily available in English households, and a thrifty addition to the lifestyle of an impoverished gallant.

Mustard and repurposed pages have a further quality in common, in contrast to mace, that illuminates one of the central features of Nashe's poetics of waste. As Joan Thirsk notes, commentators complained that 'outlandish', or imported, ingredients, used 'in place of home-grown plants', 'often arrived rotten, worm-eaten and lacking in virtue'.[87] Mustard, by contrast, would not have had far to travel from soil to pot, and so was more likely to remain fresh up to the point of consumption. Nashe claims a similar freshness for his text, ensured as it is by its rapid use: after its textual consumption, *The Vnfortunate Traueller* is imagined as being immediately transformed into waste. Quickly dismantled and either burnt, worn out, or cast onto the dunghill, Nashe depicts a situation in which his text has no time to become 'rotten, worm-eaten', or 'lacking in virtue'.[88] What the preface

[83] See, for instance, *Thomas Nashe: The Unfortunate Traveller and Other Works*, ed. by J. B. Steane (Harmondsworth: Penguin, 1972), 253.

[84] Joan Thirsk, *Food in Early Modern England: Phases, Fads, Fashions 1500-1760* (London: Hambledon Continuum, 2007), 73, 150–151, 315; Ronald Findlay, 'The Emergence of the World Economy', in *Contemporary Economic Issues: Proceedings of the Eleventh World Congress of the International Economic Association, Tunis*, vol. 3, ed. by Daniel Cohen (Basingstoke: MacMillan Press, 1998), 116–117.

[85] Nashe, *The Vnfortunate Traueller*, A3v.

[86] For an alternative interpretation of the distinction drawn between mustard and mace in the preface to *The Vnfortunate Traueller*, see Yates, *Error, Misuse, Failure*, 103–137, esp. 108. Influenced by the work of Bruno Latour, Yates argues that mace is distinct from mustard in its capacity to penetrate and transform. It 'rupture[s] the seal that ensures a clear separation between text and world, medium and message', and so draws attention to non-human agency in the processes of making and reading books. Although I am persuaded by Yates' understanding of Nashe as sensitive to the human and non-human assemblages that are created when books are dismantled, I do not agree with this particular point about mace because mustard and mace share the same penetrative and humoral qualities.

[87] Thirsk, *Food in Early Modern England*, 317.

[88] Ibid., 317.

describes, then, is a small-scale and healthy urban ecosystem. Exploding the book into a multitude of loose sheets, both the reader and the text participate in a cycle of consumption, reuse, and transformation: the uses about food and drink mean that more waste will be needed for use as toilet paper, and the use of waste as printers' napkins, or frisket sheets, contributes to the production of more texts which might, in turn, themselves be wasted. Within this fast-paced ecosystem, there is no time for a text, like a neglected foodstuff, to go stale.

In the preface to the 'dapper Mounsier Pages' of 'the Court', then, we see the culmination of Nashe's engagement with the literary and economic discourses of the early 1590s. Nashe found, in the Martin Marprelate tracts, a concern for the physical form of the book and, in this and subsequent pamphlet controversies, the rampant consumption of raw materials. Whereas for Gabriel Harvey, this consumption triggered anxieties regarding the waste of time, talent, ink, and paper, Nashe revels in the waste on which textual composition is predicated. Both Harvey and Nashe employ the waste paper commonplace in their negotiations of this excessive consumption, but only Nashe comes to celebrate it, transforming the conventional insult into an extended portrayal of impoverished but dapper living in early modern London. Highlighting the centrality of waste to a reciprocal cycle of prodigal consumption and thrifty reuse, Nashe establishes the vitality and freshness of repurposed pages. The paradoxical richness of these discarded sheets, as we will see, underpins the prodigal poetics central to Nashe's literary imagination.

2 Plague and Paper in the Summer of 1593

The preface to *The Vnfortunate Traueller* is not the only extended celebration of waste paper in Nashe's corpus. It can be placed alongside another unorthodox reference to the repurposing of a book's pages in the dedicatory epistle of *Christs Teares Ouer Ierusalem*, a lengthy and earnest sermon published in the same year as *The Vnfortunate Traueller*. Read together in this way, we can discern the way in which waste paper came to shape Nashe's imagination during this particularly productive year of his career, and how it became a vehicle with which to articulate his understanding of godly living in a time of crisis. Portraying the text as a parcel delivered to its dedicatee, Lady Elizabeth Carey, Nashe proclaims:

> Loe, for an oblation to the ritch burnish shrine of your vertue, a handfull of *Ierusalems* mummianizd earth, (in a few sheetes of wast paper enwrapped,) I heere (humiliate) offer vp at your feete.[89]

[89] Thomas Nashe, *Christs Teares Ouer Ierusalem*, STC 18366 (London: James Roberts to be sold by Andrew Wise, 1593), *2r. References are to this (the first) edition, unless stated otherwise.

As in *The Vnfortunate Traueller*, Nashe reworks the commonplace modesty trope often found in bids for patronage. As we have seen is typical in Nashe's works, the book itself—here *Christs Teares*—is imagined as an already wasted object. The pages are rendered as repurposed sheets and the textual contents are mummy—a substance glossed by Nashe in the second edition as 'mans flesh long buried and broiled in the burning sands of Arabia'.[90] Again, as in the preface to the 'Mounsier Pages', the waste object is valuable: mummy is an expensive and difficult to obtain medicine, consisting of either the preserved flesh of corpses imported from foreign deserts, as Nashe notes, or a more local product derived from the flesh of exhumed cadavers and freshly executed criminals.[91] Like the functions described in the dedication to the 'dapper Mounsier Pages' of 'the Court', this grotesque metaphor is grounded in material practice: we know that waste and brown paper was commonly used to wrap, seal, and store medicine, forming the sorts of assemblages Donne describes as 'wrap[ping] pils' to 'save a friends life' in his dedication to *Coryats Crudities*.[92] We encounter, therefore, the same sort of object in the prefaces to both *Christs Teares* and *The Vnfortunate Traueller*, though one is imagined as wrapping a costly medicine and the other, more prosaically, as stopping mustard pots, lighting tobacco, and repairing shoes. Nashe evidently considered repurposed paper an appropriate rhetorical gift for both a noble patron and a wider readership.

We have already seen that Nashe's fixation with waste paper stemmed from his engagement with Martin Marprelate, Gabriel Harvey, and contemporary discourses concerning textual waste and prodigality. Nashe's representation of waste objects in the prefaces of these two texts, both composed in 1593, is also a consequence of his material environment during this period. The years 1592–4 were busy ones for Nashe, and we can be certain of his whereabouts for at least some of this time.[93] In the winter of 1592, Nashe was a guest at Archbishop Whitgift's Croydon Palace, sheltering with the Archbishop's household from the plague that ravaged London. We know that early in 1593, Nashe spent time with his friend Robert Cotton at Conington Manor in Huntingdonshire. We cannot be certain where Nashe was in the summer that followed, but in these months Nashe seems to have completed both *The Vnfortunate Traueller* and *Christs Teares* in rapid succession. According to its colophon, *The Vnfortunate Traueller* was completed on June 27, 1593; entered into the Stationers' Register on September 17, 1593, it was

[90] Nashe, *Christs Teares* (1594), *3r.

[91] Louise Noble, *Medical Cannibalism in Early Modern English Literature and Culture* (Basingstoke: Palgrave Macmillan, 2011), 1–3.

[92] '*Incipit Joannes Donne*', in Coryate, *Coryats Crudities*, D3v.

[93] Nicholl, *A Cup of* News, esp. 135–180. 'The Thomas Nashe Project' has created an interactive tool recording Nashe's known movements. See Kate De Rycker, 'Mapping Thomas Nashe, from Lowestoft to London', *The Thomas Nashe Project*, last accessed December 8, 2021, https://research.ncl.ac.uk/thethomasnasheproject/resources/mappingnashe/

not printed until the following year.[94] *Christs Teares* was entered into the Stationers' Register nine days before *The Vnfortunate Traueller* on September 8, 1593 and was either circulating in print or in manuscript a week later, if Harvey's claim to have read the sermon and completed a response to it on September 16, 1593 are to be believed.[95]

It is possible that Nashe was living in London during these highly productive summer months: these two texts are, after all, predominantly concerned with cities suffering catastrophic plague outbreaks. In addition to fleeing from the sweating sickness that infected London in 1517, Jack Wilton travels to a Rome that, like London in 1593, is infected with plague. *Christs Teares* similarly refigures contemporary London as a Jerusalem ravaged by a disease brought on by the sinfulness of its inhabitants. One particularly transparent passage that referred directly to the immorality of London's professional classes, excised from the second edition, led to Nashe's imprisonment in the autumn of 1593.[96] Katherine Duncan-Jones has hypothesized that Gabriel Harvey brought this passage to the attention of the authorities soon after the publication of *Christs Teares*, and that his role in Nashe's imprisonment in a Newgate cell while plague raged through the city guaranteed the ongoing savagery of their quarrel. Duncan-Jones has also shown that Nashe was rescued from prison by the Carey family and brought to the safety of the Isle of Wight, perhaps, she suggests, in direct reciprocation for the 'oblation' of *Christs Teares* that Nashe dedicated to Lady Elizabeth Carey. In an ironic twist of fate, then, the text may in fact have served Nashe as a prophylactic for the diseases that ran rampant in London's prisons—not mummy, as Nashe claimed in the preface, but certainly effective in removing him from a potentially deadly situation.

Critics have long struggled to reconcile *The Vnfortunate Traueller*, described by one critic, somewhat strangely, as 'the sunniest of all his writings', with a sermon that constitutes 'the longest and least readable of Nashe's works'.[97] The latter is, it has been hypothesized, either a 'hoax' or evidence of the author's psychological 'crack-up'.[98] More thoughtfully, Philip Schwyzer has argued that there is a 'deeper affinity' between these two texts, describing them as a 'bi-textual beast', 'cooperating to such an extent' that 'one could hardly exist without the other'.[99] Though Schwyzer overstates the interconnectedness of the two texts—one aggressively irreverent and the other violently religious—it is likely that both were composed

[94] Chiaki Hanabusa, 'Notes on the Second Edition of Thomas Nashe's *The Unfortunate Traveller*', *Notes and Queries* 56 (2009): 556–559.

[95] *The Works of Thomas Nashe*, ed. by Ronald B. McKerrow and rev. by F.P. Wilson, 5 vols. (Oxford: Blackwell, 1958): 4:212.

[96] Katherine Duncan-Jones, '*Christs Teares*, Nashe's "Forsaken Extremities"', *The Review of English Studies* 49, no. 194 (1998): 167–180.

[97] Nicholl, *Cup of News*, 154, 167.

[98] Jonathan V. Crewe, *Unredeemed Rhetoric: Thomas Nashe and the Scandal of Authorship* (Baltimore: John Hopkins University Press, 1982), 59; Nicholl, *Cup of News*, 166–180.

[99] Philip Scwhyzer, 'Summer Fruit and Autumn Leaves: Thomas Nashe in 1593', *ELR* 24, no. 3 (1994): 584.

in broadly similar circumstances in the plague-ridden summer of 1593. It is not, then, a coincidence that these two texts share a network of imagery concerned with the material contexts of producing and consuming texts in a diseased, urban environment.

Now at the mid-point of his career, Nashe had lived in London on and off for four or five years. The city, 'covered in texts' in the words of Tiffany Stern, would have been by this time structurally embedded in his imagination.[100] In the early 1590s, a new category of loose sheet joined the other bills, title pages, ballads, and waste in circulation, and so slips and scraps of paper would likely have been imbued with new and pressing valences. For the first time, Weekly Bills of Mortality emerged from the presses, translating plague fatalities into paper objects. As Mark Jenner has shown, the printing press had been co-opted into civic responses to the plague since the 1570s, with sheets that read 'Lord Have Mercy Upon Us' nailed to the doors of infected houses since this earlier date.[101] Low survival rates mean that the early history of Bills of Mortality is not well documented, but 1592 may have been the first year that weekly, rather than annual, bills were printed.[102] Notably, it was John Wolfe—the printer who entered *The Vnfortunate Traueller* into the Stationers' Register in June 1593—who was granted license to print 'the billes, briefes, notes and larges gyven out for the sicknes weekely' from 1593 until his death in 1601.[103] What is more, these 'Indices of the sicknesse' were, in 1593, edited or contributed to in some way by Wolfe's printshop lodger Gabriel Harvey.[104] Nashe describes Harvey's collaboration with Wolfe in detail in *Haue vvith You*, printed several years later in 1596. He claims that 'in the ragingest furie of the last Plague', 'when there dyde above 600. a week in London', Harvey lay 'inck-squittring and printing against' him 'at *Wolfes* in *Powles Church-yard*'. In addition to producing 'empassion[ed]' and 'fiery' pamphlets, Harvey 'did for [Wolfe] that eloquent *post-script* for the Plague Bills'.[105] None of Wolfe's bills survive so we are unable to verify Nashe's claims, but Nashe was no doubt familiar with these documents. In fact, the estimate of 'above 600' deaths per week in the height of the 1593 outbreak closely matches the figures given in a 1665 broadsheet.[106] Nashe and the

100 Tiffany Stern, '"On each Wall and Corner Poast": Playbills, Title-pages, and advertising in Early Modern London', *ELR* 36, no. 1 (2006): 87.

101 Mark Jenner, 'Plague on a page: *Lord have mercy upon us* in early modern London', *The Seventeenth Century* 27, no. 3 (2012): 255–286.

102 Erin Sullivan, 'Physical and Spiritual Illness: Narrative Appropriations of the Bills of Mortality', in *Representing the Plague in Early Modern England*, ed. by Rebeccao Totaro and Ernest B. Gilman (London: Routledge, 2011), 77–79.

103 Stephen Greenberg, 'Plague, the Printing Press, and Public Health in Seventeenth-Century London', *Huntington Library Quarterly* 67, no. 4 (2004): 512.

104 Gabriel Harvey, *A Nevv Letter of Notable Contents*, STC 12902 (London: Iohn Wolfe, 1593), A2r.

105 Nashe, *Haue vvith You*, N4v–O1r.

106 Anon., *The four great years of the plague, viz. 1593. 1603. 1625. and 1636*, Wing F1658 (London: Peter Cole, 1665), broadsheet.

compiler of the broadsheet, which records the average deaths between July and September of that year as 616, likely both had access to Wolfe's now lost Bills of Mortality.

Nashe's description of an author immersed in an inky world of manuscripts and printed texts, reading Plague Bills and counting the dead whilst 'lying in the hell mouth of infection' is vivid.[107] Perhaps, like Harvey, Nashe was 'cloystred' and 'barricadoed vp' in London, unable to open his windows because the 'fat manured earth' beneath was 'the burial place of fiue parishes', blasting gusts of damp, plaguey air upwards 'at his casements'.[108] Nashe himself lodged for an unknown period prior to 1596 in the printshop of John Danter in Hosier Lane, a short distance from Wolfe's printshop at the south door of St Paul's Churchyard. Nashe's own habitation might, therefore, have informed his portrayal of Harvey's 'cloister'. Or perhaps Nashe simply dreamt up the scene, having escaped the city and its diseased air but remaining haunted by the spectre of the plague and the fatalities numerated by the Bills of Mortality. Regardless, *The Vnfortunate Traueller* and *Christ Teares* make clear that corpse-manured earth and loose sheets of paper populated Nashe's imaginative habitat for the summer of 1593. As Mark Jenner has argued, plague publications prompt us to move beyond language-based narratives of plague and to instead 'embrace a more capacious sense of the visual and the graphic'.[109] We might add to this the haptic and the kinetic experience of handling loose and vulnerable sheets that, pinned to doors and purchased for a penny before rapidly disappearing from view, enumerated fatalities and gave a new shape to the disease, the city, and the dead. The plaguey paperscape of 1593, then, is the setting for Nashe's extended renderings of repurposed pages in *The Vnfortunate Traueller* and *Christs Teares*.

In *Christs Teares*, this environment of ephemeral plague bills, miasmic air, and 'fat manured earth' is translated into a grotesquely lyrical version of the commonplace modesty trope, in which the loose sheets of paper on which Nashe writes and the book that the reader holds come to resemble the flimsy Bills of Mortality, before being imagined as a 'few sheetes of wast paper [w]rapp[ing]' the decomposing dead, or 'mummianizd earth'.[110] This image is especially fitting because the textual contents of *Christs Teares* are themselves 'fat manured': within the sermon, we are told that Jerusalem's dead are so numerous as to 'serue to mende High-wayes' and 'turne standing Quagmyres to firme ground'.[111] This earth 'ramd full of [...] cor[p]ses' resembles that beneath Harvey's window in St Paul's Churchyard, as well as the use to which Harvey's waste paper is imagined as being put

[107] Nashe, *Haue vvith You*, N4v.
[108] Ibid., N4v.
[109] Jenner, 'Plague on a Page', 261, 277.
[110] Nashe, *Christs Teares*, *2r.
[111] Ibid., G2r.

('mak[ing] mud walls [...] mend[ing] high ways, or damm[ing] vp quagmires') in *Haue vvith You*.[112] But in the preface to *Christs Teares*, this earth is positive matter: Nashe's peculiar modesty trope stakes a claim that his sermon will have the same effect on its readers as a richly fertilized portion of soil, or lump of medicinal mummy. Its attack on the sinfulness of contemporary London, Nashe hopes, will 'peirce vnawares into the marrow and reynes of' his readers, prompting them to repentance, godly living, and prayer and so move God to cleanse London of the plague.[113] The sermon is, therefore, imagined as a grossly corporeal plague cure wrapped up in paper and offered to the reader for medicinal consumption.

Though striking, this concrete metaphor is not an isolated moment in *Christs Teares* or Nashe's wider work. It conforms to what Neil Rhodes has influentially termed Nashe's 'grotesque' style, pairing as it does 'images which have a physical similarity, but which normal experience classifies quite separately'—namely, a book and a parcel of medicine.[114] Furthermore, it manifests the grotesque effect that Rhodes describes as 'rendering in solid terms what was previously a purely abstract concept, and to make that concept comically palpable', though, in *Christs Tears* the emotional effect is that of horror and fascination rather than comedy.[115] This particular instantiation of Nashe's grotesque style makes palpable the moral acts that will cleanse London of the plague, and so participates in what Ernest Gilman has labelled 'plague discourse': the strongly held belief, in the early modern period, that 'plague is itself a form of (divine) utterance, and a form of writing that inscribes itself in the natural world'.[116] Within this already textualized plague discourse, then, the idea of absorbing a cure through a waste wrapper is not so unlikely. The oblation, like the plague itself, has been, to borrow Gilman's phrasing, 'encrusted with metaphor' and is perceived as having 'a "meaning" beyond, or intrinsic to' its material form.[117]

It is not unusual for a wrapper or container to be so loaded with significance. As Lucy Razzall has demonstrated, boxes, broadly defined, have a rich rhetorical history, and Nashe's oblation might be productively considered within two such traditions.[118] The waste exterior of *Christs Teares* inverts the early modern commonplace of the apothecary's painted box that, beautifully decorated, contains 'potentially dangerous substances'.[119] The grossness of Nashe's textual object

[112] Ibid., G2r; Nashe, *Haue vvith You*, F1r.
[113] Nashe, *Christs Teares*, A1v.
[114] Rhodes, *Elizabethan Grotesque*, 13.
[115] Ibid., 25–26.
[116] Ernest B. Gilman, *Plague Writing in Early Modern England* (Chicago: University of Chicago Press, 2009), 3.
[117] Ibid., 3.
[118] Lucy Razzall, *Boxes and Books in Early Modern England: Materiality, Metaphor, Containment* (Cambridge: Cambridge University Press, 2021).
[119] Ibid., 96.

might instead deter opening, despite its curative contents. Nashe's image, therefore, conforms to the *Sileni* tradition in which an ugly exterior contains hidden truths.[120] Within this widespread formulation of the relationship between a container's interior and exterior, the box or wrapper is not a superficial or unimportant layer, functioning only as a protective cover or veil for the more important matter within. Instead, the gross outer shell is inseparable from the valuable and functional contents in the creation of a moralistic meaning. As Razzall writes, 'the box itself becomes as significant as what it might or might not contain, and the relationship between inside and outside is far from straightforward'.[121] This collaborative process of meaning making is grounded in the physical act of opening and can be usefully illuminated by Christopher Tilley's theorization of the relationship between metaphor and material culture more broadly. The 'bodily experience' of unwrapping that underpins Nashe's rhetorical oblation is one that has worked its 'way up into' an 'abstract meaning' and so manifests an 'embodied' form of 'imagination'.[122] The bodily experience of opening a waste wrapper might be usefully compared to Tilley's understanding of the experience of entering and dwelling in a house. The house has shaped the foundational metaphors of inside and outside in human culture, and so is 'an ordered and ordering device, structured and structuring fundamental principles of social life and the cosmos'.[123] For Nashe, the performance of unwrapping waste has worked its way into a similarly 'ordered' structure of thought, one that serves as a pattern for his understanding of both the rightful workings of society and the divine workings of the cosmos.

This suggestion may sound overly grand, but the kinetic experience of waste paper is central to Nashe's negotiation of a godly way of being in the world. Repurposed pages are a figure through which Nashe theorizes and manifests a philosophy that comes to fruition in his textual output in the summer of 1593. Keeping things enclosed or wrapped up, we discover in the course of *Christs Teares*, is an immoral act. The sermon dwells at length on a single sin with which Nashe was preoccupied throughout his career: miserliness. This is apparently the principal offence that has caused Jerusalem's and, by extension, London's plague. The sermon berates these sinners:

> You Vsurers and Engrossers of Corne, by your hoording vp of gold and graine, tyll it is mould, rusty, Moath-eaten, and almost infects the ayre with the stinche, you haue taught God to hoord up your iniquities and transgressions, tyll mouldinesse, putrifaction and mustinesse, enforceth hym to open them: and being opened,

[120] Ibid., 68–87. Prominent examples of this tradition, examined by Razzall, can be found in Erasmus and Rabelais.

[121] Razzall, *Boxes and Books*, 102.

[122] Tilley, *Metaphor and Material Culture*, 35.

[123] Ibid., 49.

> they so poison the ayre with theyr ill sauour, that from them proceedeth this perrilsome contagion.[124]

Tangling metaphorical and material effect, Nashe parallels the act of hoarding that, through the propagation of mould and rust, 'almost' causes infection, with the sinful act of engrossing, or hoarding, that brings divine punishment upon the city. As Peter Stallybrass notes, in an economy without banking 'hoarding tends to bring social *dis*credit': it is 'both miserly and pointless', with stored gold and grain contributing nothing to society.[125] Good citizens should instead undertake 'conspicuous consumption' and so, 'by sharing the wealth around', create social credit.[126] This economic ideology is a secular version of Nashe's poetics of waste and prodigality, in which sealed and tightly enclosed spaces and objects signify miserliness and cause social and material disease, and in which their opening and dispersal remedy these ills. Nashe goes on to explain that if 'couetise' is 'enlarged out of durance', and London's misers release their gold and grain into a cycle of conspicuous consumption, 'the infected ayre will vncongeale, and the wombes of the contagious Clowdes will be clensed'.[127] This curative opening and 'vncongeal[ing]' is modelled by the kinaesthetics of unwrapping that underpins the dedicatory 'oblation' of *Christs Teares*: the reader is prompted to open and disperse the leaves of the text, conspicuously consuming them and digesting their medicinal contents. As Jennifer Richards has argued, Nashe makes clear that 'the physical book' is not just 'an object we can hold' but 'a "thing" we can perform'.[128] This performance extends beyond the oral features that Richards explores and into the imaginatively potent kinetics of opening that is set out explicitly in several of Nashe's prefaces.

Scholars including Lorna Hutson, Alexandra Halasz, and Georgia Brown have examined the binary of miserliness and prodigality that pervades Nashe's works, and it is widely acknowledged that he finds literary profit in the prodigal expenditure of language.[129] This prodigality extends, in *Christs Teares* and elsewhere, beyond language into the physical realm of bodily health and the spiritual realm of godliness. The cause of Rome's plague in *The Vnfortunate Traueller* is, like Jerusalem's, the material consequence of miserliness: 'The clouds', we are told, 'like a number of cormorants, that keepe their corne till it stinke[s] and is mustie, kept in their stinking exhalations, till they had almost stifled all Romes inhabitants'.[130]

124 Nashe, *Christs Teares*, X2v.
125 Peter Stallybrass, 'The Value of Culture and the Disavowal of Things', in *The Culture of Capital: Property, Cities, and Knowledge in Early Modern England*, ed. by Henry S. Turner (London: Routledge, 2002), 284.
126 Ibid., 284.
127 Nashe, *Christs Teares*, Y1r.
128 Richards, *Voices and Books*, 282.
129 Hutson, *Nashe in Context*, 72–98; Halasz, *The Marketplace of Print*, 99–110; Brown, *Redefining Elizabethan Literature*, 53–99.
130 Nashe, *The Unfortunate Traveller*, K2r, M2v.

Elsewhere in Nashe's works, miserliness is personified in figures who enclose and hoard a variety of materials: in *Pierce Penilesse*, Greedinesse and his wife Dame Niggardize are allegorical embodiments of parsimony, wearing repurposed pots, pans, parchment and paper, and collecting nose drippings for candle wax.[131] Similarly, in *Summer's Last Will and Testament*, performed at Archbishop Whitgift's palace during the early months of the plague outbreak in 1592, we meet the puritanical Christmas who has 'dambd vp all' his 'chimnies' and offers no hospitality.[132] Christmas is an anti-model who teaches the dying king, Summer, a lesson: after meeting Christmas, he commands that the nobility 'keepe high dayes and solemne festivals' and let 'breathe thy rusty gold'.[133] One figure in *The Vnfortunate Traueller*, though not quite a personification of miserliness, is characterized by this vice. The physician Zachary, who is described as 'Dame Niggardize['s] sole heyre and executor', concocts a plague medicine as he 'mouldes vp a Manna' from bread crumbs and distils 'coole allom water' from his 'rheumatique eyes', mixing 'a preseruation against the plague' from the 'wormes and moathes' he picks from between the pages of his books.[134] At first glance resembling Nashe's waste oblation, the distinction between Zachary's 'preseruation' and Nashe's plague cure becomes clear when we consider the intricacies of these two acts of textual recycling. Whereas Zachary picks among his books in a miserly fashion, squeezing matter from imperceptible crumbs and the minute insects that have nibbled their pages, contributing only the water that has leaked from his eyes, Nashe, produces and bequeaths entire volumes of paper, fertile with intellectual and material value, and, in his prefaces, generously invites their dispersal.

More than 'a monument of bad taste', 'pathological' rather 'than devotional', then, *Christs Teares* is a manifestation of the macabre prodigality and virtuous unthrift that structures Nashe's poetic imagination.[135] *Christs Teares* glosses this prodigality with a paraphrase of Psalm 112: 'A good man is mercifull and lendeth, hee giueth, he disperseth, he distributeth to the poore'.[136] The mantra of all Londoners must be to 'disperse' because, Nashe continues, 'If wee should not gyue, Death wold take from vs': hoarding, the sermon makes clear, brings pestilence and death.[137] In a passage bearing striking similarities to John Donne's bibliographic

[131] Nashe, *Pierce Penilesse*, B1r–v.

[132] Thomas Nashe, *A Pleasant Comedie, called Summers Last Will and Testament*, STC 18376 (London: Simon Stafford for Walter Burre, 1600), B4v, H2r. For a discussion of the economics of this play, see Hutson, *Nashe in Context*, 155–196 and Jessica Rosenberg, 'A Digression to Hospitality: Thrift and Christmastime in Shakespeare and the Literature of Husbandry', in *Shakespeare and Hospitality: Ethics, Politics, and Exchange*, ed. by David B. Goldstein and Julia Reinhard Lupton (London: Routledge, 2016), 39–66.

[133] Nashe, *Summers Last Will and Testament*, H3r–v.

[134] Nashe, *The Unfortunate Traveller*, M2v.

[135] G. R. Hibbard, *Thomas Nashe: A Critical Introduction* (London: Routledge & Kegan Paul, 1962), 123; Nicholl, *A Cup of News*, 69.

[136] Nashe, *Christs Teares*, N2v.

[137] Ibid., N2v.

meditations on, in the words of John Carey, the 'body helplessly mouldering through its long years of dissolution', Nashe dwells on the dissipation and decomposition of 'our costly skinne-cases'.[138] This disturbing synonym for the outer shell that encloses the immortal soul suggests that skin serves the same purpose as boxes, bindings, cases, and wrappers, recalling but inverting the physical structure of Nashe's textual 'oblation'. The gross, bodily stuff (the skin) is now the wrapping, rather than the 'mummianizd' contents. Much like the wasted 'pages' of the preface to 'the dapper Mounsier Pages of Court', the textual and the corporeal are imaginatively intertwined. In the conclusion of Nashe's grotesque sermon, we are told that the fate of 'our garments (which are cases and couers for our bodies)' teach us our own fate. Though we might think that, like 'the case or couer of any thing', our skin will preserve and protect us, keeping us, as a bookbinding does a text, 'from consuming to dust', it in fact 'weare[s] away with continuance, euen as Time doth weare and fore-welke [wilt, fade] vs'.[139] No container or its contents, whether skin, paper, gold, or text, is exempt from Nashe's philosophy of prodigal dispersal. Nashe invites his audience, again sounding much like Donne, to conceive of their bodies as mundane matter that will 'wither', 'wanze [wane, waste away]', 'decay' and, eventually, 'rott' in the grave.[140] Nashe's disintegrative devotion differs from Donne's, however, in its outlook. Whereas Donne anxiously and persistently imagines the reconstitution of the dispersed body at the Resurrection, Nashe dwells on and even macabrely celebrates the earlier moment of dissipation. The carefree opening and scattering of 'our costly skinne-cases', like boxes of corn, chests of gold, and the bindings of books, is vital to good and godly living. The giving, dispersal, and distribution outlined in Psalm 112 become, for Nashe, indiscriminate acts of both curative generosity and material wasting: 'I will giue', Nashe expansively concludes, and 'be prodigal'.[141]

The pervasiveness of this poetics of wasting, centred on the kinetics of repurposed paper, is made visible in the works that Nashe composed in his quarrel with Harvey after this plague year. It is perhaps inevitable that the simultaneously profligate and productive dismemberment of books, an idea developed partly in response to Harvey's own writings as well as the broader discourse of usefulness and the material environment of the city in the time of plague, would come to define Nashe's representation of Harvey's and his own acts of composition.

3 Pickling Waste Paper for Posterity

The binary of miserliness and prodigality, expressed throughout Nashe's works in concrete metaphors of enclosure and release, helps us to understand an apparent contradiction in Nashe's use of the waste paper trope. Nashe consistently hurls

[138] Carey, *John Donne*, 226. See p. 91 above. Nashe, *Christs Teares*, T1v.
[139] Ibid., T1v.
[140] Ibid., T1v.
[141] Ibid., V4v.

waste paper insults at Gabriel Harvey throughout his career. How is it, then, that waste paper can embody both that which is positive and good and that which is derided within Nashe's worldview? Though this is in part a result of the elasticity of the trope, adapted by Nashe for different circumstances, the distinction between Harvey's and Nashe's waste reveals the complexities of his poetics of dispersal: not all waste is created equal. Dame Niggardize's 'apron made of almanacs out of date' and the 'purgations and vomits wrapt vppe in wast paper' on sale in the bookshops of St Paul's are not the same as the pages bequeathed 'to the dapper Mounsier Pages' and the waste oblation gifted to Elizabeth Carey.[142] The difference, we will see, is a matter of timing and how skilful an author is in managing and responding to the organic life cycle of their text.

Nashe's fixation with waste paper stems from his sensitivity to the biography of the book: its passage from the discarded matter of rags to pages inked and read, to matter potentially repurposed in the form of waste, before disintegrating or being, once again, discarded. Harvey, Nashe claims in *Haue vvith You*, 'hath laid waste paper in pickle, and published some rags of treatises against Master *Lilly* and mee'.[143] Here, Nashe telescopes the projected fate and the past life of Harvey's printed text into a polychronic object that is simultaneously rags, text, and waste. Nashe is, therefore, sensitive to what Jonathan Gil Harris terms the 'untimely' nature of matter, manifested in the capacity of objects to 'collate many different moments' and prompt their users to dwell on their past and future, as well as their present forms.[144] All books have the potential to end up as waste, but the precise biography of individual texts can differ, just as the fate of gold or corn quickly consumed differs from that which, hoarded and unused, grows mouldy and corrupt. Rotten texts, Nashe suggests throughout his works, will have the same material effect as the plague clouds resulting from the miserliness of Rome's, Jerusalem's, and London's inhabitants: many books on sale in St Paul's are, Nashe claims in *The Anatomie of Absurditie*, composed of 'waste paper [...] fraught with nought els saue dogge daies effects', loaded with the disease that accompanies the hottest days of the year.[145] Some texts, Nashe makes clear, begin their lives rotten: 'Looke to it, you Booksellers and Stationers', Nashe writes, 'and let not your shops be infected with any such goose gyblets or stinking garbadge, as the jygs of newsmongers'.[146] Harvey's texts are, as we would expect, lumped in with this category of putrid and prematurely festering text: because of Harvey's sluggish and stolid style, Nashe claims, he writes with a 'superficial slime of poison' that 'congeal[s] into clodderd garbage'.[147] Here, in *Strange Newes*, Nashe uses the same language to describe the inky matter of Harvey's words as he does the plague clouds that need

[142] Nashe, *Pierce Penilesse*, B1r, I2r.
[143] Nashe, *Haue vvith You*, E2v.
[144] Harris, *Untimely Matter*, 2–4.
[145] Thomas Nashe, *The Anatomie of Absurditie*, STC 18364 (London: by J. Charlewood for Thomas Hacket, 1589), B2r.
[146] Nashe, *Pierce Penilesse*, I2v.
[147] Nashe, *Strange Newes*, H4r.

to be 'vncongealed' in *Christs Teares*.[148] The lack of movement and freshness in Harvey's prose means that his texts are, from their birth, stale and infectious: they share the corruptive properties of corn and gold that does not circulate. In contrast, good writing, and so implicitly Nashe's writing, flows freely from the inkhorn and 'leap[s] like a cup of neat wine new powred out'.[149] It is important to note that Nashe does not claim his own texts escape the organic trajectory from freshness to corruption entirely, but rather that they are, at least in the beginning, 'new'. Implicitly, if it remains undrunk, a newly poured cup of wine will also grow stale. This is why, as we have seen, Nashe always imagines his works as being rapidly and prodigally consumed, and so having no time to turn bad.

The dangers of consuming one of Harvey's texts is made clear in the description, in *Haue vvith You*, of his *Four Letters* as a '*Gargantuan* bag-pudding', containing 'nothing [...] but dog-tripes, swines liuers, oxe galls, and sheepes guts'.[150] Listing a gross hotchpotch of animal innards, some foul, some edible, Nashe alludes to the birth of Rabelais' Gargantua, prematurely induced by his mother, Gargamelle, after she had eaten 'a huge deale of tripes' ('mangea grand planté de tripe').[151] Gargamelle's tripe were consumed to prevent them from going off, because 'in a very short while they would have stunk, which had been an undecent thing' ('en ce que possible n'estoit longuement les réserver, car ells feussent pourrie. Ce que sembloit indécent'). This 'shitten stuff' ('matière fécale'), swiftly consumed to save it from putrefying, makes Gargamelle 'a little unweil in her lower parts' ('commença se porter mal du bas'). The midwives arrive and, groping about, pull out 'a certaine filthy stuffe' ('assez de maulvais goust') that, despite resembling a child, is in fact semi-digested tripe that had 'slipt out' of 'her *straight intrall* [entrail], which you call the bum-gut' ('mais c'estoit le fondement qui luy escapoit, à la mollification du droict intestine lequel vous appellez le bouyau cullier').[152] In alluding to this confusion of 'shitten' animal organs, foodstuff, and excrement, Nashe suggests that Harvey's text is, like Rabelais' tripe, both a monstrous birth and untimely stuff likely to make its consumers queasy at best.

148 Nashe, *Christs Teares*, Y1r.

149 Nashe, *Strange Newes*, H4r.

150 Nashe, *Haue vvith You*, F1r.

151 This translation is taken from Thomas Urquhart's *The First Book of the Workes of Mr. Francis Rabelais Doctor in Physick*, Wing R105 (London: [Thomas Ratcliffe and Edward Mottershead] for Richard Baddeley, 1653), 22–23; for the original French, see François Rabelais, *Oeuvres Completes*, ed. by Jacques Boulenger (Paris: Gallimard, 1955), 14–15. We do not know for certain whether Nashe read Rabelais, and there was no English translation of *Gargantua and Pantagruel* until Urquhart's in 1653. However, as Anne Prescott argues, a 'Rabelaisian energy field' permeated the literary circles that Nashe moved in and Nashe's collaborator, Ben Jonson, owned and annotated a copy of Rabelais' *Oeuvres* (Anne Lake Prescott, *Imagining Rabelais in Renaissance England* (New Haven: Yale University Press, 1998), viii, 53–55, 196–206). I believe that this passage shows Nashe's relatively detailed familiarity with the French author's work.

152 Urquhart, *Rabelais*, 30–31; Rabelais, *Oeuvres*, 22–23.

In *Haue vvith You*, Nashe portrays a Harvey who remains comically oblivious to the rapid putrefaction of his works: despite the fact his books 'ly[e] dead & neuer selling' and are consequently 'turne[d] to wast paper', Harvey assumes that they will survive indefinitely.[153] Importuno, the buffoonish Harvey figure in *Haue vvith You*, outlines a theory of textual endurance, claiming that Harvey's printed attacks on Nashe will 'sticke by [him] for euer'.[154] '[W]hile printing lasts', Importuno prophecies, 'thy disgrace may last, & the Printer (whose Copie it is) may leaue thy infamie in Legacie to his heyres, and his heyres to their next heyres, successiuely to the thirteenth and fourteenth generation'.[155] Passed down through the generations of a publishing dynasty like the rights to a valuable steady seller, Harvey's attacks will be printed, Importuno assumes, '*in sempiternum*', or eternally.[156] The qualifying clause, 'while Printing lasts' suggests, though, a degree of uncertainty, and the epistrophic listing of the printer's many heirs becomes absurd. This was after all, as Adam Smyth has demonstrated, a period in which the 'prevailing culture of textual transience' upset 'any direct coupling of print with textual permanence'.[157] Though far from a new technology, print was not yet close to being thirteen or fourteen generations old, and, as Kate De Rycker notes, hindsight allows us to 'see how tenuous a hold writers such as Nashe had over the survival of their work'.[158] Nashe and Jonson's play, *The Isle of Dogs* (1597), for instance, did not survive its censorship, and the 1599 Bishops' Ban ordered, admittedly ineffectively, the destruction of all of Nashe's and Harvey's works.[159] Because of the uncertainty of this material environment, Nashe's description of what he terms elsewhere in *Haue vvith You* 'the Immortality of the Print', might be read ironically.[160] It pre-empts, as Steve Mentz has noted, the prominent debate between Elizabeth Eisenstein and Adrian Johns concerning the perceived fixity and longevity of print in the early modern period, with Johns critiquing Eisenstein's emphasis on the capacity of print to serve as a reliable and stable 'agent of change'.[161] It is, in *Haue vvith You*, the Harvey-like figure, Importuno, who communicates a proto-Eisensteinian faith in the capacity of print to transform textual culture and so to last '*in sempiternum*'. The 'cultural fantasy' of the printing press

153 Nashe, *Haue vvith You*, T3v.
154 Ibid., E1v.
155 Ibid., E1v.
156 Ibid., E1v.
157 Smyth, *Material Texts*, 57.
158 Kate De Rycker, 'Commodifying the Author: The Mediation of Aretino's Fame in the Harvey-Nashe Pamphlet War', *ELR* 49, no. 2 (2019): 145–171, 169. On this uncertainty of textual endurance, see also Calhoun, *Nature of the Page*, 138–150.
159 De Rycker, 'Commodifying the Author', 169.
160 Nashe, *Haue vvith You*, E3v.
161 Mentz, 'Day Labor', 31. The initial publications that sparked this long-lasting debate are Elizabeth Eisenstein, *The Printing Press as an Agent of Change: Communications and Cultural Transformations in Early Modern Europe*, 2 vols (Cambridge: Cambridge University Press, 1979) and Adrian Johns, *The Nature of the Book: Print and Knowledge in the Making* (Chicago: University of Chicago Press, 1998).

as an 'agent of change' and a preservative technology is not, then, Nashe's own, as Mentz argues, but something portrayed ironically and attributed to the buffoonish Importuno.[162]

Instead, Nashe's portrayal of Harvey's attempts at textual preservation conform to what Katie Kadue has termed the 'domestic georgic'—the way in which 'the imaginations of male authors' in the sixteenth and seventeenth centuries 'repeatedly returned to the rhythms of domestic labor', revealing a sense of authorship as 'only temporary in its effectiveness' and often wholly unsuccessful.[163] Several pages prior to Importuno's Eisensteinian articulation in *Haue vvith You*, Nashe describes Harvey's bid for textual permanence in domestic terms, framing writing as kitchen work that attempts to stave off the natural rhythms of rot and decay. The claim, cited above, that Harvey 'hath laid waste paper in pickle' suggests that Harvey is a bad householder who attempts to prevent the decomposition of an object that is already past its best.[164] The 'treatise' that Harvey pickles is not even new, Nashe claims. Rather than composing fresh texts, Harvey recycles old ones that 'haue lyne by him euer since the great matches of bowling and shotting on the *Thames* vpon the ice', a reference to the frost fair of 1565 that halted the course of the Thames.[165] Seeking to similarly halt the temporal progression of his own books, Harvey will inevitably fail to achieve longevity through his texts because they have always been and so will always remain rotten.

In publishing already old and 'congeal[ed]' texts, Harvey's 'treatise' resembles the belatedly preserved wares described in Nashe's final work: the three smoked herrings that we meet in his 1599 *Lenten Stuffe* that, already 'worn [...] stale throughout England', are transported from Great Yarmouth to Rome.[166] '[E]ndungeond' in his 'pocket a twelvemonth', a witty English hawker travels to the papal court and succeeds in selling the Pope his stale wares.[167] The Pope's cooks serve up the fish as part of a banquet and once 'uncovered' at the table, the herring releases 'such an ayre' that no 'hostler was able to endure', prompting comical acts of exorcism and desperate prayer in the papal court.[168] In addition to providing an aggressively anti-Catholic burlesque, this narrative contains a pattern that we have now encountered many times in Nashe's writing: 'endungeond' or tightly enclosed matter, kept past its best, turns rotten and releases corrupted 'ayre'.[169] In *Lenten Stuffe*, as elsewhere, Nashe displays a lack of faith in the capacity of preservative

[162] Mentz, 'Day Labor', 31.

[163] Katie Kadue, *Domestic Georgic: Labors of Preservation from Rabelais to Milton* (Chicago: The University of Chicago Press, 2021), 9.

[164] Nashe, *Haue vvith You*, E2v.

[165] Ibid., E2v.

[166] Thomas Nashe, *Lenten Stuffe*, STC 18360 (London: [Thomas Judson and Valentine Simmes] for N[icholas] L[ing] and C[uthbert] B[urby], 1599), H3r.

[167] Ibid., H4v.

[168] Ibid., I1r.

[169] Ibid., H4v, I1r.

practices such as pickling and smoking to delay the inevitable decomposition of an organic object, or to undo the effects of time that has already passed. Resembling this rotten herring and the plague clouds of *The Vnfortunate Traueller* and *Christs Teares*, Harvey's texts, along with all of the other putrid works Nashe describes throughout his corpus, are implicitly at risk of releasing corruption and poison once opened. Paradoxically, the cure to this rottenness is the timely, and so rapid, dissolution and dispersal of the book's pages.

As Lorna Hutson has argued, this is a seasonal pattern of ripening and rotting that belongs to the festive world of the carnivalesque, in contrast to the 'stale accumulation of exhausted scraps of discourse' that Nashe locates in the work of Gabriel Harvey and humanist culture more broadly.[170] Nashe, as Hutson suggests, 'sets about to restore vitality' by 'breaking' up Harvey's text, 'dismembering' it and 'burying' it within his own works in a process of composting not unlike Hester Pulter's 'dunghill poetics'.[171] We can add to this the rhetorical dismemberment of Harvey's books in the form of repeated waste paper attacks: once situated within Nashe's broader poetics of dispersal, these violent motions are revealed as curative rather than straightforwardly destructive. The claims that Harvey's books have been 'torne' up to 'wrappe the excrements of huswifery' or, in the form of toilet paper, 'bequeath[ed] [...] to the Priuie, leafe by leafe', should in fact be read as the safe disposal—or, more accurately, repurposing and dissolution—of hazardous waste.[172]

We also uncover in Nashe's waste paper insults a commentary on the competing bibliographic discourses available in the 1590s. In Nashe, we find another counternarrative to the figuration of texts as enduring monuments, with its source in Horace and its most famous expression in the late sixteenth and early seventeenth centuries in Shakespeare's *Sonnets*. Pickling sits in comic opposition to the culinary practices evoked in Shakespeare's poetry—the distillation that Wendy Wall describes as being 'like preserving, a chemical transformation that enabled a substance to take on a new and potentially more enduring incarnation'.[173] At stake in distillation, Wall continues, 'was nothing short of the elimination and transcendence of elements that corrupted matter'.[174] In Nashe's account of Harvey's pickling, there is no such transcendence of corrupted matter. Instead, texts either remain whole and become rotten, or are dispersed in the form of useful and so paradoxically wholesome fragments. Nashe does not claim for himself what Wall

170 Hutson, *Nashe in Context*, 110.

171 Ibid., 110. On Dolan, 'Dunghill Poetics', see pp. 32–33 and 135 above, as well as Stenner's description of how Nashe 'envisages printed texts not as a means to longevity for their author but as fully disposable and capable of being recycled and fed back into the economy' (*The Typographic Imaginary*, 181).

172 Nashe, *Haue vvith You*, C1r; Nashe, *Strange Newes*, G3v; Nashe, *Pierce Penilesse*, E1v.

173 Wall, *Recipes for Thought*, 187.

174 Ibid., 187.

terms the 'monumental and grand ambitions' of a poetry that endures.[175] Instead, it is Nashe's Harvey who harbours these hopes: much like Importuno in *Haue vvith You*, Harvey is described in *Strange Newes* as 'a forestaller of the market of fame' and 'an ingrosser of glorie', a bibliographic miser in the pattern of the 'Vsurers and Engrossers of Corne' in *Christs Teares* who bring plague upon ancient Jerusalem and contemporary London.[176] Misunderstanding the organic rhythms central to the life cycle of the material text, Harvey publishes already rotten pages that, 'founde' and shut up only 'in his owne Deske', threaten to release dangerous corruption if, after such a lengthy period of closure, the drawer or the book's cover ever chance to be opened.[177]

By contrast, Nashe embraces the bibliographic life cycle of consumption and the kinetics of opening and disintegration, inviting it in the prefaces to *Christs Teares* and *The Vnfortunate Traueller*. Although our modern-day understanding of the material text may be more in line with Importuno's '*in sempiternum*', and although many of both Nashe's and Harvey's texts do survive, we should not underestimate the credibility of Nashe's waste paper poetics in the period in which he wrote. As Adam Smyth writes, the 'willingness to cut'—and we might add, disperse into waste—'reflects the fact that the coherent, bound, unannotated, "complete" printed book, with which modern bibliographical culture has been fixated, was not yet the dominant mode for conveying text'.[178] Nashe's poetics tells us about more, though, than the textual culture of the period. The bodily experience of waste paper was not isolated to the realm of the bibliographic and the literary. Instead, encoded with meaning through their repeated reuse, the fragmentary nature of early modern books transformed into a concrete metaphor that, in turn, constituted Nashe's understanding of how one might exist within society and the world. For Nashe, then, waste paper was an 'ordering device' that structured the fundamental patterns of literary, economic, medicinal, and spiritual health.[179]

Just as *Christs Teares* teaches the healthful effect of prodigally releasing our corn, gold, and 'skinne cases' into the cycles of use and decay, Nashe concludes his parody of 'the immortality of the Print' in *Haue vvith You* with characteristically cheerful morbidity. He scoffs at the 'vrging of posteritie and after ages whose cradle makers are not yet begot', rejecting thoughts of fame and the distant future along with the foolish belief that 'there is anie thing so eternall and permanent, that consumes and dies not'.[180] All matter, whether book or body, as *Christs Teares* has shown us, 'consumes' and 'dies'.[181] Nashe is, therefore, unconcerned as to whether

175 Ibid., 190.
176 Nashe, *Strange Newes*, B2r; Nashe, *Christs Teares*, X2v. On the representation of Harvey as a miser, see Halasz, *The Marketplace of Print*, 99–100 and Viviers, 'Unprofitable Satire', 433.
177 Nashe, *Strange Newes*, B2r.
178 Smyth, *Material Texts*, 26.
179 Tilley, *Metaphor and Material Culture*, 49.
180 Nashe, *Haue vvith You*, G1r, E4r.
181 Ibid., G1r.

Harvey's printer and his progeny will 'baffull and infamize' his name for as long as print lasts.[182] Nashe, according to Nashe, will by then be 'in heauen, & shall never feele it', having adhered to the philosophy of prodigality outlined in *Christs Teares* and that runs, in his poetics of waste, throughout his works.[183] He has lived, as he says every poet should, 'in vnthrifts consistory' and has sung '*George Gascoignes* Counter-tenor', having spent time in debtor's prison.[184] Far from hoarding gold and the materials of writing until they turn mouldy, Nashe has lived an exceptionally prodigal, and therefore godly, life. Nashe's concrete metaphor of repurposed pages, then, teaches us that to get to heaven, to avoid the plague, and to write good books, we should bequeath that which we hold dear generously and willingly, scattering it, 'vnthrift[ily]', into waste and dust.

We will now turn to an author who was an avid reader of Nashe and whose imagination was similarly shaped by the kinetics and visible biographies of repurposed books: Thomas Middleton.[185] But whereas Nashe is primarily concerned with the transformation of his and Harvey's texts into waste, Middleton, writing in the early decades of the seventeenth century, was drawn to a particular subset of repurposed books: the discarded almanacs that, turning rapidly out of date, populated the environment particularly thickly at certain times of the year.

182 Ibid., E3v.

183 Ibid., E3v.

184 Thomas Nashe, *The Terrors of the Night or, A Discourse of Apparitions*, STC 18379 (London: Iohn Danter for William Iones, 1594), C1v; *Strange Newes*, I1r.

185 On the influence of Nashe on Middleton, see Rhodes, *Elizabethan Grotesque*, esp. 86.

5
Out of Date Almanacs and Middleton's 'Mouldy Stuff'

> An apology must be made that these notes contain little or no technical or critical bibliography, but they are only intended to show that the Almanack was a very human book.
>
> Eustace Bosanquet, 'English Seventeenth-Century Almanacks'[1]

In her 2010 *Vibrant Matter: A Political Ecology of Things*, Jane Bennett describes an encounter between herself and the debris that had accumulated on 'the grate over the storm drain' on 'a sunny Tuesday morning' in Baltimore.[2] Bennett catalogues the debris and describes its effect on her:

> Glove, pollen, rat, cap, stick. As I encountered these items, they shimmied back and forth between debris and thing—between, on the one hand, stuff to ignore, except insofar as it betokened human activity [...] and, on the other hand, stuff that commanded attention in its own right, as existents in excess of their association with human meanings, habits, or projects. In the second moment, stuff exhibited its thing-power: it issued a call, even if I did not quite understand what it was saying.[3]

For Bennett, debris is a particularly expressive form of matter. Removed from the anthropocentric culture of use, discarded 'objects appeared as things, that is, as vivid entities not entirely reducible to the contexts in which (human) subjects set them'.[4] Bennett goes on to explain that the 'call' the debris issues concerns the way in which 'human being and thinghood overlap', and so constitutes an announcement that 'we are also non-human and that things, too, are vital players in the world'.[5] An understanding of this ecology is 'political', for Bennett, because in developing a sensitivity to our 'strange and incomplete commonality with the outside', or the non-human, we will be prompted to live and think more ethically and

[1] Eustace Bosanquet, 'English Seventeenth-Century Almanacks', *Library* 4, no. 10 (1930): 397.
[2] Bennett, *Vibrant Matter*, 4.
[3] Ibid., 4.
[4] Ibid., 5.
[5] Ibid., 4.

Waste Paper in Early Modern England. Anna Reynolds, Oxford University Press. © Anna Reynolds (2024).
DOI: 10.1093/oso/9780198882701.003.0006

to behave more responsibly toward both the human and non-human inhabitants of our world.[6] This vibrant sensitivity characterized some 'premodern' cultures, according to Bennett, and the loss of this way of inhabiting the world is connected in part to our relationship to the debris, junk, rubbish, and waste that surrounds us.[7] A consumerist society is necessarily anthropocentric because the 'sheer volume of commodities, and the hyperconsumptive necessity of junking them to make room for new ones, conceals the vitality of matter'.[8] Bennett's suggestion that, in order to live 'more carefully, more ecologically, more strategically', we should cultivate 'a capacity for naivete' and become 'temporarily infected' by the 'discredited philosophies of nature' is, then, inseparable from how we should live with and think about our waste.[9]

This chapter is concerned with one such discredited philosophy of nature: the relationship between the human and the world as it is communicated in early modern almanacs, within the science of which all earthly bodies are understood to be constituted of corresponding matter, influenced in the same way by the celestial bodies above. The chapter is also about the early modern ecology of waste that is grounded in these flimsy and ephemeral publications. It explores the call that almanacs issued through their material biography, when they form, as discarded things, an early modern version of the debris caught in Bennett's storm drain. Relevant for only one year and published in their hundreds of thousands, almanacs were the type of text most often turned to waste, and perhaps the most repurposed item in the sixteenth and seventeenth centuries.

After overviewing the ecology that underpins the almanac's textual contents, I consider the additional meanings that almanacs accrued once discarded. Lingering in the environment in the form of repurposed paper, almanacs occupied what William Viney has termed 'waste-time', a temporality of disuse that 'makes time an explicit, tangible thing of thought'.[10] Waste, as Viney argues, 'gives witness to how discarded things are temporally and spatially dispersed', prompting us to structure broader narratives around the material biography of these objects.[11] These waste narratives overlap in productive ways with the intellectual content of almanacs, and so these little books, in various states of use and disuse, shaped a pervasive early modern understanding of the interconnected place of the human within the environment. The chapter concludes with an exploration of the representation of greasy and discarded almanacs in the works of Thomas Middleton: though always sceptical about their accuracy and use, Middleton's works manifest his discomfort with the way that almanacs blur the distinction between the

[6] Ibid., 18.
[7] Ibid., 18.
[8] Ibid., 5.
[9] Ibid., 17.
[10] Viney, *Waste*, 3, 9–14.
[11] Ibid., 29.

human and the non-human. His repeated attempts to extricate the human subject from an environmental assemblage of non-human things, however, reveals the hold that almanacs had over his and the wider cultural imagination. Almanac bodies—assemblages made up of the human and the pages of these little books—recur throughout Middleton's *Masque of Heroes* and the play *No Wit/Help Like a Woman*, and they also underpin the philosophy of his mock- *Owl's Almanac*. Upon close inspection, these assemblages reveal an alternative mode of thought grounded in the widespread intimacy with a specific sub-category of repurposed books. Not yet fully discredited in this early stage of modernity, we might term this philosophy ecosystemic rather than anthropocentric.

1 Almanac Ecology

It isn't entirely clear what Eustace Bosanquet was suggesting when, in the passage quoted in the epigraph to this chapter, he described the almanac as 'a very human book'.[12] He is perhaps referring to the fact that almanacs were the most widely owned book in the early modern period: printed in their hundreds of thousands in the sixteenth century and in their millions in the seventeenth, almanacs were the most profitable part of the Stationers' Company's trade.[13] Costing one or two pence and made up of between two and six sheets, folded into octavo or duodecimo format, almanacs were easily slipped into the pocket and so lived close to the human body, as well as pervading the wider material environment.[14] The almanac's 'human[ity]' might also stem, for Bosanquet, from what Adam Smyth has termed their 'unpredicted social lives': almanacs contained an annual calendar, and those printed with additional blank space—and so known as blanks—were often filled with extensive annotations comprising a form of 'self-accounting' or 'life-writing'.[15] The almanac, Smyth argues, encourages its user to 'turn back on herself, in the first-person', and to write her 'individuality through the adoption of shared, public templates'.[16] It is these 'shared, public templates' that are the focus of this chapter, and the primary way in which almanacs were, I suggest, human.

[12] Bosanquet, 'English Seventeenth-Century Almanacks', 397.

[13] On the history of the almanac see Cyprian Bladgen, 'The Distribution of Almanacks in the Second Half of the Seventeenth Century', *Studies in Bibliography* 11 (1958): 107–116; Eustace Bosanquet, *English Printed Almanacks and Prognostications, A Bibliographical History to the year 1600* (London: the Bibliographical Society at the Chiswick Press, 1917); and Bernard Capp, *Astrology and the Popular Press: English Almanacs, 1500–1800* (London: Faber and Faber, 1979).

[14] An extant copy of *Allestree 1639* provides evidence for how almanacs were stored and carried. The crease of a vertical fold is visible down the centre of the page. See Fig. 5.1.

[15] Adam Smyth, 'Almanacs and Ideas of Popularity', in *The Elizabethan Top Ten: Defining Print Popularity in Early Modern England*, ed. by Andy Kesson and Emma Smith (Farnham: Ashgate, 2013), 125–133, 130; Adam Smyth, 'Almanacs, Annotators, and Life-Writing in Early Modern England, *ELR* 38, no. 2 (2008): 200–244, 204.

[16] Smyth, 'Almanacs, Annotators, and Life-Writing', 239.

Crucially, as Smyth argues, these 'shared [...] templates' were unrelated to the 'birth' of the 'detached' and 'coherent' modern individual that is often located in these centuries.[17] Instead, almanacs, both used and disused, were a vehicle through which individuals learnt to negotiate their place within the wider environment and to inhabit its complex network of human and non-human bodies. Bosanquet is certainly right, then, that the almanac was 'a very human book', but a closer examination of the intellectual content and physical format of these little books will prompt us to reconsider what it meant to be human in the early modern period. It will also reveal the inseparability of this expansive sense of the human from the matter of waste.

Almanacs have two parts—a calendar and a prognostication—each with a separate title page and signatures but printed and stitched together in one thin volume.[18] The first section contains preliminary material outlining instructions for use and key information; brief rules of thumb regarding 'Physick' and husbandry; an 'Anatomical' or 'Zodiac Man' that showed, in woodcut form, which part of the human body each astrological sign influenced (see Fig. 5.3); and an ever-expanding chronology of history, or 'Computation of Time', counting down from the Creation to the present year. This is followed by a month-by-month calendar containing information for each day, with columns noting special dates; the exact minute of sunrise, sunset and the length of the day; the sign and degree of the moon at noon; a symbol specifying 'good' and 'bad' days for prescribed activities (when best and worst to let blood, bathe, cut hair, sow seeds, and so on); and the phases of the moon through the course of the month. The prognostication expands on the contents of the calendar. It outlines the year's eclipses of the sun and moon and details the 'disposition' of the four seasons, describing the zodiacal movements and their effects on the weather and any consequent disease and dearth. What follows are even more detailed descriptions of the 'disposition' of each month and its weather, sometimes down to the minute, and advice regarding health and husbandry. By the turn of the seventeenth century, almanacs had accumulated an additional section at the rear containing information such as the time and location of England's fairs; the distances between major cities; and tables of weights, measures, and tides.

The structure and content of almanacs remained stable from their rise in the second half of the sixteenth century until the end of their golden age in the late seventeenth, though some almanacs became increasingly specialized. Some were calculated according to the meridian of certain cities or were aimed at certain professions (merchants, seamen, doctors and lawyers, etc.), and others contained especially large amounts of medical or agricultural material. Some

[17] Ibid., 242.

[18] For a comprehensive outline and explanation of the contents of early modern almanacs, see Phebe Jensen, *Astrology, Almanacs, and the Early Modern English Calendar* (Abingdon: Routledge, 2021), 57–73.

almanac compilers were, or claimed to be, physicians or mathematicians, and some were particularly religious. Almanac users, it seems, had their favoured publications, be they Securis or Dade, for medical advice, Johnson for agriculture, or Browne for religion. The fundamental science behind the almanac remained, though, constant. As Thomas Blount summarized in 1656, 'Astrology is a Science which tells the Reasons of the Stars and Planets motions'.[19] It 'professeth to discover the influence and domination of the superior Globe over the inferior, and therefore may be tearmed a kind of natural divination'.[20] This definition is almost identical to that given more than a century earlier in a 1537 almanac, which states that 'all thynges happenynge on the erthe dependeth on natural causes of the bodyes aboue folowynge'.[21] In short, almanac science outlined how the movements of the planets and stars altered the air which, in turn, influenced the humoral make-up of all mundane bodies, be they human, animal, vegetable, or mineral. This meant, in the words of a 1555 almanac, that by 'knowynge the disposicion of the celestial bodies, by the which all ementate [emanant, or created] bodyes be ruled', it is possible to 'prognosticat[e] [...] thynges to come, as mutacions of the ayre, pestilencis, & al other infirmities [...] which depende on the heavens, as second causes of God'.[22] Knowing what alterations in weather, health, and humoral balance to expect, almanac readers were better informed when best to undertake any number of activities, ranging from bathing, paring nails and cutting hair, to letting blood, chopping down trees, or sowing, harvesting, and preserving particular crops. Armed with an almanac's knowledge, early moderns were able to manage and mitigate the celestial impact on their own bodies, and on the non-human bodies around them.

As Todd Borlik argues, almanac science roots the human within a 'biological and cosmic cycle', situating them in an 'animistic' and 'vitalistic' world in which 'objects' are not 'bandied about by external forces' but 'possess an innate tendency for motion that amounts to a quasi-agency'.[23] The human is decentred within this environment. As Katherine Walker puts it, 'a narrative of human experience' is established that is 'more interactive than typically recognized by early modern scholars of the body and environment'.[24] Within this ecology, human and non-human bodies correspond to one another, made up as they are of homologous

[19] 'Astrology', in Thomas Blount *Glossographia: or, a Dictionary, Interpreting all such Hard Words*, Wing B3334 (London: Tho. Newcomb to be sold by Humphrey Moseley, 1656).

[20] Ibid.

[21] George Seyfridt, [*Almanake and Prognostication for the Yeare of our Lorde MCCCCC, and XXXVII*], STC 513 (Antwerp: Widow of C. Ruremond?, [1537]), 1 sheet.

[22] Antonius de Montulmo, *An Almanacke and Prognostication for the Yere of Our Lord God D.CCCCC.LV.* [*sic*], STC 483.14 (London: Thomas Marche, [1555]), A2r-A2v.

[23] Todd A. Borlik, *Ecocriticism and Early Modern English Literature* (Abingdon: Routledge, 2011), 13, 118.

[24] Borlik, *Ecocriticism and Early Modern English Literature*, 117; Katherine Walker, 'Early Modern Almanacs and the Witch of Edmonton', *Early Modern Literary Studies* 18, no. 1 (2015): 1–25, 2.

matter and influenced by the celestial bodies in exactly the same way.[25] Certain bodies should not come into contact with one another at certain times: 'No part of man's body ought to be touched with any chirurgical instrument', John Securis tells his readers, 'when the sunne or moone [...] is in the same signe that ruleth that part of manne's body'.[26] The same rules apply to the human, the animal, and the vegetable: if it is a good day for shearing your sheep, it is also a safe day to cut your hair; similarly, the pores of both the earth and the body open and close with the passing of the seasons, so that 'the humors begin to moove', according to Johnson 1604, in the warmth of April.[27] At this time, the human body is 'very apt to be purged', and plants begin to grow.[28] The human and the organic, almanacs taught their users, move according to the same rhythms because of their shared material constitution and their shared situation beneath the stars.

It is difficult to determine how seriously almanac users engaged with the scientific content of their calendars and prognostications. We should be wary, though, of the assumption that almanac readers were predominantly rural and uneducated, and that almanac science did not represent a widespread early modern understanding of how the world worked.[29] Grounded in the gradual relegation of almanacs and their astrological contents to the realm of superstition and rural tradition in the late seventeenth and eighteenth centuries, this assumption obscures the vibrancy of almanacs as a locus of thought in the century and a half prior to their eventual decline.[30] In this earlier period, as the work of scholars including Bernard Capp, Ann Geneva, and Patrick Curry has demonstrated, almanacs were read and compiled across the socio-economic spectrum: they were often written by university men, such as Gabriel Harvey's brothers John and Richard, graduates of the University of Cambridge and much-maligned by Thomas Nashe, and the mathematician Leonard Digges.[31] Almanacs themselves were read by royals, the poor, and everyone in between, although the majority of surviving books belonged

[25] The homology of the material makeup of the human and the environment is the foundation of early modern humoral theory. See Michael Schoenfeldt, *Bodies and Selves in Early Modern England: Physiology and Inwardness in Spenser, Shakespeare, Herbert, and Milton* (Cambridge: Cambridge University Press, 1999), 3, and Gail Kern Paster, *Humoring the Body: Emotions and the Shakespearean Stage* (Chicago: University of Chicago Press, 2004), 4.

[26] John Securis, *1574. An Almanacke and Prognostication*, STC 512 (London: Richarde Watkins and James Robertes, 1574), A2r.

[27] *Johnson. 1604 an Almanack and Prognostication*, STC 466.7 (London: E. Allde for the Company of Stationers, [1604]), B6r.

[28] Ibid., B6r.

[29] Maureen Perkins, for instance, states that 'the more astrological the content of an almanac, the more plebeian its intended audience' (*Visions of the Future: Almanacs, Time, and Cultural Change, 1775–1870* (Oxford: Clarendon Press, 1996), 3).

[30] On this broader cultural 'decline' of superstition, see Keith Thomas, *Religion and the Decline of Magic: Studies in Popular Beliefs in Sixteenth and Seventeenth Century England* (London: Weidenfeld and Nicolson, 1971), 355–357.

[31] Patrick Curry, *Prophecy and Power: Astrology in Early Modern England* (Oxford: Polity Press, 1989), 23–40; Ann Geneva, *Astrology and the Seventeenth-Century Mind: William Lilly and the Language of the Stars* (Manchester: Manchester University Press, 1995), 55–71.

to yeomen, husbandmen and artisans.[32] Educated and non-sceptical use is visible in the margins and blanks of many extant almanacs, constituting cryptic astrological calculations as well as the more common daily accounts of the middling sort.[33]

There is also a risk of taking the satirical representation of almanacs and almanac readers in early modern literature too much at face value: foolish almanac readers were a stock figure on the early modern stage, with examples including Sordido in Ben Jonson's *Every Man out of his Humour* and Cuddy Banks in *The Witch of Edmonton*, as well as Thomas Middleton's Weatherwise, in *No Wit/Help Like a Woman* who we will meet again later in this chapter. Almanacs also make fleeting appearances in *A Midsummer Night's Dream* (Bottom: 'A calendar, a calendar! Look in the almanac—find out moonshine, find out moonshine!') and a number of works by Francis Beaumont and John Fletcher (*Philaster, The Chances*, and *A Wife for a Month*, to name a few).[34] Mock-almanacs and prognostications were also popular in the period, published by writers such as the pseudonymous Adam Foulweather (sometimes identified as Thomas Nashe), Thomas Dekker, and Middleton.[35] But parody and scepticism do not preclude serious almanac use: just as Jonson's farcical portrayal of alchemy and astrology in *The Alchemist* does not mean that he rejected the foundations of all alchemical science, those who portrayed naïve almanac readers who are over-reliant on their good and bad days shared with them a fundamental understanding of the world.[36] This fundamental understanding was communicated, in its most widespread and accessible form, in cheap, ephemeral almanacs. Jonson, as Margaret Healy points out, seems to have known a vast amount about the alchemical science he derides, and the same can be said, for instance, of Middleton and the contents of the almanacs he often portrayed sceptically: his works reveal that he regularly purchased and read almanacs throughout his career, in particular, *Bretnor*, *Dade*, and *Neve*.[37]

[32] Capp, *Astrology and the Popular Press*, 60; Thomas, *Religion and the Decline of Magic*, 379.

[33] Smyth, 'Almanacs, Annotators, and Life-Writing', 200–244. The British Library's copy of Allestree 1639 contains the signature 'Herman ye Earl of Arundalls Secretary' (another page of this almanac is reproduced at Fig. 5.1), and the Bodleian Library's *Pond 1629* (Rawl. Alm. 83 (1)) contains astrological annotations.

[34] *A Midsummer Night's Dream* in *The Oxford Shakespeare: Complete Works*, 2nd ed., ed. by John Jowett et al (Oxford: Clarendon Press, 2005), III.1.48–49.

[35] Adam Foulweather, *A Wonderfull, Strange and Miraculous, Astrologicall Prognostication*, STC 11209 (London: [Thomas Scarlet], 1591); Thomas Dekker, *The Ravens Almanacke*, STC 6519.2 (London: E[dward] Allde and another for Thomas Archer, 1609); Thomas Middleton, *The Owles Almanacke*, STC 6515 (London: E[dward] G[riffin] for Lawrence Lisle, 1618). See also F. P. Wilson's 'Some English Mock-Prognostications', *The Library* 4, no. 19 (1938): 6–43.

[36] Margaret Healy, 'Alchemy, Magic, and the Sciences', in *Ben Jonson in Context*, ed. by Julie Sanders (Cambridge: Cambridge University Press, 2010), 322; see also William Eamon, 'Astrology and Society', in *A Companion to Astrology in the Renaissance*, ed. by Brendan Dooley (Leiden: Brill, 2013), 190.

[37] Healy, 'Alchemy, Magic, and the Sciences', 325. Middleton quotes directly from Bretnor 1611, Dade 1611, and Neve 1611 in *No Wit/No Help Like a Woman's; Or, The Almanac.* He also quotes from Bretnor 1618 in *The Owles Almanacke* and Bretnor 1619 in *The Masque of Heroes*. See David George, 'Weather-Wise's Almanac and the Date of Middleton's No Wit No Help Like a Woman's', *Notes and Queries* 13,

An almanac's science of celestial 'influence and domination' was, after all, a remarkably capacious one, repeatedly modified rather than rejected by a surprising array of practitioners including Tycho Brahe, Johanes Kepler, and Francis Bacon.[38] As the work of scholars including Gail Kern Paster, Mary Floyd-Wilson, and Katherine Rowe has demonstrated, early moderns conceived of themselves as participants in an ecology grounded in the 'interpenetrability' and 'transactionality of body and environment', with a correspondence rooted in 'the shared materiality of body and world'.[39] Paster has, for instance, demonstrated the centrality of the humors and the passions to the drama of the period, and so has radically altered our understanding of the relationship between early modern science and literature.[40] There is less scholarship, however, on the importance of almanacs within both literary and scientific thought, though recent work by Abigail Shinn and Katherine Walker has made a persuasive case for the importance of the little books in the work of Edmund Spenser and in *The Witch of Edmonton*, respectively.[41] As Shinn notes, almanacs underpin the agricultural rhythms of *The Sheapeardes Calendar* and demonstrate 'the universal link between microcosm and macrocosm'.[42] Spenser's poem is, therefore, embedded in an understanding of a 'shared astrological compulsion as man and his environment both incline towards the path plotted for them by the stars'.[43] Similarly, Katherine Walker argues that almanacs offered their readers 'a heterogeneous understanding of the environment', intermingling 'popular and scientific discourses' and providing a 'valuable means for interpreting the cosmos'.[44] Walker demonstrates that Cuddy Banks, the stock comic almanac reader of *The Witch of Edmonton*, is empowered by his popular reading material: he is able to control the relationship between his body and the cosmos, closing it off against malicious influence by using his insight into the rhythms of seasonal time.[45]

As Walker notes, although much work has been done on the 'circulation and medical content' of almanacs, 'literary studies has yet to explore how prognostications and ephemeral texts influence our understanding of the early modern notion

no. 8 (1966): 297–301 and *No Wit, No Help Like a Woman's*, ed. by Lowell E. Johnson (Lincoln, NE: University of Nebraska Press, 1976), xii.

[38] Darrel H. Rutkin, 'Astrology', in *The Cambridge History of Science*, ed. by Katharine Park and Lorraine Daston (Cambridge: Cambridge University Press, 2006), 548–552.

[39] This quotation is taken from Mary Floyd-Wilson and Garret A. Sullivan's summary of the field in the introduction to *Geographies of Embodiment in Early Modern England* (Oxford: Oxford University Press, 2020), 4.

[40] See Paster, *Humoring the Body*, passim.

[41] Walker, 'Early Modern Almanacs'; Abigail Shinn, '"Extraordinary discourses of vnnecessarie matter": Spenser's *Shepheardes Calendar* and the Almanac Tradition', in *Literature and Popular Culture in Early Modern England*, ed. by Matthew Dimmock and Andrew Hadfield (Farnham: Ashgate, 2009), 137–149.

[42] Shinn, 'Extraordinary discourses', 144.

[43] Ibid., 144.

[44] Walker, 'Early Modern Almanacs', 5.

[45] Ibid., 1–25.

of the environment'.[46] This chapter works, in part, to fill this critical gap, concluding as it does in an examination of the importance of discarded almanacs in the works of Middleton. Before we arrive at Middleton, however, we need to gain a better understanding of how these little books communicate the interconnectedness of the body and the world, articulating as they do a particular mode of being in the world through both their textual contents and their material form. Crucially, the way in which almanacs were handled, used, dismantled, and discarded collaborates with the intellectual matter of the calendar and prognostication, shaping an ecological understanding of the environment and the place of the human within it. In addition to inviting annotation, almanacs cultivated other kinetic and dexterous forms of material use. As Julian Yates suggests, the portability of almanacs 'stitch[ed] their human bearer into the information networks they condensed', 'secreted', as they often were, in pockets 'like some prosthetic skin'.[47] The limp parchment wrapper of an extant copy of Allestree 1639, held at the British Library, bears material evidence of this common practice: the little book has been folded firmly length-ways and tied with a parchment cord, suggesting that it was kept with regularity in a pocket (see Fig. 5.1). In addition to this habitual proximity of an almanac and its user, the woodcuts, diagrams, and tables invited active engagement, prompting users to fill in blank spaces with self-accounting and to project their own body onto the shape of the 'Zodiac Man' and the network of corporal-celestial relations it mapped.

A number of almanacs offered an even more hands-on experience: *White 1638*, for instance, includes a double-sided 'briefe and easie Kalender for this Yeere 1638. which being cut out, is fit to be placed *into any Booke of Accompts, Table-booke, or* other, conteyning the whole Almanacke, in a short method'.[48] Inviting disassembly, as Adam Smyth has shown, almanacs often directed their readers to cut and paste their pages onto walls and into other books.[49] Others, such as Allestree 1639, contain engravings of astrological instruments that, more than a printed set of instructions, were themselves to be cut out and pasted onto 'a smooth square little board that will not skellow', or warp (see Fig. 5.1).[50] With holes pierced for the sun to shine through and the addition of a silken thread, lead plummet, and moveable pearl, the cut-out page becomes an instrument that allows *Allestree*'s reader to tell the hour of the day and 'the height of any thing standing Perpendicular'.[51]

[46] Walker, 'Early Modern Almanacs', 3.

[47] Julian Yates, 'Thomas Middleton's Shelf Life', in *The Oxford Handbook of Thomas Middleton*, ed. by Gary Taylor and Trish Thomas Henley (Oxford: Oxford University Press, 2012), 16–31, 21.

[48] *White. 1638 a New Almanacke and Prognostication*, STC 527.28 (London: F.K. for the Company of Stationers, [1638]), C4r–v.

[49] Smyth, 'Almanacs and Ideas of Popularity', 131.

[50] *Allestree. 1639 a New Almanacke and Prognostication*, STC 407.22 (London: T. Cotes for the Company of Stationers, [1639]), C6v–7v.

[51] Ibid., C7r.

Comparable to the volvelles found in more expensive astrological publications, these technologies make use of the capacity of paper 'to fold, crease, slice and bow', as Helen Smith writes, as well as to be easily pierced, pasted, and spun.[52] Providing a tool with which to 'materialis[e] problems' and perform calculations, the dismantled pages of almanacs were a physical, as well as intellectual, instrument with which to negotiate the wider environment.[53]

These almanacs-turned-instruments also materialize the way in which the human body is embedded within a network of other non-human bodies. As Adam Smyth's work on cutting up books has demonstrated, texts were 'often imagined as an assemblage of pieces, which might be crumbled, or shredded, into its constituent parts'.[54] The cutting up and handling of almanacs as instruments suggests that the environment is, as Jane Bennett argues, a similarly 'heterogeneous assemblage'.[55] Almanacs, after all, encourage their users to conceive of the world as a collage in which human and non-human bodies are intermingled, responding in identical ways to the influence of the celestial bodies above. In cutting up an almanac and transforming it into an instrument, its users made an instruction manual for negotiating the environment a prosthetic with which to do so. In these moments, the fragmented almanac becomes a tool that extends the human body further into the world, creating a microcosmic assemblage of human and paper body that, in its response to the air and stars, forms a miniature version of the wider environmental macrocosm.

What is more, in the form of cut up fragments, almanacs further communicate the temporal rhythms contained within their calendars, just as they do the interconnected ecology of their scientific content. Cutting is, in many ways, analogous to the transformation of a text into waste, and the dismemberment of almanacs in the form of instruments and detachable tables closely resembles the habitual discarding and repurposing of the little books at the end of the year. Almanacs were wasted like clockwork, as it were. In light of Juliet Fleming's work on cutting, we might conceive of these apparent acts of destruction as 'open[ings]' in that they 'introduc[e] the future', with acts of wasting adding new potency to the intellectual and ecological content of almanacs, rather than rendering them inert and meaningless.[56] Waste almanacs, once cast off and transformed, continue to communicate the place of the human in the world just as they continue to mark the passage of time. Only now, we will go on to see, they do so with the vibrancy of discarded things.

[52] Smith, 'A unique instance of art', para. 39–43.
[53] Ibid., para 43, para 41.
[54] Smyth, *Material Texts*, 26.
[55] Bennett, *Vibrant Matter*, 23.
[56] Juliet Fleming, *Cultural Graphology: Writing after Derrida* (Chicago: University of Chicago Press, 2016), 99.

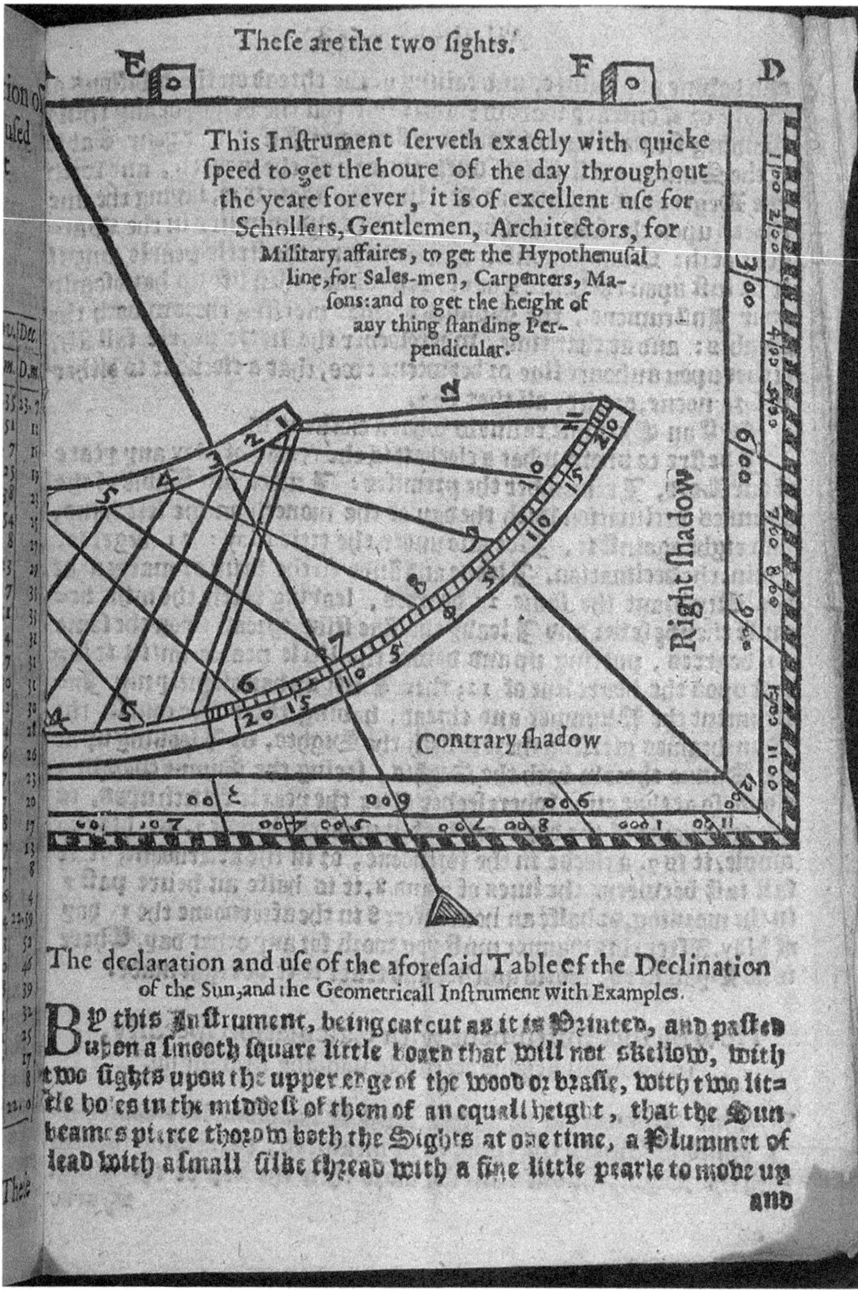

Theſe are the two ſights.

This Inſtrument ſerveth exactly with quicke ſpeed to get the houre of the day throughout the yeare for ever, it is of excellent uſe for Schollers, Gentlemen, Architectors, for Military affaires, to get the Hypothenuſal line, for Sales-men, Carpenters, Maſons: and to get the height of any thing ſtanding Perpendicular.

Right ſhadow

Contrary ſhadow

The declaration and uſe of the aforeſaid Table of the Declination of the Sun, and the Geometricall Inſtrument with Examples.

By this Inſtrument, being cut out as it is Printed, and paſted upon a ſmooth ſquare little board that will not ſkellow, with two ſights upon the upper edge of the wood or braſſe, with two little holes in the middeſt of them of an equall height, that the Sun beames pierce thorow both the Sights at one time, a Plummet of lead with a ſmall ſilke thread with a fine little pearle to move up and

Fig. 5.1 'This Instrument serveth exactly with quicke speed to get the houre of the day throughout the yeare for ever', in *Allestree 1639*, STC 407.22 (London: T. Cotes, for the Company of Stationers, [1639]), C7r. British Library, C.194.a.337.

2 Almanac Waste-Time

In addition to teaching their users how best to inhabit the world, almanacs marked the passage of time. These two primary functions of early modern almanacs are, after all, intertwined, with the ecology of celestial influence dependent on the movements of the stars through the course of the year. Managing this inexorable influence was a matter of good timing, with the 'good and bad days' marked in the calendar, along with the more general rules for self-application in the prognostication, helping almanac readers to time particular tasks and behaviours safely and effectively. Furthermore, in addition to telling their users how to manage their time, almanacs were a technology that both reflected and shaped the cultural understanding of time itself. The early modern period, as Kristen Poole and Owen Williams have argued, was a transitional one regarding 'temporal epistemologies'.[57] They suggest that 'time was not always, or even primarily, hegemonic and teleological', and that there was 'tremendous diversity in terms of temporal models and habits of thought concerning one's place in time', with 'normative linear depictions' of 'chronology' not stabilizing until the eighteenth century.[58]

This multiplicity of temporal models is nowhere more apparent than in the early modern almanac, depicting as it does the cyclical return of the seasons and calendrical months in conjunction with the linear computation or chronology of time. Almanacs are grounded, therefore, in both the age-old agricultural rhythms of annual renewal and in the teleological understanding of the progression of events beginning in the creation and ending in the present year. The computation suggests a conception of time as a line rather than a circle, and so, implicitly, as something that will eventually come to an end. This entropic linearity, as we have already seen in relation to John Bale, underpins the widespread religious worldview in which time and the world, progressing through a series of ages, are understood to be irrevocably decaying.[59] The cyclical idea of renewal is, in contrast, grounded in non-religious, seasonal rhythms, as well as in the understanding of matter that underpins the science of almanacs: composed of the same fundamental material, human and non-human bodies dissolve into minute fragments, or atoms, that endlessly rearrange into new forms and, again and again, disintegrate.[60] Almanacs, then, simultaneously present two competing narratives—one

[57] Kristen Poole and Owen Williams, 'Introduction', in *Early Modern Histories of Time: The Periodizations of Sixteenth- and Seventeenth- Century England*, ed. by Kristen Poole and Owen Williams (Philadelphia: University of Pennsylvania Press, 2019), 1–17, 9.

[58] Ibid., 9, 11.

[59] See pp. 112–113 above.

[60] Early modern almanacs rarely delved into the details of scientific complexities and controversies. They are, however, a product of the wider scientific and philosophical culture which, in the sixteenth and seventeenth centuries, grappled with the implications of atomistic theory. On the relationship between atomic philosophy and temporal epistemology, see Victor Harris, *All Coherence Gone: A Study of the Seventeenth Century Controversy Over Disorder and Decay in the Universe* (London: Frank Cass & Co. Ltd., 1966), 6, 104, 164, and 192–193.

of renewal, the other of entropy—with neither prioritized above the other. This intellectual content is, once again, reiterated by the physical biography of the little books, with the duality of renewal and decay, of circle and line, encapsulated in the annual act of discarding an old almanac and its replacement with a new one.

Discarded almanacs were, in early modern England, potent emblems for communicating these contradictory temporal epistemologies. We know that almanac waste was both pervasive and conspicuous in the early modern environment because of the frequent references to repurposing the little books, and these references constitute a sizeable sub-category of the waste paper trope. If a book 'seeme[d] of lesse use than an old Almanack', it would most likely be 'ma[d]e wast paper'.[61] A sub-par book would, within a year, 'hang up with old Almanacks or sell for waste Paper', or, 'in a short time [...] be out of Date, and but as wast Paper, or an old Almanack'.[62] If the contents of a book were of passing relevance, they were 'like old Almanacks out of date, fit for nothing but to cover Mustard pots'.[63] Almanac users would have habitually repurposed their own almanacs at the turn of the year, most likely repurposing them domestically. The majority of almanac waste would, though, have been made up of the bookseller's unsold stock, and so have been sold in bulk to bookbinders or grocers partway through the year. In addition to one's own almanac, repurposed in the kitchen or privy, early moderns would have regularly encountered the little books in the form of binding waste, box linings, and food and spice wrappers—an occurrence borne out by extant bindings and still life paintings (see Fig. C.1).

The Stationer's Company records bear witness to the likelihood of vast quantities of almanacs becoming waste at one time, describing as they do several instances when the market for almanacs was flooded by unlicensed publications at a considerable loss to the company. In 1637, for instance, Mr Stansby was punished for printing 'the Midleborough Psalmes and Almanacks' without licence.[64] He was fined £10 to cover the 'Paper made waste' and the subsequent 'Losse' to the Company.[65] Similarly, in 1666, Francis Mawborne of York printed four thousand almanacs, as well as an unspecified number of sheet almanacs, without a licence.[66] As a result, the Company's own almanacs for 1666 went unsold and were 'turnd only into wast paper'.[67] Mawborne, like Stansby, was fined heavily to recuperate these losses. In addition to these exceptional cases, we can assume that not

[61] Lewes Roberts, *The Merchants Mappe of Commerce*, STC 21904 (London: R. O[ulton, Eliot's Court Press?] Thomas Harper, and Felix Kingston] for Ralph Mabb, 1638), 1.

[62] Francis Howgill, *The Mouth of the Pit Stopped*, Wing H3172 (London: for Thomas Simmons, 1659), 6; George Whitehead, *The He-Goats Horn Broken*, Wing W1933 (London: for Robert Wilson, 1660), 32.

[63] Bramhall, *The Serpent Salve*, 105.

[64] *Records of the Court of the Stationers Company*, ed. by Jackson, 281.

[65] Ibid., 281.

[66] *A Chronology and Calendar of Documents*, ed. by McKenzie and Bell, 562.

[67] Ibid., 562.

insignificant quantities of almanacs would have been surplus to demand every year. Alongside the proportion of almanacs found in extant binding waste, these records allow us to surmise that almanacs were the category of text most likely to be wasted in the period, and that they populated the urban and domestic environment particularly thickly at certain times of the year. Perhaps the flood of almanacs entering the grocers' and binders' shops was an urban version of seasonal markers, such as geese flying overhead and the falling of leaves, of the encroaching end of the year.

The capacity of a discarded almanac to mark the linear passage of time, like that computed in its chronology, is reflected in the language used to describe them in the sixteenth and seventeenth centuries. Almanacs were, in the early modern period, a by-word for ephemerality. It became commonplace in the late sixteenth and early seventeenth centuries, to describe a discarded thing as having been 'laid aside as an Almanacke out of date', or 'like an old Almanack laid aside as useless'.[68] The phrase 'out of date' was coined in the late sixteenth century and it became rapidly associated with the expiry of an almanac: an early use is in Richard Hakluyt's 1582 *Divers Voyages*, and a decade later Thomas Nashe described Dame Niggardize as wearing an outfit of offcuts, trimmings, and repurposed things, including 'an apron made of Almanackes out of date'.[69] In the same year, in *Greenes Groatsworth of Wit*, a 'country author' describes how literary tastes have changed and so his figurative 'Almanack is out of date'.[70] This phrase solidified into an aphoristic simile, 'like an Almanacke out of date', early in the seventeenth century.[71]

Almanacs, then, materialized the passage of a year in both their calendars and in their transformation into waste in the autumn or winter months. It was a particularly expressive marker of the passage and loss of time because, in its calendar, it counted down the months and days until its own expiry. As Joseph Andrewes modestly suggests in the preface to a 1621 sermon, once it 'hath burdened the *Stationers Stall* the 12. part of the date of *an Almanack*, it may serve well enough for waste *Paper*'.[72] Here Andrewes, like so many other early modern writers who refer tropically to an out of date almanac, illuminates what William Viney has argued of waste more broadly: '[w]aste', Viney writes, 'frequently requires a sense of how time has somehow passed, paused, or is no longer available to use through the things that surround us'.[73] In the case of almanac waste, this 'sense' of time passing

[68] William Troughton, *Saints in England Under a Cloud*, Wing T2319 (London: Mathew Simmons for Hannah Allen, 1648), *5v; Anon., *Bradshaw's Ghost*, Wing B4165([London: s.n.], 1659), 11.

[69] Richard Hakluyt, *Divers Voyages Touching the Discoverie of America*, STC 12624 (London: [Thomas Dawson] for Thomas Woodcocke, 1582), ¶1r; Thomas Nashe, *Pierce Penilesse*, STC 18373 (London: Abell Jeffes for J. B[usby], 1592), B1r.

[70] *Greenes, Groats-Worth of Witte*, STC 12245 (London: [J. Wolfe and J. Danter] for William Wright, 1592), E1r. Authorship of this text, printed shortly after Greene's death, is uncertain.

[71] See Dawes' description of how godly behaviour is 'like an Almanacke out of date', *Gods Mercies*, H7v.

[72] Andrewes, *The Brazen Serpent*, A4r.

[73] Viney, *Waste*, 3.

and lost is made all the more explicit through its textual content, and so this subcategory of repurposed paper serves as a particularly potent concrete metaphor for any object or idea that will soon expire, as well as for the passage of time itself: it stands in for both an unread sermon and 'the 12. part [...] date of *an Almanacke*'.[74]

Peeking out of the corners of bindings and from the crumpled edges of food, spice, and tobacco wrappers, wasted almanacs were unique in that they had expired at a particular moment and that they were emblazoned with markers of this expiry. Usually, in the early modern period, materials were worn, used, or read indefinitely until they wore out or, if esculent, until they rotted or perished. Almanacs, however, expired on the final day of the year, and this year was printed in large type on its title page and running headers and so typically remained visible even after it had been repurposed. The date of an almanac, along with the years listed in its computation, might have taken on particular significance when encountered as an untimely, wasted object: what, for instance, might have crossed the mind of the Royalist Bishop of Durham, John Cosin, when he caught glimpses of an almanac for 1638 in the binding of his edition of Henry Spelman's 1639 *Concilia, decreta, leges, constitutiones* (see Fig. 5.2)? Cosin might have made out snippets of the 'Chronologie or Computation of yeeres to this present yeare. 1638'; the running header 'White. 1638' across alternate pages; the disposition of the months and seasons; and a list of Kings in the gutter of the folio. King Charles I's reign, beginning in 1625 was, by White's calculations, '[a]fter which, writing 14-yeares'. The almanac's chronology may have taken on additional significance with the outbreak of Civil War in 1642 or with the execution of the King in 1649. Cosin was by this time in exile with few possessions but his books. Partially concealed by its use as guards, the incomplete events, 'Since his Majesties return [...]', 'Since his Majesties procl[...]', and 'Since Prince Charles was [...]' might have prompted either hope or despair in the Royalist bishop, if he chanced to glance on them. The wasting of an almanac and the execution of a monarch might both have encouraged a particularly pessimistic, even eschatological, understanding of the decay of the state of the world. If Cosin happened to use his copy of Spelman after 1660, though, when both he and Charles II had returned triumphantly to England, the waste fragments may have triggered happier thoughts: in this year, the Royalist rhythms that structured the majority of early modern almanacs were restored.

The sense that things were turning out of date extended far beyond the rapid expiry of fleeting books and fashions. As this critical imagining of Cosin's encounter with binding waste suggests, grand historical events and narratives might become intertwined with the chronology and expiry of an almanac. There is a consensus that, in the course of the early modern period, the conceptualization of time shifted from an older, more traditional cyclical model to what Poole

[74] Andrewes, *Brazen Serpent*, A4r.

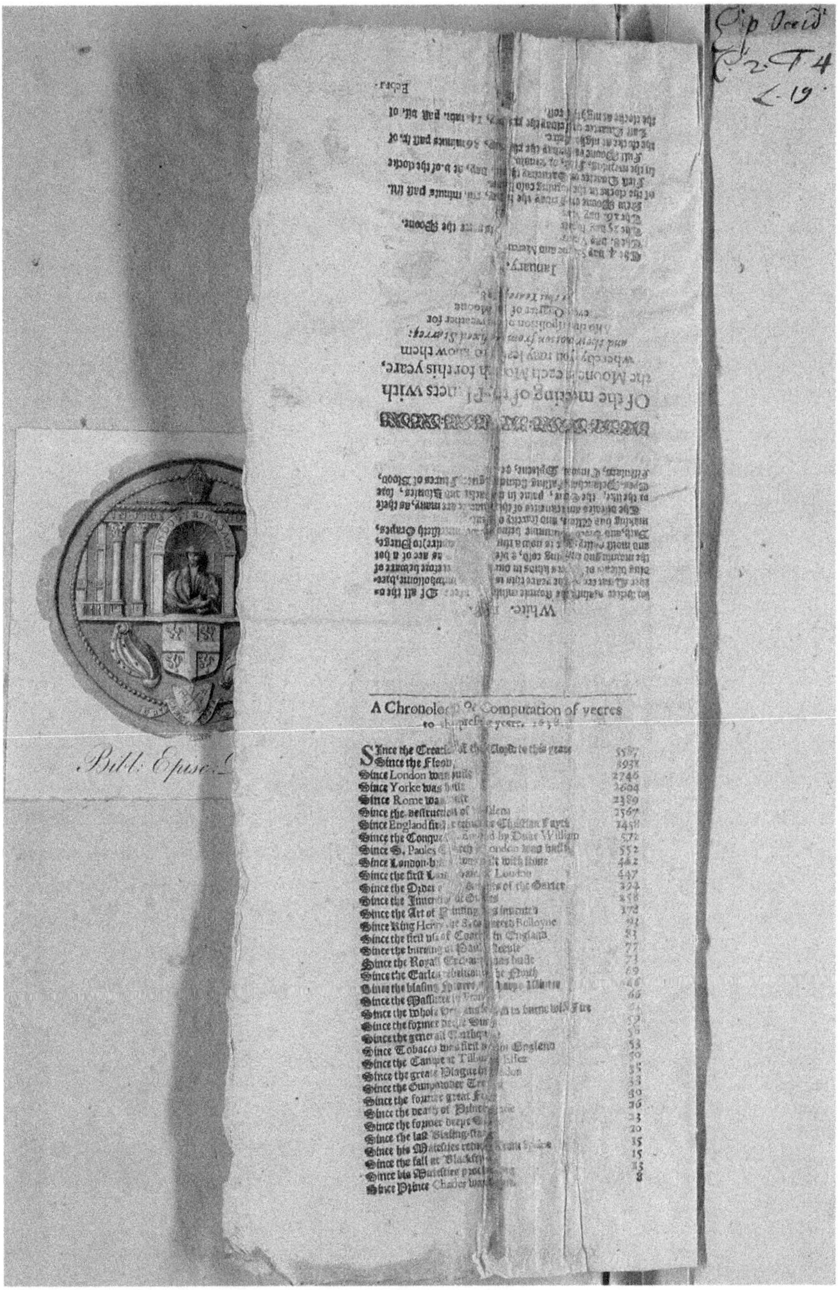

Fig. 5.2 Fragments of *White 1638*, STC 527.28 (London: F.K. for the Company of Stationers, 1638) used as guards and now removed, repaired, and tipped into the binding. Found in Henry Spelman, *Concilia, decreta, leges, constitutione*, STC 23,066 (Londini: Richardus Badger for Ph. Stephani, & Ch. Meredith, and Aureo Leone [1639]), Cosin D.1.1.

and Williams term 'normative linear depictions'.[75] Almanacs, it has been suggested, marked and contributed to this shift, along with other time-telling devices such as watches. Neil Rhodes, for instance, notes how in this period, time shifted to become 'artifactual'.[76] No longer ingrained in the physical experience of the days, nights, and seasons, it came to be understood as a remote and abstract concept, more a product of officialdom than a part of nature.[77] Similarly, in her study of almanacs and English Protestantism, Alison Chapman argues that the decreasing cost of books and the annual nature of almanacs meant that 'textual calendars [were] more disposable', and that 'time itself came to seem more temporary'.[78]

Furthermore, as Victor Harris has demonstrated, astrological texts became increasingly concerned with the idea of nature corrupting between the years of 1570 and 1630, and so the world came increasingly to be perceived as progressing along an entropic trajectory.[79] The sense of inevitable degeneration was a dominant one, as we have seen in relation to John Bale's eschatological thought, but prior to the late sixteenth century the heavens were typically conceived as being made up of a fifth element—a 'quintessence'—that was not subject to decay.[80] A number of late sixteenth- and seventeenth-century astrologers and natural philosophers began to query the nature of this celestial matter, and when Tycho Brahe discovered a new star in 1572 he furthered the sense that, like the mundane world, the heavenly spheres were wearing out.[81] 'The signes and revolutions of the heaven, are changed and remooved from the olde accustomed places', astrologers noted, because the sun and the pole star had declined by a number of degrees.[82] Rainbows seemed fainter and the comet of 1577 suggested that the substance of the heavens was beginning to alter. Time was coming to be understood, in the words of a late sixteenth-century thinker, as 'the generall rust of the world, which weareth, eateth, consumeth, and perforateth all thynges'.[83] Although almanacs were not typically concerned with such recent discoveries and the intricacies of astrological science, particularly devout almanacs did often dwell on the degenerative trajectory that had been set in motion by the Fall, an event recorded early on in

[75] Poole and Williams, 'Introduction', in *Early Modern Histories of Time*, 11.

[76] Neil Rhodes, 'Time', in *The Ashgate Research Companion to Popular Culture in Early Modern England*, ed. by Andrew Hadfield, Matthew Dimmock, and Abigail Shinn (Farnham: Ashgate, 2014), 283–294: 285.

[77] Ibid., 285.

[78] Alison A. Chapman, 'Marking Time: Astrology, Almanacs, and English Protestantism', *Renaissance Quarterly* 60, no. 4 (2007): 1270.

[79] Harris, *All Coherence Gone*, esp. 86–172.

[80] Ibid., 77; Kathleen Crowther, 'The Scientific Revolution', in *The Oxford Handbook of Early Modern European History, 1350–1750*, vol. 2, *Cultures & Power*, ed. by Hamish Scott (Oxford: Oxford University Press, 2015), 65.

[81] Herschel C. Baker, *The Wars of Truth: Studies in the Decay of Christian Humanism in the Earlier Seventeenth Century* (Eugene: Wipf and Stock, 2006), 72; Harris, *All Coherence Gone*, 137–138.

[82] Robert Pond, *A Newe Treatise of the Right Reckoning of Yeares*, STC 20104 (Edinburgh: Robert Waldegrave, 1599), 2, quoted in Harris, *All Coherence Gone*, 115.

[83] John Banister, *The Historie of Man*, STC 1359 (London: John Day, [to be sold by R. Day], 1578), B2r, quoted in Harris, *All Coherence Gone*, 104.

most computations of time. A prognostication for 1536, for instance, highlights the incongruity of its own cyclical calendar and the linearity of a religious understanding of time: astronomers, the compiler notes, 'teche to descryve [to give a written description or interpretation] | devyde or discerne one tyme from an other | and to observe dayes | monethes | & yeres | of this time that continually decayeth'.[84] Although the calendar divides 'one tyme from an other' and the prognostication teaches us how to manage our lives within these cycles, we should not forget, the compiler suggests, the gradual senescence of the world.

The imaginative leap from the decay of an almanac to that of the world was, therefore, an easy one, with the annual dismemberment and disintegration of a text mirroring the progressive degeneration of the world. This trajectory might also be mapped onto that of the human body: as Bernard Capp suggests, the little books might serve as *memento mori*, reminding their users that, in the words of an anonymous writer in 1671, 'every year thy almanac thou buyest | Th'art one year nearer to the year thou diest'.[85] Less piously, jests in the works of Thomas Dekker and Thomas Middleton indicate the way in which almanacs and ageing (typically female) bodies kept imaginative company. A popular joke in Dekker's 1607 *Jests to Make You Merie*, for instance, describes:

> A Gentlewoman comming to one that stood at a window reading a booke, Sir (said she) I would I were your booke, (because she loved the Gentleman,) So would I quoth he, I wish you were. But what booke would you have me to bee (said the other) if I were to be so? Mary, an Almanacke (quoth the Gentleman) because I would change every yeare.[86]

Middleton reworks this jest in *No Wit/Help Like a Woman* when a character claims, 'When wives are like almanacs, we may have every year a new one. Then I'll bestow my money on 'em; in the mean time, I'll give 'em over and ne'er trouble my almanac about 'em'.[87] Elsewhere, in his *An/The Old Law*, Middleton's clown tells his wife, 'Go, go thy ways, thou old almanac—at the twenty-eight day of December e'en almost out-of-date!'.[88] An ageing woman, according to Middleton's clown, is like an almanac on the cusp of irrelevance, and so he hopes to lay her aside just as easily. Similarly, for Dekker's gentleman almanacs signify a rapid cycle of consumption and disposal: if only, he wishes, women were *more* like almanacs.

[84] Otto Brunfels, *A Very True Pronosticacio[n], with a Kalender* [...] *for the Yere of our Lorde M. CCCCC. Xxxvi*, STC 421.17 (London: Joh[a]n Byddell, 1536), B3r.

[85] Capp, *Astrology and the Popular Press*, 147, 285; *An Almanac But for One Day* (Glasgow: Robert Sanders, 1671), quoted in Capp, 147.

[86] Thomas Dekker, *Jests to Make you Merie*, STC 6541 (London: N[icholas] O[kes] for Nathaniell Butler, 1607), 8.

[87] *No Wit/No Help Like a Woman*, ed. by John Jowett, in *Thomas Middleton: The Collected Works*, ed. by Gary Taylor and John Lavagnino (Oxford: Clarendon Press, 2007), 9.539–542.

[88] *An/The Old Law*, ed. by Jeffrey Masten, in *Thomas Middleton, The Collected Works*, IV.1.149–150.

The early modern perception of the shared trajectory of human and paper matter is also suggested by the frequent descriptions of almanacs as notably mouldy objects. As Dekker notes in his 1609 *Guls Horne-Booke*, many readers follow 'moth-eaten fashion' and 'mouldy custom'.[89] He claims to have a much more discerning nose, and to 'have smelt out the musty sheetes of an old Almanacke'.[90] Similarly, a religious writer lamented in 1657 that 'Bibles are like old Almanacks moulding in corners', and in 1659 a controversialist described how a book with a false title page, and so feigning newness, attempts to avoid 'grow[ing] old [o]r stick[ing] upon th[e] hands, as such other mouldy stuff hath done'.[91] It will, however, 'hang up with old Almanacks or sell for waste Paper' when the potential buyer sees through this cunning sales ploy. The association of almanacs with mould is a further indication of the rapidity with which they go out of date, and so are understood to decay and rot in the same manner as foodstuffs and other organic matter past its best. At the same time, 'mould' refers to the matter of the earth itself, in its clods of soil and in the sense of the 'earthly mould'.[92] Almanacs, therefore, are imagined as tending, in their disintegration and decay, toward the originative substance of the earth.

This material biography shares the same pattern as that of the human body which, as the Bible tells its readers, disintegrates into clay, dust, and earth after its entry into the grave. This biographical parallel is further heightened by an examination of the etymology of 'mould', which, in some early modern translations of the Bible, is explicitly associated with human matter. The 1535 Coverdale Bible contains the apocryphal verse 'Thou maydest Adam of the moulde of the earth' (Tobit 8:6), and a gloss of 1 Corinthians 15:47, rendered in English in 1570 as 'The first man is of the earth earthy', considers the etymological valences of 'earth'.[93] The verse refers, the gloss notes, 't[o] the originall word Adam, which signifieth earth or clay, whereof he was made, like as also among the Latines man is called *Homo* of *Humus*, which signifieth the earth'.[94] The Latin for mould or earth, the gloss notes, is 'humus', which shares the root of 'homo', or human, meaning 'of the earth', a resonance also present in a 1639 paraphrase of Job 19:26, which reads: 'Though wormes devoure me, though I turne to mold'.[95] This is also the language

[89] Thomas Dekker, *The Guls Horne-Booke*, STC 6500 (London: [Nicholas Okes] for R. S[ergier?], 1609), 1.

[90] Ibid., 1.

[91] Christopher Love, *The Dejected Soules Cure*, Wing L3151 (London, John Rothwell: 1657) 134; Howgill, *The Mouth of the Pit Stopped*, 6.

[92] See *OED* 'mould, n. 1', 2.a. and 5.

[93] *The Coverdale Bible*, STC 2063 ([Cologne?: by E. Cervicornus and J. Soter?, 1535]); David Chytraeus, *A Postil or Orderly Disposing of Certeine Epistles*, trans. by Arthur Golding, STC 5263 (London: Henry Bynneman for Lucas Harrison and George Bishop, 1570), 161.

[94] Ibid., 161.

[95] See 'Mold' in Peter Levens, *Manipulus Vocabulorum*, STC 15532 (London: Henrie Bynneman for Iohn Waley, 1570) and in Anglicus Galfridus, *Incipit Liber Q[ui] Dicitur Promptorium Parvulorum Siue Clericorum*, STC 20434 (London: Richard Pynson for Frederic Egmondt and Petrus Post pascha,

with which man is described in John Milton's *Paradise Lost*: he is 'Adam, earth's hallowed mould, | Of God inspired'—or more literally, a lump of earth that God has breathed into—and later, after the Fall, Adam puns on the verbal and noun forms of mould: 'Did I request thee, Maker, from my clay | To mould me Man?'[96] Both almanac and man, this network of literary tropes and religious glosses suggests, originate in the same earth, soil, mould, and dust, and will both return to it. The fragmentation of an almanac at the end of the year, persistently imagined as mouldy as it disintegrates into earthy fragments, might model its user's own eventual disintegration and decay. The material trajectory of an almanac, therefore, encountered and enacted by its user year on year, teaches the same lessons as its intellectual contents: time passes and all bodies, whether human or non-human, pursue the same biography of fragmentation and dissolution.

The biography of matter that returns to earth and dust, whether human or bookish, does not, though, have to be conceived of as exclusively linear or entropic. This biography might instead be understood as sharing the cyclical and seasonal structure that is also contained within the almanac's calendar and prognostication. Frances E. Dolan, in her work on the 'dunghill poetics' of Hester Pulter, has argued that in the early modern period, soil was understood as having a 'transformative power' in which it 'decomposes and then re-matters' the human body after death.[97] Highlighting the 'intimate connections of human and humus', Dolan describes how this 'composting' process points to 'the intermeshment of life forms' and 'the erosion of the human/non-human boundary'.[98] The religious and literary poetics of the dunghill that Dolan finds in Pulter's verse chimes with the intellectual and embodied knowledge that was propagated by almanacs: the new year signals the rebirth of the almanac in a new edition, as well as the 'composting' of the old one, not on the dunghill, but in diverse and disintegrative ways in households, privies, and shops. Rather than entropic decay, this repurposing might be conceived of as the book returning, at the end of its biography, to its original condition as dirt and dust. A waste almanac might, therefore, be perceived as nourishing future growth in the manner of compost.

This cycle could be conceived of as a narrow and specific one, with the almanac waste nourishing future flax seeds which will, in turn, become linen, rags, paper sheets, and, once again, waste; or it might be conceived as an entry into a much broader network of interconnected matter, like that modelled in the science of the almanac. Perhaps the composted, or recycled, waste will be 're-matter[ed]' in a more meandering fashion, in a process akin to that described by Hamlet as he

1499); George Sandys, *A Paraphrase upon the Divine Poems*, STC 21725 (London: [John Legatt] to be sold by [Andrew Hebb], 1638), 26.

[96] John Milton, *Paradise Lost*, ed. by John Leonard (Harmondsworth: Penguin, 2000), V, 321–22 and X, 743–744.

[97] Dolan, 'Hester Pulter's Dunghill Poetics', 21.

[98] Ibid., 29.

meditates on death and decomposition in the graveyard.[99] 'Alexander died', Hamlet muses, and 'Alexander was buried, Alexander returneth into dust, the dust is earth, of earth we make loam, and why of that loam whereto he was converted might they not stop a beer-barrel?'[100] Considering the mouldy matter to which both almanacs and the human return, Hamlet sees in it a microcosm of the heterogeneous assemblage within which all human and non-human bodies dwell. When the scattered fragments that were once human are 're-matter[ed]', they might end up in any material form because, as almanacs teach, all matter is interchangeable. Why then, Hamlet asks, might an emperor not end up 'stop[ping] a hole' or 'patch[ing] a wall'?[101] Strikingly, Hamlet imagines the human body transformed into materially indistinct stoppers and bungs, functions that, as we have seen, waste paper and parchment often performed in the period.[102] Although the Danish prince is not talking specifically, here, about almanacs, he conveys an understanding of the biography of matter grounded in their ecology: for Hamlet, as for many attentive early modern almanac users, human matter shares the biography and physical basis of non-human matter—its origins in the earthly mould, its response to the influence of the stars, and its return to the earthly mould. After death, the correspondence between the human and the non-human becomes even more apparent, as fragments of one readily transform, on the level of atoms and dust, into the other.

The biography of a waste almanac, then, might shape two distinct temporal epistemologies, with time conceptualized as either linear or cyclical, just as the calendar and computation of the almanac offers a confusion of the two. Once wasted, therefore, an almanac becomes more, rather than less meaningful. In the process of being discarded, the calendar and prognostication gain new ways of marking the passage of time and materializing the interconnected trajectory of human and non-human bodies. The way in which a discarded almanac demands a particular mode of attention can be usefully illuminated by Martin Heidegger's theorization of the 'broken tool'.[103] Heidegger notes that when objects serve their typical, utilitarian purpose (*gebrauchender Umgang*) they remain inconspicuous, but this changes when they break and are no longer useful. 'Un-ready-to-hand' (*unzuhanden*), they are obtrusive, disrupting the normal state of subject-object relations. The waste almanac is not, though, quite 'un-ready-to-hand' or fully broken because it continues to serve a wide range of functions and remains particularly

[99] Dolan, 21.
[100] *Hamlet* in *The Oxford Shakespeare*, 5.1.204–207.
[101] Ibid., 5.1.163–164.
[102] See pp. 117 and 48 above for evidence of wasted texts bunging beer barrels and stopping mustard pots.
[103] Martin Heidegger, *Being and Time*, trans. by John Macquarrie and Edward Robinson (Oxford: Blackwell, 1962, repr. 2001), 102–107. For a comprehensive account of this element of Heidegger's philosophy, see Graham Harman, *Tool-Being: Heidegger and the Metaphysics of Objects* (Chicago: Open Court, 2011).

prone to frequent handling. Heidegger's understanding of the obstinacy and obtrusiveness of objects that have stopped performing their primary purpose is, though, useful, particularly when it is paired with William Viney's definition of what he terms 'waste-time'. '[T]his temporal condition', Viney writes, is

> an anachronic tense, one that commingles pasts and presents and leaves things lingering, loitering and persisting above and beyond the time of its functionality. It is the persistence of what has passed that makes waste an articulate thing of time.[104]

Here, Viney modifies Heidegger's theory of the broken tool by focusing on temporal effect rather than a lack of utility: waste is particularly articulate and communicative because it lasts longer than we expect it to. In lingering past the end of the year, therefore, a waste almanac prompts its users to dwell on the nature of time, be it cyclical or entropic. This capacity is shared, as Viney argues, by all waste objects, but the temporal articulation of a waste almanac is heightened by the way in which its textual contents mark the passage of time and offer a scientific account of time's effect on the body and the environment.

We might describe waste almanacs, therefore, as particularly vibrant objects. Although they are only meaningful because of the significances they have been granted within their cultural context, and because of what Jane Bennett terms the 'certain anticipatory readiness', or 'perceptual style open to the appearance of thing-power', that their users bring to them, waste almanacs were effective and agentic things in the early modern period.[105] Commanding attention and communicating an ecology not unlike Bennett's 'complicated web of dissonant connections between bodies', waste almanacs prompted a particularly embodied knowledge of one's place in the world.[106] Heightening the degree to which the almanac already, in its pre-wasted form, removed the human from the centre of things and situated them within a broader, environmental assemblage, waste almanacs prompt us to reconfigure Bosanquet's description of almanacs as 'very human book[s]'.[107] Although, according to the almanac's ecology, the little books were very human, their science of corresponding matter and their shared material biographies also taught that humans were strikingly almanac-like. The remainder of this chapter will demonstrate that the humanity of the almanac, or rather, the almanac-like nature of the human, is explored nowhere more extensively than in the work of the avid but always sceptical almanac reader, Thomas Middleton.

[104] Ibid., 178.
[105] Bennett, *Vibrant Matter*, 5.
[106] Ibid., 4.
[107] Bosanquet, 'English Seventeenth-Century Almanacks', 397.

3 Thomas Middleton's Almanac Bodies

Active between 1597 and the year of his death in 1627, Thomas Middleton has been heralded as the only early modern writer other than Shakespeare 'who created plays still considered masterpieces in four major dramatic genres: comedy, history, tragedy, and tragicomedy'.[108] We can add to this impressive range Middleton's talents in producing masques, civic pageants, pamphlets, and poems. This broad commercial and civic success prompted T. S. Eliot, in a 1927 essay, to claim that Middleton had neither a 'point of view' nor a 'peculiar personality', and it is certainly difficult to find a consistency of voice and theme across his more than six dozen single and collaboratively authored works.[109] There are, though, two definable features of Middleton's dramatic and prose writings that are pertinent to this study: the first is the considerable influence of Thomas Nashe on Middleton's early pamphlets and plays, indicating that he was an avid and careful reader of the older writer's works; the second is Middleton's persistent preoccupation with almanacs throughout his three-decade long career.

Middleton imitated Nashe most explicitly in *The Black Book* (1604), a continuation of *Pierce Penilesse*, but, as Neil Rhodes has demonstrated, Middleton frequently borrowed Nashe's diffuse 'grotesque imagery', particularly the frequent 'base comparison' of bodies with absurd and piecemeal materials.[110] Rhodes argues that Middleton's 'breathless' and so Nashean 'cramming together of physical detail in prose' is 'more or less abandon[ed]' after his earlier works, but it is intriguing to note that we uncover glimpses of this grotesque style in Middleton's career-long fixation with almanacs.[111] Always on the cusp of becoming waste, almanacs are, after all, a particularly Nashean object, and Middleton's preoccupation with their material and temporal biographies might have stemmed in part from his formative reading in the older writer's corpus. As we will go on to see, Middleton does not appear to have been persuaded by Nashe's poetics of wholesome and godly waste, instead conceiving of discarded almanacs within the secular and environmental framework outlined in their textual ecology. Middleton's almanacs do, however, retain a certain Nashean grotesqueness, and we find the grotesquely 'base comparison' of body and miscellaneous matter in Middleton's frequent staging of almanac readers and human–almanac hybrids.[112]

Middleton's references to almanacs, then, extend far beyond the misogynistic jests cited in the section above. In addition to authoring the mock-almanacs *Plato's*

[108] Gary Taylor, 'Thomas Middleton: Lives and Afterlives', in *Thomas Middleton, The Collected Works*, 25.

[109] T. S. Eliot, 'Thomas Middleton' (1627), quoted in Taylor 'Thomas Middleton', 55.

[110] Rhodes, *Elizabethan Grotesque*, 19–22. Rhodes also observes that the title and idea for *The Owl's Almanac* is borrowed from Nashe's anti-Marprelate tracts. See the introduction to *The Owl's Almanac*, ed. by Neil Rhodes in *Thomas Middleton: The Collected Works*, 1272.

[111] Rhodes, *Elizabethan Grotesque*, 86.

[112] Ibid., 86.

Cap (1604) and *The Owl's Almanac* (1618), the little books play a central role in *No Wit/Help Like a Woman's*, performed at Whitehall and at the Fortune in 1611, and in *The Masque of Heroes*, performed at the Inns of Court in 1619. As critics including Meredith Molly Hand, Katherine Walker, and Julian Yates have demonstrated, Middleton's attitude towards almanacs is first and foremost a sceptical and 'derisive' one.[113] Weatherwise in *No Wit/Help* is a stock almanac reader who 'woos by the almanac', living a life dictated by its good and bad days (1, 112). In this play and elsewhere, Walker argues, Middleton satirizes 'astrological determinism' and parodies the idea that all knowledge can 'be housed in a single, directive text'.[114] In doing so, Walker concludes, Middleton promotes 'rationalized ways' of negotiating one's place in society and the environment.[115] Similarly for Hand, *The Owl's Almanac* prompts readers to 'reconsider their own ways of apprehending themselves, their assumptions and perceptions, and their positions in the world' by fashioning 'a savvy, sceptical reader', and for Yates, Middleton's *Owl* observes a contemporary shift in reading practices. By creating an almanac that has already been filled in, Yates argues, Middleton critiques 'the transformation of reading into a practice which sells itself to book users as a one-way technology of use rather than a two-way technology of mutual transformation'.[116] The majority of readings of Middleton's almanacs, therefore, focus on his rejection of their deterministic philosophy and typically situate the little books within the emergent capitalist system that transforms readers into passive, quasi-objects. This consensus in scholarship on Middleton is encapsulated by Katharina Boehm's suggestion that, because almanacs were a highly commercialized textual form, Middleton portrays through them 'a commercial world in which humans are objectified while commodities are anthropomorphized and fetishized'.[117]

My understanding of Middleton's almanacs is indebted to each of these readings, but it dwells further on the contradictions at the heart of the author's apparent preoccupation. Middleton, I suggest, is not able to reject an almanac's ecology so straightforwardly, and his attempts to negotiate the distinction between the human and the non-human, the agentic subject and the passive object, underpins his frequent return to the little books. In addition to situating Middleton's almanacs within the context of emergent capitalism, we need to consider their place within the environmental framework that was communicated by the calendars and

[113] Meredith Molly Hand, '"More Lies Than True Tales": Scepticism in Middleton's Mock-Almanacs', in the *Oxford Handbook of Thomas Middleton*, 312–328; Katherine Walker '"Daring to Pry into the Privy Chamber of Heaven": Early Modern Mock-Almanacs and the Virtues of Ignorance', *Studies in Philology* 115, no. 1 (2018): 129–153; Yates, 'Thomas Middleton's Shelf Life', pp. 16–31. Hand describes 'Middleton's treatment of almanacs [a]s keenly and wittily derisive', 318.

[114] Walker, 'Daring to Pry', 131.

[115] Ibid., 152.

[116] Hand, 'More Lies', 316, 315; Yates, 'Thomas Middleton's Shelf Life', 30.

[117] Katharina Boehm, 'The Economies of Intellectual Labor in Thomas Middleton's City Comedies', *Studies in English Literature, 1500–1900* 56, no. 2 (2016): 351–372, 364.

prognostications themselves. Middleton's representation of almanacs reveals that he was sensitive to the temporal and material rhythms contained within both their intellectual contents and their physical biography. He was attuned to the waste-time of a discarded almanac because of both his careful reading of Nashe's waste-filled works and his avid consumption and probable reuse of the ephemeral, little books. As a result, Middleton would have approached almanacs with the 'certain anticipatory readiness' that Jane Bennett terms the 'perceptual style open to the appearance of thing-power'.[118] This 'thing-power' is manifest in Middleton's works where he situates stock almanac readers alongside more complex figurations of the correspondence between the human and the non-human. In his masques and plays, Middleton stages the ecological assemblage communicated by an almanac's scientific content. These assemblages take the form of human–almanac hybrids, and so Middleton materializes the almanac's ecology by blending its material make-up with the human in the form of an actor wearing or imitating its pages. This is part of Middleton's persistent attempts, identified in the scholarship cited above, to untangle the human subject from the wider network of non-human things. These attempts are not, though, ever entirely successful, and so Middleton's works reveal a vexed and ongoing negotiation of the relationship between the self and the world as it is articulated in the cheap and ephemeral little books.

This tension is apparent in *Masque of Heroes*, a text that complicates any straightforward reading of Middleton's sceptical parody of almanacs and their readers.[119] The onstage director and narrator of the play's action is Doctor Almanac, a physician who, instructed by the New Year and assisted by Time, resolves the conflict between the days of fasting and the days of feasting and so creates a harmonious calendar grounded in the traditional rhythms of the almanac. This masque reiterates the argument of James I's 1618 *Declaration of Sports*, but it also, seemingly contradictorily, mocks almanac readers: the second antemasque that threatens to disrupt the calendrical harmony is composed of eight boys 'habited' as the 'Good, Bad, and Indifferent Days' (26–29). The Good Days are 'attired all in white garments, sitting close to their bodies, their inscriptions on their breasts'; the Bad Days 'all in black garments, their faces black' and also bearing inscriptions; and the Indifferent Days wear 'garments half white, half black, their faces seamed with that parti-colour' (266.3–13). Their 'ridiculous figures' purchase 'a smile from the cheeks of many a beauty' before they 'vanish, proud of that treasure' (266.16–18). In this dance, the 'pleasures of low births and natures' (267) are parodied for the enjoyment of the educated audience. The faces and bodies of the boys materialize the pages of an almanac's calendar, with white and black textiles or paint transforming them into the typographical

[118] Bennett, *Vibrant Matter*, 5.

[119] Thomas Middleton, *Masque of Heroes. Or, The Inner Temple Masque*, ed. by Jerzy Limon in *Thomas Middleton: The Collected Works*, 1320–1330. Line numbers are given in the text.

symbols that mark the prescribed days for particular activities.[120] This literal version of what Miles Grier terms 'inkface' serves to associate the 'lowly' classes who rely too heavily on almanacs with the 'elastic racial category' of blackness, and so others them from the audience.[121] But just as the 'parti-colour' faces destabilize these racial categories, the human is blurred with the non-human in the embodiment of the almanac's pages. Although the boys raise smiles and vanish, they demonstrate how readily the human body can transform into something akin to mobile and ephemeral sheets of paper. *Heroes* contains, as James Knowles argues, apocalyptic 'undertones', with the restoration of the cyclical, festive calendar situated uncomfortably alongside the death of the Old Year and the decline of Master Kersmas.[122] Performed in the first month of the year, when many members of the audience would have recently discarded, dismantled, and repurposed their own old almanacs, Middleton's masque contains an underlying narrative of dissolution and termination. *Heroes*, therefore, is a sensitive representation of both the cyclical and the linear temporalities contained within an almanac's pages, as well as, in its antemasque, a manifestation of its heterogeneous ecology.

In this 1619 masque, Middleton returns to the same concerns that he staged and explored in the 1611 play *No Wit/Help Like a Woman.* In an early scene the 'gull' (4, 1) Weatherwise transforms the stage into a woodcut of the Zodiac Man—the most widespread representation of the human body in the period, found in the opening pages of most almanacs. Attempting to woo the play's wealthy widow, Weatherwise hosts an astrologically themed banquet: a table is brought on stage bearing twelve trenchers, one for each month of the year, and each containing a versified posie lifted directly from that year's almanac (from Bretnor, Neve, or Dade; 4, 1). Next, a banquet appears 'with the twelve signs' of the Zodiac 'made like banqueting-stuff' (4, 95.1–8)—presumably pasteboard props imitating marchpanes and other delicate foodstuffs which, in turn, imitate the astrological symbols inked on the page of the almanac. It is perhaps unlikely that, as David George suggests, the guests sit 'round the figure of a man cut out in paper', but the allusion to the almanac's Zodiac Man would have been immediately apparent to the audience:[123] in a sense, the Zodiac Man is turned inside out on the stage, with the actors seated around the astrological signs (made of pasteboard 'banqueting-stuff'), rather, as in the almanac's woodcut, the astrological signs encircling the human figure within, connected by lines denoting their sympathy and influence (see Fig. 5.3).

[120] 'Seamed' (266.12) may be metaphorical, but it suggests that the Bad Day's black ink may have been represented by a thin, fabric vizard rather than with paint. See Ian Smith on the history of the representation of black skin with cloth in early modern performance: 'Othello's Black Handkerchief', *Shakespeare Quarterly* 64, no. 1 (2013): 1–25. This also raises the intriguing question of whether the whiteness—or absence of ink—of an almanac's Good Days were represented by white cloth or by the white skin of the performers.

[121] Grier, 'Inkface', 207.

[122] James Knowles, 'Introduction' to *Masque of Heroes*, in *Thomas Middleton, The Collected Works*, 1322–1323.

[123] George, 'Weather-wise's Almanac', 297–301.

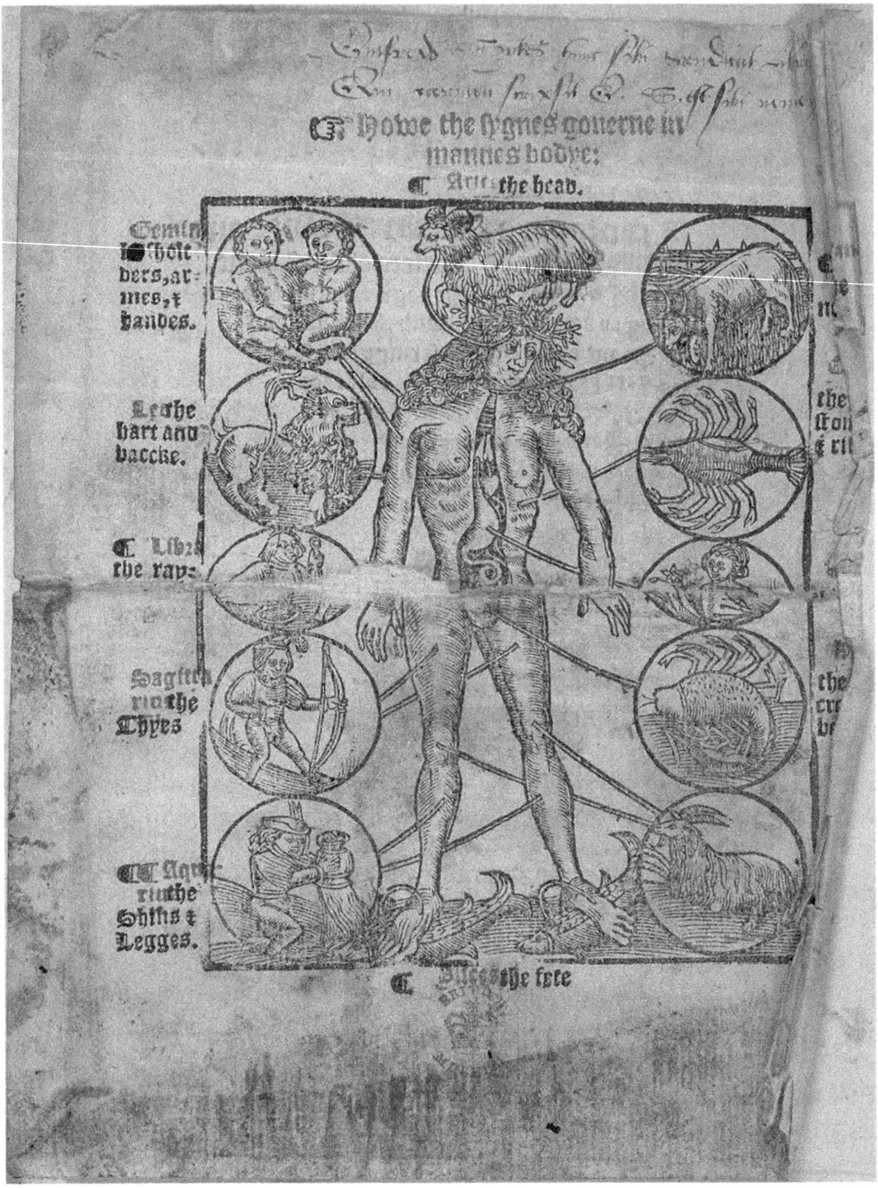

Fig. 5.3 Fragment of an almanac containing the woodcut of a Zodiac Man, removed from an unidentified binding by John Bagford, Harley 5937, item 54. The British Library.

This set-piece leads to lewd wordplay regarding rams and bulls and concludes with the staging of the sun's annual passage through the zodiacal regions in the form of an alcoholic 'sun-cup' passed between Weatherwise's guests (4, 249–306). As with the Good and Bad days in *Heroes*, the audience are prompted to laugh at the almanac reader's expense; but as is the case with *Heroes*' Good and Bad Days, we are also presented with a scene that intermingles the material form of the almanac and the human body. Whereas *Heroes*' Good and Bad Days scatter and disappear from the stage, the figure of the Zodiac Man at the centre of Weatherwise's banqueting table is turned inside out, with the inky border that separates the human body from the wider environment removed. In the same way that an almanac that has been cut up and turned into an instrument serves a prosthetic function, Middleton's materialization of the Zodiac Man extends the human out into the world. At the same time as parodying the naivety of Weatherwise, who believes that such an elaborate device will help him to woo the play's widow, the inverted Zodiac Man suggests the correspondence between the inside and the outside, the human and the non-human. Weatherwise's banquet intermingles the human with the wider environment visually on stage, materializing and dissolving the pages of the almanac through the actors' bodies just as he did in *Heroes*' antemasque. In these two moments, then, Middleton embodies the science of the almanac at the same time as satirizing its naïve usage.

The Zodiacal banquet is not the only almanac–human hybrid that appears on stage in *No Wit*. As the alternative title the play was performed under at Whitehall, *The Almanac*, suggests, the little books were the primary source of the play's spectacle and theatrical effect.[124] Described as a 'most necessary property' (4, 32), Weatherwise's own body is repeatedly objectified throughout the play. Figuratively blurred with the flimsy book—most likely an actual almanac—that Weatherwise carries throughout the performance, he is described as 'waste lumps of almanac stuff | Kned with May-Butter' (1, 118–119).[125] This gross, sticky metaphor is worth unpacking as, in these lines, Weatherwise's identity as a human subject is replaced by a concoction of materials and temporalities. He is akin to a dismembered and discarded old almanac, rubbed with, or wrapping, rancid butter that has been left out in the sun for the month of May (a common cure-all in the period).[126] The

[124] See John Jowett, 'Middleton's *No Wit* at the Fortune', *Renaissance Drama* 22 (1991): 191–208.

[125] The first edition of *No Wit* has 'waste lumps', whereas the Oxford *Thomas Middleton, The Collected Works* has 'masty lumps of almanac stuff'. I would argue that 'waste lumps' is likely to be authorial as it is in keeping with the network of waste imagery explored in this chapter. 'Masty' (meaning fattened) does, though, create much the same effect as 'waste', here. Thomas Middleton, *No Wit, Help Like a Womans*, Wing M1985 (London: [Thomas Newcombe] for Humphrey Moseley, 1657), 9 and *Thomas Middleton and Early Modern Textual Culture: A Companion to the Collected Works*, ed. by Gary Taylor and John Lavagnino (Oxford: Oxford University Press, 2007), 1152, I.118.

[126] 'May-butter' was the product of setting butter out in the sun for the month of May until it became 'exceeding soveraigne and medicinable for wounds, straines, aches, and such like grievances', and, presumably, rancid. See Gervase Markham, *The English House-Wife*, STC 17353 (London: Nicholas Okes for Iohn Harison, 1631), 199.

image reminds us of an almanac's origins in the 'stuff' or pulp made from soaked and beaten rags, macerated to form the raw material of the page. At the same time, we are reminded of paper's likely afterlife as waste wrapping and as a vehicle for the application of medicines, such as May butter, in the household.[127] The paper blurs with the foodstuffs it touches: as well as paper's pulp, 'stuff' can refer to pie-filling, and so Weatherwise is metaphorically transformed into both paper and a pie that, we are told, will be 'spoil[ed] in the baking' (1, 120). In this densely packed image, Middleton intertwines the sense of almanacs as often mouldy, or spoiled, and out of date with the common kitchen practice of lining pie dishes with waste paper which might, as when wrapping butter, become greasy. Weatherwise's body, then, congeals into shapeless gobbets of rancid matter and greasy, heterogeneous 'stuff'. We are presented with a gross confusion: something part-man, part-almanac, and wholly past its best.

In these lines, Middleton alludes to the widely perceived greasiness of almanacs. Thomas Nash (no relative of Thomas Nashe) compliments clothes that are not 'greasy, and like an *Almanacke* out of date', and Dryden's translation of Juvenal describes a superstitious woman, 'By whom a greasie Almanack is born | With often handling, like chaft Amber, worn'.[128] In Dryden's image, the pages of the woman's prognostication are threadbare from over-handling, gaining a greasy sheen from being repeatedly rubbed as if they were a magical charm. Among the most handled objects in the sixteenth and seventeenth centuries, most almanacs, not just those owned by the excessively superstitious, would have been thoroughly polished, soiled, and stained by the natural greases left behind by human skin.[129] In describing Weatherwise as a confusion of oils—human, paper, and animal—Middleton recognizes what Joshua Calhoun has described as 'the presence of animals on and within plant-based paper' and its consequent 'hybridity' and 'heterogeneity'.[130] The page emerges as a site at which multiple sorts of matter combine, with the grease of flax imagined as blending with animal size made from boiled animal skins, leather offcuts, and waste parchment, as well as the grease left behind by hands after sustained use and, after its reuse, the grease of the

[127] On the medicinal use of brown paper see Smith, 'A unique instance of art', paras. 34–38, and Leong, 'Papering the Household', 32–45. Waste paper would have been used in much the same way as brown paper. See p. 49 above.

[128] Thomas Nash, *Quaternio, or a Fourfold Way to a Happie Life*, STC 18382 (London: John Dawson, 1633), 260; *The Satires of Decimus Junius Juvenalis*, trans. by John Dryden, Wing J1288 (London: for Jacob Tonson, 1693), 118.

[129] Kathryn M. Rudy has quantified this process in her analysis of late medieval prayer books. Using a 'densitometer', Rudy has traced patterns of use over time. She argues that the accumulation of dirt in specific areas reveals the most intensely used sections of a manuscript, and might also indicate the unique way in which an individual held their codex, 'Dirty Books: Quantifying Patterns of Use in Medieval Manuscripts Using a Densitometer', *Journal of Historians of Netherlandish Art* 2, no. 1–2 (2010), http://www.jhna.org/index.php/past-issues/volume-2-issue-1-2/129-dirty-books.

[130] Calhoun, *The Nature of the Page*, 100.

foodstuffs and medicines the pages will wrap.[131] Reduced to grease, the description of Weatherwise suggests, all bodies begin to look and feel the same.

But the transformation of Weatherwise's body into that of an almanac is not only figurative in *No Wit*. Later in the play, Weatherwise reprises his astrological banquet in the form of a wedding masque. At this point, the widow is betrothed to the play's cross-dressing heroine and so Weatherwise plots a nuptial sabotage. Casting himself as the Air that his 'almanac speaks on' (7, 307), the vehicle for astrological influence, the other failed suitors are given the parts of earth, wind, and fire. In the starring role, Weatherwise wears 'a coat made like an almanac, all the twelve moons set in it, and the four quarters, Winter, Spring, Summer, and Autumn, with change of weathers, rain, lightning, and tempest, etc.' (9, 62.6–11). Emerging from above a 'thing like a globe' which 'opens of one side o'th'stage and flashes out fire' (9, 62.1–2), Weatherwise wore a coat that was perhaps made from the sheets of several almanacs, cut up and fixed directly onto his garment. This would, after all, have been the cheapest and easiest way to make a coat *like* an almanac, as well as the visual form in which audience members would most readily recognize the twelve moons, or months; the four quarters of the year; and the alterations of weather, as outlined in the stage directions. It would have been much easier to make out the distinctive rubrication, woodcuts, and layout of an almanac from a distant part of a theatre or hall than to decipher more obscure and unfamiliar painted symbols and emblems. Furthermore, almanacs were likely a common stage property because of their cheapness and the relative frequency with which they appear in plays. Part-way through the masque, the widow's failed suitors go off-script and transform into a disruptive antemasque, but this is rapidly dispelled by the arrival of the four winds. Here, the suitors' costumes, among them Weatherwise's almanac coat, 'seem to yield and almost fall of themselves' (9, 170.10–11). At this moment, the actor playing Weatherwise would have shed his coat and so discarded the pages of the almanac on the stage as, 'discovered' (9, 170.13), the suitors exited the scene. Weatherwise's masque might, then, have staged the dissolution of the almanac in two ways at once: firstly, in the fragmented, material make-up of the coat itself, made from almanacs, and secondly, in the antemasque's dispersal by the winds and the subsequent onstage discarding of the almanac coat.

However this scene was staged, in it, Weatherwise is materially transformed into the 'waste lumps of almanac stuff' (1, 118–119) he was characterized as earlier in the play. For Julian Yates, Middleton's 'systematic exploration of the comedic

[131] Tim Barrett describes the process in detail, explaining that '[t]here was typically more gelatin in historical European papers than any other ingredient other than the cellulose fiber itself', and that 'old well-worn rags' needed to be sized to obtain 'a supple, pliant, yet durable and long-lasting paper'. See 'Sizing' in 'Paper Through Time: Non-Destructive Analysis of 14th Through 19th Century Papers'. The University of Iowa, 2012, last accessed August 19, 2020, http://paper.lib.uiowa.edu/european.php#sizing. For the boiling of old manuscripts to make size, see Yale, *Sociable Knowledge*, 216.

possibilities afforded by the yoking together of a human person with this slim book-cum-list of key days, events and predictions' is a parody of the way that the suitors 'have given themselves over so completely to a book' that 'they embody a walking fetish'.[132] Similarly, for Hand the 'inversion of human and book, the objectification of the former and the fetishization of the latter' is both 'a troubled reflection upon the effects of emergent capitalism' and a correction to the passivity of the 'undiscerning reader'.[133] But there is more at work in these hybrid figures, found in *Heroes* as well as *No Wit*, than a commentary on the reading practices of the naïve and gullible. In these moments, Middleton is thinking through the ecology of the almanac. Though he certainly mocks those who read almanacs so literally as Weatherwise, he evidently cannot escape the sense that almanacs—through their intellectual content and the material biography that he stages—communicate an unavoidable truth about the human condition.

This unavoidable truth might be glossed by several passages in Middleton's 1618 mock-prognostication *The Owl's Almanac*. This prose work is, as Neil Rhodes writes, 'an elaborately designed parody of a printed document, in which the typography and page layout have a comic function'.[134] Middleton was evidently fascinated by the typographical dimensions of almanacs, persistently representing their symbols and structures as he did the Good and Bad Days in *Heroes* and the Zodiac Man and astrological masque in *No Wit*. Offering a 'general calendar for the common motion of the moon in all the months of the year' (1159–1160), Middleton replaces lunar descriptions of waxing and waning with mundane and urban ones, pairing them with the typographical symbols for the four quarters of the moon. The 'Last quarter' of the moon is described as corresponding to 'thieves at Newgate two or three days after the sessions. With sick persons when the bell rings out for them. With my almanac when 'tis put under pie-crust' (1167–1170). For Middleton, hanged bodies and the plague dead share the temporal rhythms of a repurposed almanac: all three resemble the waning of the moon as they turn out of date at specific moments, and so are reduced to greasy, indistinguishable, and disintegrating matter. According to the Owl who narrates Middleton's pamphlet, the almanac is, like the battered clothes and shoes she goes on to describe, 'an excellent memento (make but use of it) of your mortality' (2317–2318), and so serves as a concrete metaphor for the way that 'the world is out at the elbows' (1708). Almanac users, *Owl* suggests, found in the waste-time of their little books an uncomfortable truth about their own material biographies, as well as the material biography of the world around them.

But a *memento mori* does not quite do justice to the complexity of the representation of almanacs within Middleton's works. Waste almanacs are more than

[132] Yates, 'Thomas Middleton's Shelf Life', 22.

[133] Hand, 'More Lies Than True Tales', 322.

[134] Neil Rhodes 'Introduction' to *The Owl's Almanac*, in *Thomas Middleton, The Collected Works*, 1272.

emblems for the eventual decay of the body. Instead, they are descriptions of and guides for living within a world in which the human is not at the centre. As Meredith Molly Hand argues in her particularly sensitive reading of *Owl*, a 'crucial "truth"' that 'Middleton subverts is the idea of humans' essential primacy in the world'.[135] By giving animals such as Owl a voice, Middleton 'destabilizes humans' privileged place in the natural order', removing the 'clear distinction between humans and animals' and as such imagining an 'alternative world'.[136] A 'world' that lacks a vertical hierarchy between the human and the non-human is not, though, one that Middleton had to imagine or fabricate: it was one available within the ecology of the almanac, and one that expands beyond the animals that Hand writes of to the wider network of human and non-human bodies. Middleton, then, neither objectifies the human nor anthropomorphizes owls or books in his plays, masques, and mock-almanacs. Instead, he presents them as akin to the assemblages that are found within the pages of an almanac. We might, then, term him an ecological writer, not in the sense of his being concerned with the impact of the human on the world and the consumption of its finite resources, but in his attentiveness to what Mary Floyd-Wilson and Garret A. Sullivan have termed 'the *interpenetrability* and the *transactionality* of body and environment'.[137] It is important to note, though, that Middleton remains uncomfortable with the ecology that he repeatedly stages. In his conventional mockery of naïve almanac readers, over reliant on the little books' good and bad days, he works again and again to distinguish the human from the rest of the world and to confirm their agency. But in Middleton's almanac hybrids, the human is always absorbed back into the indistinguishable grease of the broader world. Middleton was likely unable to escape this uncomfortable truth because, year on year, he was interrupted, in the form of bookbindings, food wrappers, and privy tokens, by the vibrant articulations of almanac waste.

*

By attending to repurposed almanacs and their representation in literature, we discover a significant facet of early modern book culture as well as implications that extend beyond the textual and the bibliographic. Almanac waste is a particularly potent articulator of the place of the human in a wider ecosystem of non-human but fundamentally analogous bodies. Once discarded, these little books become even more visible nodes within this network, marking the greasy and fragmentary point at which skin and paper, like all other mundane matter, overlap. Middleton, as we have seen, reads almanacs 'ecosystemically', sensitive to not just the 'non-human elements' that make up early modern books, as Calhoun has argued is

135 Hand, 'More Lies Than True Tales', 316.
136 Ibid., 326, 322.
137 Floyd-Wilson and Sullivan, 'Introduction', in *Geographies of Embodiment in Early Modern*, 4

the case of Renaissance writers more broadly, but the way in which the human spills over and into the pages of an almanac.[138] Middleton, like so many of his contemporaries, both the naïve and the sceptical, found in the discarded sheets of out-of-date almanacs a mirror for his own place in the world.

We might, then, ask what our own waste has to say to us if we were to approach it with the 'anticipatory readiness' early moderns such as Middleton brought to their discarded almanacs.[139] It is difficult for us to decipher the vitality of our waste, not simply because there is so much of it, as Jane Bennett suggests, but because we rarely spend time in its presence.[140] When we discard something, it does not typically remain in our household or appear in our local grocers' or booksellers' shops, as sheets of waste paper did in the sixteenth and seventeenth centuries. Instead, it disappears out of sight, and so, typically, out of mind. When we do dwell on it, does our waste tell us of the cyclical rhythms of the world, as seems to be promised in the term 'recycling'? Do we imagine our rubbish as being almost magically transformed into a previous stage in its material biography, and so its readiness for future reuse? Or, does it tell us of a linear version of temporality, something much more pessimistic and akin to *Owl's* world wearing itself out at the elbows? The answer, as in the case of early modern waste, is not straightforward, and this is in part because of the radical shift in the material make up of much waste matter in the past century. Crucially, waste no longer decentres the human, as almanac waste did its early modern users. Instead, our modern-day waste articulates the destruction that we have wrought on the environmental assemblage outlined by the pages of Middleton's ephemeral, little books.

This is sensitively expressed in Kathleen Jamie's 2005 account of wandering amongst a plastic 'midden', or dunghill, on the coast of a Hebridean island.[141] Swept on land by the Atlantic currents, Jamie describes the 'fascination' held by the 'hundreds of marker buoys', 'the shampoo and milk cartons, the toilet-cleaner bottles we could turn over with our feet' (59). Seeking out the remnants of a plane wreck, a whale carcass, and bird bones to clean and take home, Jamie becomes side-tracked by the vitality of waste. It is familiar stuff—'we knew their shapes, had seen them ranked in supermarkets and hardware stores'—but, out of place, it discomforts her (59). Though she chooses to take home a 'gannet's shank' and 'the whale's vertebra', it is the rubbish that lingers in her imagination, embodying as it does the peculiar temporal mode of waste-time. She later wishes she had 'brought home the doll's head' that had looked up at her from the beach (69). Meditating on the temporalities contained within the various objects populating the Hebridean 'midden', Jamie notes that we typically find value 'not [in] the things that endured, but those

[138] Calhoun, *The Nature of the Page*, 150, 12.
[139] Bennett, *Vibrant Matter*, 5.
[140] Ibid., 5.
[141] Kathleen Jamie, *Findings* (London: Sort Of Books, 2005), 49–69: 66. Further references are included in the text in parentheses.

that had been transformed by death or weather', such as animal bone (66). Unlike naturally worn animal bones, plastic waste discomforts us because it resists transformation and decay. Neither cyclical nor linear in its biographical rhythms, it 'issue[s] a call', not of the 'thing-power' stemming from its existence in 'excess of [its] association with human meanings, habits, or projects', as almanac waste does, but of its place within an anthropocentric system.[142] It articulates the disruption and damage that has resulted from our placing ourselves at the centre, rather than as one component in the interconnected environmental assemblage. Our waste, then, remains as vital and expressive as early modern waste; the message that it communicates is simply a different one, and we are significantly less attuned to its voice.

[142] Bennett, *Vibrant Matter*, 4.

Coda

Living with Waste

Let us imagine, for a moment, a relatively normal day in the life of an early modern city-dweller: a housewife, perhaps, of comfortable means, at the turn of the seventeenth century. I suggest a housewife not only because the pages of this study have been dominated by male writers, but because waste paper would have been encountered by men and women across the social scale: by elite and middling women in their books, lining their chests and boxes, and aiding their needlework, as well as by their lower-class counterparts in shops and in kitchens.

Our housewife begins her day by reading a sermon. An old favourite, the suggestion in its opening pages that it will probably be turned to waste paper raises a wry smile. She does, as she lays it aside, dwell on its parchment wrapper: an old breviary. An earlier reader has traced its illuminated letters with spidery lines of black ink (see Fig. 1.2); perhaps, she thinks, they were practising their letters, or perhaps they did it out of admiration for the shimmering, colourful shapes—remnants of a past belief system at odds with the contents of the sermon. Later, she visits the grocer's shop. She buys a few sheets of paper, some flax, and a cornet of pepper. Glancing at the wrapper, she sees lines of red intermixed with the black, and the words 'Frende. 1602' along its fore edge: it is autumn, and this year's unsold almanacs are already circulating as waste. Back at home, she sets to work on a pie. Seasoning the meat, she regrets not buying a twist of mace, but it is too expensive, and the herbs she has gathered from her garden will do the job. After tipping the mixture into the pastry coffin, she reaches over to take the top leaf from a stack of unwanted papers gathered in the corner. A draft of an old letter, she is reminded of the paper she bought from the grocer's shop that morning and the letter she needs to finish. In the same moment, she deftly lines the dish beneath the pie. After a pause, she takes another leaf and covers the top of the pie to stop it from burning, as it did last time.[1]

We cannot know whether this precise series of events ever happened, but we can be certain that waste paper was common stuff in early modern England. References to waste paper are neither wholly rhetorical nor a symptom of a paper-short environment. Instead, waste practices were part of both an economy of thrift

[1] For instructions on the use of paper in the kitchen, see Anon, *A Booke of Cookerie*, STC 3299 ([London: E. Allde, 1597], 18.

Waste Paper in Early Modern England. Anna Reynolds, Oxford University Press. © Anna Reynolds (2024).
DOI: 10.1093/oso/9780198882701.003.0007

and frugality and an indication of the plenitude of paper in the period. Letters, almanacs, poetry, drama, sermons, Bibles, controversial texts, along with medieval manuscripts, recently redundant legal documents, and handwritten papers, circulated in manifold forms: in addition to whole books and loose sheets, they were regularly encountered as binding waste, grocery wrappers, tobacco-lighters, stoppers for mustard pots, linings for boxes, pies, and textiles, and as toilet paper. Such things, as Daniel Miller writes, 'direct our footsteps and are the landscapes of our imagination, as well as the cultural environment to which we adapt'.[2] They are, therefore, far more than the basis for a throwaway joke, and we might instead describe the fragments that thickly populated the early modern landscape as a scaffold for thought. We might also describe these objects as vibrant and articulate, prompting varied and generative thought through their textures, their parameters of use (how they were torn, cut, folded, and pasted), and their visible life cycle (the material traces of having transformed from flax, to rags, to text, to waste). But waste paper did not do this humbly or invisibly, as Miller suggests 'stuff' does more broadly.[3] Instead, early moderns were sensitive to a particular set of meanings encoded within repurposed pages, and these meanings can be uncovered in a wide range of religious, historical, and literary texts. As Christopher Tilley writes, 'structures of bodily experience work their way up into abstract meanings', and so form an 'embodied imagination'.[4] Metaphor is made up of the shifting interactions between things, language, and thought. Crucially, the haptic and kinetic experience of handling waste paper, in this historical moment, was intertwined with a re-emerging classical trope and a longstanding concern regarding literary immortality and textual survival. As a result, waste paper structured a thing-based way of thinking, grounded in its kinaesthetic and biographical contexts: whether encountered as an object in everyday life or in a carefully crafted passage of text, waste paper readily became a material emblem, a concrete metaphor, for the passage of time and the passage of things through time.

Waste paper, then, offered multilayered narratives that we, like our early modern counterparts, might endeavour to read. A waste fragment tells its own individual biography—how it was once principally valued as a text, was then either read and discarded or remained unread and unsold, before being repurposed. Those examples that survive to tell their stories are, more often than not, binding waste. Trimmed and pasted into other books, some waste speaks of extraction, collection, and archiving, having been removed and stored elsewhere, whereas others remain partially concealed as guards, pasteboards, and flyleaves. These fragments also tell broader cultural and historical narratives: of the rhythms of the printing house and the book trade; of the dissolution of the monasteries and the violence

[2] Miller, *Stuff*, 53.
[3] Ibid., 50.
[4] Tilley, *Metaphor and Material Culture*, 35.

of the Reformation; of the conceptualization and practice of waste and prodigality; of an understanding of the relationship between body and world; and of the development of antiquarian practice and bibliography. We can tell from the insults, tropes, poems, and meditations on and about waste paper that, in the early modern period, these narratives were often emblematic of the failed and the lost. They make visible the supersession of out-of-date texts, unsold or no longer wanted, and the erasure of past devotional practices and ways of understanding the world. But the past contained within and represented by waste paper might also erupt into the present, triggering nostalgia or a desire to recuperate that which has been lost.

The past and the present might also co-exist, serving, perhaps above all else, as a *memento mori*, a reminder of the eventual senescence and decay of all organic things. Our imagined housewife, as she glances across her kitchen table, might notice the cornet of almanac waste sitting alongside her pie, lined as it is in now greasy paper and oozing its meaty innards. Perhaps it also sits next to other remnants and offcuts of food—crusts of bread, worm-eaten apples, emptied oyster shells, and half-peeled fruit. This is similar to a scene depicted in Pieter Claesz's *Still Life with a Turkey Pie*, a breakfast still life painting, or *ontbijtjes*, representative of the genre (see Fig. C.1).[5] It is worth dwelling on the visual syntax of this painting for a moment as it renders waste paper at once emblematically and literally, and so shares the temporal framework of the tropes and literary representations explored within this volume. These breakfast pieces are closely related to the mid-seventeenth-century genre of *trompe l'oeil* and the hyper-realism of their carefully positioned and often paper objects, as well as the older tradition of *vanitas*.[6] In addition to depicting the muted opulence of the dinner tables of the wealthy in the Dutch Golden Age, paintings such as Claesz's share with the *vanitas* tradition a moral intent: they display 'interrupted' meals, and warn against gluttony and 'dissolute' behaviour.[7] The dissolution of the objects, in various states of rupture and decay, invoke a precariousness, characterized as they are by a pattern of broken exteriors and vulnerable, exposed interiors. Opened shells and peeled skin reveal fleshy oysters and lemon pith, and a sliced, crusty loaf suggests the texture of bread gradually turning stale. Crumpled waste paper is part of this language: it is a container that has progressed far along its life cycle that, in the moment depicted by Claesz, spills out pepper or spice. This moment in the life of a text recurs throughout paintings of this genre: the waste cornet in William Claesz Heda's *Breakfast with a Lobster* (1648, The Hermitage), for instance, visually echoes the curve of

[5] For an introduction to the conventions and history of this genre, see Hans Vlieghe, *Flemish Art and Architecture, 1585–1700* (New Haven: Yale University Press, 1998), 219.

[6] See, for instance, Samuel van Hoogstraten, *Trompe-l'oeil Still-Life* (oil on canvas, Dordrechts Museum: Dordrecht, 1664) and Heather Wolfe, 'Filing, seventeenth-century style', *The Collation* (blog), March 7, 2016, last accessed August 19, 2020, http://collation.folger.edu/2013/03/filing-seventeenth-century-style/.

[7] Kathryn Calley Galitz, *The Metropolitan Museum of Art: Masterpiece Paintings* (New York: Skira Rizzoli, 2016), 286.

Fig. C.1 Pieter Claesz. 1627. *Still Life with a Turkey Pie.* Oil on panel. Amsterdam: Rijksmuseum.

Reproduced by kind permission of the museum.

the dismembered crab leg above and the spiralling lemon peel below, as well as the creased fabric of the luxurious tablecloth it sits on. Waste paper, in these paintings, is one element of a complex whole that, hyper-realistically, portrays the perishability and vulnerability of emphatically organic matter. These artworks, therefore, contain narratives of decay and mortality that are also present in the works of Donne, Bale, Aubrey, Nashe, and Middleton. A careful onlooker such as our housewife might be reminded that she too will, like her kitchen-stuff, waste away.

But as well as being emblems of ephemerality, these fragments are also survivors—flotsam of the shipwreck of time. Their untimeliness might transform into something of an aura: these fragments were (and are) often fetishized, and that which was valued as a text before being turned into waste might, once again, be valued as a historical or aesthetic object. Readers of this book may be familiar with the thrill that one feels when handling a book bound in the crumpled, discoloured, and disjointed pages of another book—one that an early modern user has decided to throw away and that has, seemingly against the odds, survived. There are perhaps a multitude of reasons that we, in our own cultural moment, are drawn to these paper remnants, and that we find in their creases, stains, tears, and colours something attractive and affecting. Perhaps we find in discarded paper something that feels inexplicably real and authentic because of the ever-increasing dominance of the digital, and so of glowing, glassy screens, in our everyday lives. Or perhaps because we are so often faced with the permanence and the synthetic feel of plastic, we have become nostalgic for the coarse textures and visible vulnerability of rag-based paper. Or perhaps there is something universally human in our attraction to fragments and waste. Waste paper is akin, in many ways, to the ruins that are a more strikingly visible and widely studied form of material leftover.[8] Ruins have provoked a range of emotional and intellectual responses throughout the centuries, and, like waste paper, ruins make visible the passage of time and seem to offer a connection to the otherwise inaccessible past.

Where waste paper differs from ruins, however, is in its scale and its material form: waste paper suggests intimacy and secrecy. Held in the hands and often concealed between covers, waste paper prompts us to think about the other hands that have touched it, either reading it or writing on it, or that have torn, folded, stitched, and pasted it into its current form. What is more, we are highly conscious that this object was not intended to survive: whereas the makers of now-ruined stone architecture and marble statues evidently hoped that their handiwork would survive, waste paper is paradoxically preserved in the moment it is discarded. This peculiar situation is perhaps part of its affective pull: when we open a book and unveil waste, we are encountering an inadvertent survival—something unlikely and therefore precious, and something that may have gone unnoticed and overlooked for centuries. We might, then, conceive of waste paper as being 'privy' in a multitude of senses: as well as being the punchline of a common scatological joke because it might be used as toilet paper, waste paper serves as 'privy tokens' because it often survives secretly, tucked out of sight in boxes or bindings. In it, we might find the hidden treasure of manuscripts and printed texts unknown, rare, or thought lost. Much waste paper, though, sat and circulated in plain sight, and most does not contain rare or singular texts. But wherever it is or was encountered, and whatever text it contains, waste paper is always something that both we and its early users are privy to, in the sense of our being 'intimately acquainted with' it.[9] Grasped, turned, folded, wiped, and crumpled, waste paper served as such a potent emblem, or scaffold for thought, because of the hand-held scale of this intimacy.

These repeated, privy encounters are the root of what I have termed the widespread early modern sensitivity to repurposed pages, and I have used the term sensitivity because of its capacity to capture the importance of both the material and the metaphorical, or the sensory and the intellectual, in this paper-based intimacy and knowledge. Sensitivity encompasses both the corporeal 'function of sensation or sense perception' and the cerebral susceptibility or openness to 'emotional or aesthetic impressions' or, more broadly, the exercise of a 'delicate and profound appreciation' for a given thing.[10] Though there is little that is delicate about the sensitivity examined in these pages, it is something grounded at

[8] See, for instance, Hui, *The Poetics of Ruins.*
[9] 'Privy', adj. I.4.a, *OED.*
[10] 'Sensitive', adj. 1.a and 4.c, *OED.*

once in the senses—particularly, the haptic—and in the intellectual openness to the narratives and meanings suggested by these objects.

It is in large part my own affective response to waste paper, experienced while sitting in archives and rare book libraries, that has inspired this study: did early moderns feel something similar, I wondered, when looking at and handling waste? The answer is, of course, that in the most part they did not. Common and everyday stuff, waste and fragments were encountered in an entirely different setting and were interpreted within a historical framework distanced by centuries from our own. Further research might pursue the life cycle of waste paper, both newly created and enduring, through the intervening centuries: paper continued to be repurposed, though it was used for different purposes and within shifting intellectual frameworks, with the continued rise of cheap, domestic white paper in the eighteenth century, and the invention of wood-pulp paper in the nineteenth. An extended history of waste paper might also consider in more detail how waste fragments have been extracted and collected, re-cut, re-pasted, and catalogued by bibliophiles, librarians, and archivists, and how these fluctuating collections constitute a digressive but unbroken line of connection between past centuries and our own. Such a history, decipherable in catalogues and extant objects, would be inseparable from the gradually shifting affective responses to waste across time. It would provide an account of both past and present systems of preservation and valuation and reveal what, at various moments in time, has been considered worth keeping, and what has been thrown away.

But what I have been most struck by in this examination of early modern accounts and descriptions of waste paper, are the differences made visible between our own understanding and relationship with waste and those of this earlier period. Waste, as we have seen, has not always meant 'refuse', the wholly discarded, the useless and the used-up, the 'unserviceable', and 'unsaleable' stuff left behind after a process of manufacture or use.[11] In a world where very little went to waste, where even heretical texts might be used as wrappers rather than be destroyed, waste had economic value and extensive utility. What is more, the continuing biographies of these objects as they progressed between stages of use led to an accumulation of figurative value. Both literally and imaginatively productive, waste paper was something that might be repopulated and reconfigured again and again. This is in stark contrast to our own experience of waste today in the developed world. We dispose of our waste and it is removed from our homes and our lives. We assume that we will not meet it again, although we are aware that, more often than not, it will sit indefinitely in landfills or float in the ocean, and perhaps, every now and again, wash up on a distant shore. Periodically, news stories tell us that this is also the fate of much of our recycling (a relatively recent sub-category of waste), but, in most instances, we choose to forget the fate of our rubbish—recyclable or

[11] 'Waste', n. III.11.a, *OED*.

otherwise—once it has been discarded, only conceiving of it intermittently and in the abstract.[12] There is a large gap, for instance, in our knowledge of the life cycle of a plastic milk carton. We are unaware what occurs between the moment we throw it into the recycling bin and when we buy a new bottle in packaging that tells us it has been made in part from old ones. This is very different from the early modern experience of reusing one's own waste in the household or buying books and groceries wrapped in a neighbour's discards. In this world, that which was discarded did not disappear from view. Instead, early moderns lived *with* their waste: made of organic stuff, often rapidly disintegrating but persisting for a time, it cohabited with the human. Because of this difference, we do not typically approach or conceive of our own waste with the sensitivity of an early modern handling repurposed paper. Waste does, though, still contain narratives and meanings for those who are willing to look for them. Admittedly, these are very different from the early modern poetics and philosophies of waste uncovered in this book. Instead, they articulate waste's inability to disintegrate or decompose, and of environmental contamination—in other words, of the Anthropocene, and the devastating effect of the human on the world. Because we do not cohabit with our waste, however, these narratives often remain far from our minds.

Admittedly, in the years since I began this study, sensitivity to waste has been on the rise. Waste and its effects are becoming increasingly unavoidable as the volume of discarded matter, unable to decompose at any meaningful rate, grows, and as the damage wrought by the broader by-products of modern living finally begin to affect our daily lives in the developed world. Extreme weather events now regularly occur close to home, no longer limited to distant parts of the globe. I wonder if we might still find value, though, in something as seemingly trivial as early modern waste paper, or rather, if there is something that we might learn from a different way of understanding that which we throw away. Our waste is not going anywhere, and so we need to learn to live with it. And perhaps we would produce less of it if we cultivated the sort of sensitivity this book has uncovered. It could only be a good thing, after all, if we became more intimate with, or privy to, the entire life cycle and the potential narratives and meanings contained within our waste.

[12] See, for instance, Oliver Franklin-Wallis, 'Plastic recycling is a myth: what really happens to your rubbish', *The Guardian*, August 17, 2019, last accessed March 5, 2022, https://www.theguardian.com/environment/2019/aug/17/plastic-recycling-myth-what-really-happens-your-rubbish.

APPENDIX

Waste Cited

British Library

British Library, Harley 5937, item 54. A page of an unidentified almanac removed from a binding and stored in an album by John Bagford, http://explore.bl.uk/BLVU1:LSCOP-ALL:BLL01020556659

Cambridge University Library

Syn 8.63.181. Fols.121–122 and 143–144 of *Ovidii Nasonis Metamorphoseon Libri XV*, STC 18954 (Cambridge: *Ex Academiae celeberimmae typographeo*, 1631), used as flyleaves in the contemporary binding of Thomas Phillips, *Book of Lamentations*, STC 19878.5 (Cambridge: J. D[awson] for P[eter] Cole, 1639). Possibly recovered waste, https://idiscover.lib.cam.ac.uk/permalink/f/t9gok8/44cam_alma21336423670003606

Durham University Library

Cosin AA.3.28. Annotated sheet (fol. Hh3) of Thomas Elyot's *Bibliotheca Eliotae*, STC 7661 (London: Thomas Bethelet, 1548) as front flyleaf in the seventeenth-century binding of Robert Burton's *The Anatomy of Melancholy*, STC 4161 (Oxford: John Lichfield for Henry Cripps, 1628), http://library.dur.ac.uk/record=b1697611~S1

Cosin BB.5.27. Pages of a seventeenth-century letter, folded to form the guards in the seventeenth-century binding of *De usuris liber, Claudio Salmasio auctore*, USTC 1027926 (Lugd. Batavor.: ex officina Elseviriorum, 1638), https://library.dur.ac.uk/record=b1697882~S1

Cosin D.1.1. Fragments from Sheet B of *White 1638*, STC 527.28 (London: F.K. for the Company of Stationers, 1638) used as guards in Henry Spelman, *Concilia, decreta, leges, constitutione*, STC 23066 (Londini: Richardus Badger for Ph. Stephani, & Ch. Meredith, and Aureo Leone [1639]). Most likely unsold waste, now removed, silked and slipped loose into the binding, http://library.dur.ac.uk/record=b1695211~S1

Cosin K.2.14. Fragments from a Geneva Bible used as guards and reinforced with strips of a fourteenth-century manuscript, in the contemporary binding of *The Works of the Very Learned J. Jewell*, STC 14580 (London: [Eliot's Court Press for J. Norton], 1611), http://library.dur.ac.uk/record=b1696508~S1

Harry Ransom Center

BV 4615 B647 1635. Fragments of seventeenth-century rough accounts used as guards in the seventeenth-century binding of Robert Bolton, *Instructions for a right comforting afflicted consciences*, STC 3239 (London: T[homas] H[arper] for

Thomas Weaver, 1635), https://search.lib.utexas.edu/permalink/01UTAU_INST/9e1640/alma991009904739706011

BV 4817 B45 1657. Manuscript waste paper containing clothier's accounts used as rear flyleaf in the seventeenth-century binding of William Bell, *Incomparable Company-Keeping*, Wing B1813 (London: M. S. for George Eversden, 1657 [i.e. 1656]), https://search.lib.utexas.edu/permalink/01UTAU_INST/9e1640/alma991013126439706011

D 158 F95 1647. Fragments recovered from a seventeenth-century notebook of rough accounts (a booksellers?), used as guards along with printed waste from a large-format Dutch almanac, in the contemporary binding of Thomas Fuller, *The historie of the holy vvarre*, Wing F2438 (Cambridge: Roger Daniel for John Williams, 1647), https://search.lib.utexas.edu/permalink/01UTAU_INST/9e1640/alma991010065209706011

DA 330 B12 1629. Fragments of accounts used as guards in the seventeenth-century binding of Francis Bacon, *The historie of the reigne of King Henry the Seuenth*, STC 1161 (London: by I. H[aviland] and R. Y[oung] for Philemon Stephens and Christopher Meredith, 1629), https://search.lib.utexas.edu/permalink/01UTAU_INST/9e1640/alma991007614049706011

Henry E. Huntington Library

12963. Parchment fragments of a Catholic service-book used as guards in the binding of John Bale, *The First Two Partes of the Acts or Vnchaste Examples of the Englyshe Votaryes*, STC 1274 (London: Iohn Tysdale, 1560), https://catalog.huntington.org/record=b1495758

13101v.1. A fifteenth-century manuscript of St. Anselm used as a wrapper for Thomas Morley's *Canzonets. Or Little Short Songs to Three Voyces*, STC 18121 ([London]: Tho: Est, [1593]), https://catalog.huntington.org/record=b1826523

327725. Fragments of a Catholic service-book used as guards in the binding of Théodore de Bèze, *Confessio Christianae fidei*, STC 2007 (London: Richard Serll, [1565?]]), https://catalog.huntington.org/record=b1496438

438880. A medieval, monastic manuscript used as flyleaves in the binding of John Hooper, *A Funeral Oration*, STC 13755(London: by Thomas Raynalde, [1550]), https://catalog.huntington.org/record=b1497631

492913. Medieval manuscript service book wrapper of Sebastian Münster, *A Briefe Collection and Compendious Extract of Straunge and Memorable Thinges*, STC 18243 (London: Thomas Marshe, 1574), https://catalog.huntington.org/record=b1043778

69747. Fragments of an illuminated manuscript used as guards, with leaves of Matthew's Gospel as flyleaves, in the contemporary binding of Polydore Vergil, *De inventoribus rerum*, STC 24644 (London: VVithin the precincte of the late dissolued house of the Grey Friars, by Richard Grafton [1546]), https://catalog.huntington.org/record=b1504038

The Metropolitan Museum of Art

28.220.7,.8. Scraps of pattern books and receipts used as embroidery support for a pair of gloves made in Britain ca. 1600, 311×159 mm, https://www.metmuseum.org/art/collection/search/222229

The National Archives

C 107/110. Sheets from several late sixteenth- and seventeenth-century texts, including unsold sheets from *Magna Charta Cum Statutis*, STC 9283 (London: [Adam Islip] in aedibus Thomas Wight, 1602) used to line a seventeenth-century deed box, https://discovery.nationalarchives.gov.uk/details/r/C7028719

Shakespeare's Birthplace Trust

SBT: 1992–1994. Printed waste paper dated 1597 (text and edition not specified in catalogue) lining a circular pasteboard box containing sycamore posy trenchers, sixteenth century, 3×130 mm, https://collections.shakespeare.org.uk/search/museum/strst-sbt-1992-4/view_as/grid/search/everywhere:1992-4-140638/page/1

The Victoria and Albert Museum

W.17:1-1910. Proof sheets of William Camden, *Remaines of a Greater Worke, Concerning Britaine* (1615), edition not specified in catalogue, used as the lining of a box made in Southwark between 1600 and 1625, various woods, 255×760×545 mm, https://collections.vam.ac.uk/item/O7888/box-unknown/

W.51-1926. Decorated and waste sheets including calendars for 1680 (text and edition not specified) lining a box on stand made in England between 1600 and 1680, oak, 750×770×385 mm, https://collections.vam.ac.uk/item/O130436/box-on-stand-unknown/

Bibliography

Primary Sources

Manuscript Sources

London, British Library

Add. MS 27343.
Egerton MS 2982

Printed Sources

Allestree, Richard. *Allestree. 1639. A New Almanack and Prognostication, for the Yeere of Our Lord God, 1639.* STC 407.22. London: T. Cotes for the Company of Stationers, [1639].

Andrewes, John. *The Brazen Serpent: or, the Copie of a Sermon Preached at Pauls Crosse, Decemb. 31. 1620.* STC 591. London: G[eorge] P[urslowe] for Thomas Thorp, 1621.

Anthony, Francis. *The Apologies, or Defence of a Verity Heretofore Published Concerning a Medicine Called Aurum Potabile, that is, the Pure Substance of Gold, Prepared, and Made Potable and Medicinable Without Corrosiues.* STC 667. London: Iohn Legett, 1616.

Aske, James. *Elizabetha Triumphans. Conteyning the Damned Practizes, that the Diuelish Popes of Rome Haue Vsed Euer Sithence her Highnesse First Comming to the Crowne, by Mouing her Wicked and Traiterous Subjects to Rebellion and Conspiracies, Thereby to Bereaue her Maiestie Both of He[R] Lawfull Seate, and Happy Life.* STC 847. London: Thomas Orwin for Thomas Gubbins and Thomas Newman, 1588.

Aubrey, John. *Brief Lives with An Apparatus for the Lives of our English Mathematical Writers*, 2 vols., edited by Kate Bennett. Oxford: Oxford University Press, 2015.

Bacon, Francis. *The Tvvoo Bookes of Francis Bacon. of the Proficience and Aduancement of Learning, Diuine and Humane.* STC 1164. London: Thomas Purfoot and Thomas Creed, 1605.

Bacon, Francis. *The Works of Francis Bacon*, vol. 2, *Philosophical Works*, edited by James Spedding, Robert Leslie Ellis, and Douglas Denon Heath. Cambridge: Cambridge University Press, 2011.

Bale, John and John Leland. *The Laboryouse Journey & serche of John Leylande, for Englandes Antiquities.* STC 15445. London: S. Mierdman, 1549.

Baret, John. *An Aluearie or Triple Dictionarie.* STC 1410. London: by Henry Denham, 1574.

Banister, *The Historie of Man.* STC 1359. London: John Day, [to be sold by R. Day], 1578.

Blount, Thomas. *Glossographia: or A dictionary, interpreting all such hard vvords, whether Hebrew, Greek, Latin, Italian, Spanish, French, Teutonick, Belgick, British or Saxon; as are now used in our refined English tongue.* Wing B3334. London: Tho. Newcomb to be sold by Humphrey Moseley, 1656.

A booke of cookerie. otherwise [sic] called: The good huswiues handmaid for the kitchin. STC 3299. [London: E. Allde, 1597].

Bradshaw's ghost: being a dialogue between the said ghost, and an apparition of the late King Charles. Wing B4165. [London: s.n., 1659].

Bramhall, John. *The Serpent Salve, or, A remedie for the biting of an aspe.* Wing B4236. [S.L: S.N], 1643.

Brinsley, John. *Ludus Literarius: or, the Grammar Schoole.* STC 3768. London: [Humphrey Lownes] for Thomas Man, 1612.

Brome, Alexander. *Bumm-fodder, or VVaste-Paper Proper to Wipe the Nation's Rump With, or Your Own.* Wing B4864A. [London: s.n., 1660].

Browne, Thomas. *Hydriotaphia, = urne-buriall, or, a Discourse of the Sepulchrall Urnes Lately Found in Norfolk.* STC B5154. London: for Hen. Brome, 1658.

Brunfels, Otto. *A Very True Pronosticacio[n], with a Kalender* [...] *for the yere of our Lorde M. CCCCC. Xxxvj.* STC 421.17. London: Ioh[a]n Byddell, 1536.

Burton, Robert. *The Anatomy of Melancholy, vvhat it is. VVith all the Kindes, Causes, Symptomes, Prognostickes, and Seuerall Cures of it.* STC 4159. Oxford: John Lichfield and James Short for Henry Cripps, 1621.

Camden, William, *Remaines of a Greater Worke, Concerning Britaine, the Inhabitants Thereof, their Languages, Names, Surnames, Empreses, Wise Speeches, Poësies, and Epitaphes.* STC 4521. London: G[eorge] E[ld] for Simon Waterson, 1605.

Catullus, edited and translated by G. P. Goold. London: Duckworth, 1983.

Cawdry, Robert. *A Table Alphabeticall, conteyning and teaching the true vvriting, and vnderstanding of hard vsuall English wordes, borrowed from the Hebrew, Greeke, Latine, or French. &c.* STC 4884. London: J[ames] R[oberts] for Edmund Weaver, 1604.

Chamberlain, Robert. *Chamberlain's Arithmetick: being a plain and easie explanation of the most useful and necessary art of arithmetick in whole numbers and fractions, that the meanest capacity may obtain the knowledge thereof in a very short time.* Wing C1813. London: for John Clark, 1679.

Charas, Moyse. *The Royal Pharmacopoeia, Galenical and Chymical, According to the Practice of the Most Eminent and Learned Physitians of France, and Publish'd with their Several Approbations.* Wing C2040. London: for John Starkey, 1678.

Chillingworth, William. *The Religion of Protestants a Safe Way to Salvation.* STC 5138. Oxford: by Leonard Lichfield for John Clarke, 1638.

Chytraeus, David. *A Postil or Orderly Disposing of Certeine Epistles vsually red in the Church of God, vppon the Sundayes and Holydayes Throughout the Whole Yeere*, translated by Arthur Golding. STC 5263. London: Henry Bynneman for Lucas Harrison and George Bishop, 1570.

Churchyard, Thomas. *A Sparke of Frendship and Warme Goodwill, that shewest the effect of true Affection and vnfoldes the Finenesse of this World.* STC 5257. London: [T. Orwin], 1588.

Clark, Andrew (ed.). *The Life and Times of Anthony Wood*, 5 vols. Oxford, 1891–1895.

Cockeram, Henry. *The English Dictionarie: or, An Interpreter of Hard English vvords.* STC 5461. London: [Eliot's Court Press] for Edmund Weauer, 1623.

Cooper, Thomas. *Thesaurus Linguae Romanae & Britannicae.* STC 5686. London: Henry Wykes, 1565.

Cornwallis, William. *Essayes.* STC 5775. [London]: [S. Stafford and R. Read] for Edmund Mattes, 1600–1601.

Coryate, Thomas. *Coryats Crudities: Hastily Gobled Vp in Five Moneths Trauells in France, Sauoy, Italy, Rhetia Co[M]Monly Called the Grisons Country, Heluetia Aliàs Switzerland, Some Parts of High Germany, and the Netherlands; Newly Digested in the Hungry Aire of Odcombe in the County of Somerset, & Now Dispersed to the Nourishment of the Trauelling

Members of this Kingdome. STC 5808. London: W[illiam] S[tansby for the author], 1611.

The Coverdale Bible. STC 2063. [Cologne?: by E. Cervicornus and J. Soter?, 1535].

Culverwel, Nathaneal. *An Elegant and Learned Discourse of the Light of Nature, with Several Other Treatises*. Wing 7569. London: T. R[atcliffe] and E. M[ottershed] for John Rothwell, 1652.

Dafforne, Richard. *The Merchants Mirrour. or, Directions for the Perfect Ordering and Keeping of his Accounts. ... As Likewise, A Waste-Book, With a Compleat Journall, and Leager Thereunto Appertaining; unto the Which I Have Annexed two Other Waste-Bookes for Exercise of the Studious Famous Art*. Wing D102. London: J. L. for Nicholas Bourn, 1651.

Davies, John. *The Poems of Sir John Davies*, edited by Robert Krueger and Ruby Nemser. Oxford: Clarendon Press, 1975.

Dawes, Lancelot. *Gods Mercies and Ierusalems Miseries. A Sermon Preached at Pauls Crosse, The 25. of Iune. 1609*. STC 638. London: [John Windet] for Cle. Knight, 1609.

Dekker, Thomas. *Nevves from Graues-End: Sent to Nobody*. STC 12199, London: T[homas] C[reede] for Thomas Archer, 1604.

Dekker, Thomas. *Jests to Make you Merie: with the Coniuring vp of Cock VVatt, (the Walking Spirit of Newgate) to tell tales*. STC 6541. London: N[icholas] O[kes] for Nathaniell Butler, 1607.

Dekker, Thomas. *The Guls Horn-Booke*. STC 6500. London: [Nicholas Okes] for R. S[ergier?], 1609.

Dekker, Thomas. *The Rauens Almanacke foretelling of a Plague, Famine, and Ciuill Warre*. STC 6519.2. London: E[dward] Allde and another for Thomas Archer, 1609.

Dekker, Thomas. *Lanthorne and Candle-Light*. STC 6486. London: [by Edward Allde] for John Busby, 1609.

Dickens, Charles. *Bleak House*. Oxford: Oxford University Press, 1996.

Digges, Dudley. *The Defence of Trade. In a letter to Sir Thomas Smith Knight, gouernour of the East-India Companie, &c. From one of that societie*. STC 6845. London: William Stansby for Iohn Barnes, 1615.

Donne, John. *An Anatomy of the World. Wherein, by Occasion of the Vntimely Death of Mistris Elizabeth Drury the Frailty and the Decay of this Whole World is Represented*. STC 7022. London: William Stansby for Samuel Macham, 1611.

Donne, John. '*Incipit Joannes Donne*'. In Thomas Coryat. *Coryats Crudities*. STC 5808. London: W[illiam] S[tansby for the author], 1611.

Donne, John. *Devotions Upon Emergent Occasions*, edited by Anthony Raspa. Oxford: Oxford University Press, 1987.

E., B. *New Dictionary of the Terms Ancient and Modern of the Canting Crew, in its several tribes, of gypsies, beggers [sic], thieves, cheats, &c. with an addition of some proverbs, phrases, figurative speeches, &c*. Wing E5. London: for W. Hawes, P. Gilbourne and W. Davis, 1699.

Everard, John. *The Gospel-Treasury Opened*. Wing E3531. London: by John Owsley for Rapha Harford, 1657.

Featley, Daniel. *The Grand Sacrilege of the Church of Rome in taking away the sacred cup from the laiety at the Lords Table*. STC 10733. London: Felix Kyngston for Robert Milbourne, 1630.

Fisher, Samuel. *Rusticus ad academicos in exercitationibus expostulatoriis, apologeticis quatuor*. Wing F1056. London: for Robert Wilson, 1660.

Fitzgeffrey, Henry. *Satyres: and Satyricall Epigrams with certaine obseruations at Black-Fryers*. STC 10945. London: Edw[arde] Allde for Miles Patrich, 1617.

Florio, John. *A Worlde of Wordes, or Most copious, and exact dictionarie in Italian and English.* Wing 11098. London: Arnold Hatfield for Edw. Blount, 1598.

Foulweather, Adam. *A vvonderfull, strange and miraculous, astrologicall prognostication for this yeer of our Lord God. 1591.* STC 11209. London: [Thomas Scarlet], 1591.

The four great years of the plague, viz. 1593. 1603. 1625. and 1636. Wing F1658. London: Peter Cole, 1665.

Galfridus, Anglicus. *Incipit liber q[ui] dicitur Promptorium parvulorum siue clericorum.* STC 20434. London: Richard Pynson for Frederic Egmondt and Petrus Post pascha, 1499.

Gayton, Edmund. *Pleasant Notes Upon Don Quixot.* Wing G415. London: by William Hunt, 1654.

Goslicki, Wawrzyniec. *The Sage Senator Delineated: or, A discourse of the qualifications, endowments, parts, external and internal, office, duty and dignity of a perfect politician.* Wing G2027. London: Ja: Cottrel for Sam. Speed, 1660.

Greenes, Groats-VVorth of Witte, bought with a million of repentance. Describing the follie of youth, the falshood of makeshifte flatterers, the miserie of the negligent, and mischiefes of deceiuing courtezans. STC 12245. London: [J. Wolfe and J. Danter] for William Wright, 1592.

Hakluyt, Richard. *Diuers Voyages Touching the Discouerie of America and the ilands adiacent vnto the same, made first of all by our Englishmen, and afterward by the Frenchmen and Britons.* STC 12624. London: [Thomas Dawson] for Thomas VVoodcocke, 1582.

Hall, John. *A Serious Epistle to Mr. William Prynne, wherein is interwoven an answer to a late book of his, the title whereof is inserted in the next leafe.* Wing H359A. London: for John Place, 1649.

Harvey, Gabriel. *Foure Letters, and Certaine Sonnets: especially touching Robert Greene, and other parties, by him abused: but incidently of diuers excellent persons, and some matters of note.* STC 12900.5. London: Iohn Wolfe, 1592.

Harvey, Gabriel. *A Nevv Letter of Notable Contents. With a straunge sonet, intituled Gorgon, or the wonderfull yeare.* STC 12902. London: Iohn Wolfe, 1593.

Harvey, Gabriel. *Pierces Supererogation, or, a New prayse of the old asse.* STC 12903. London: John Wolfe, 1593.

Hollyband, Claudius, *A Dictionarie French and English.* STC 6737. London: T[homas] O[rwin] for Thomas Woodcock, 1593.

Horace. *A Medicinable Morall, that is, the Two Bookes of Horace his Satyres, Englyshed Accordyng to the Prescription of saint Hierome*, translated by Thomas Drant. STC 13805. London: Thomas Marshe, [1566].

Horace. *Horace his Arte of Poetrie, Pistles, and Satyrs, Englished*, translated by Thomas Drant. STC 13797. London: Thomas Marshe, 1567.

Horace. *Q. Horatii Flacci Venusini.* STC 13784. [London]: [J. Kingston] for [Will]. Norton, 1574.

Horace. *Horace; Satires, Epistles and Ars Poetica*, translated by H. Rushton Fairclough. Cambridge, MA: Harvard University Press, 1942.

Horace. *Horace, Odes and Epode*, edited and translated by Niall Rudd. Cambridge, MA: Harvard University Press, 2004.

Howgill, Francis. *The Mouth of the Pit Stopped, and the Smoke that Hath Arisen out of it Scattered by the Breath of Truth.* Wing H3172. London: for Thomas Simmons, 1659.

Jeanes, Henry. *A Second Part of The Mixture of Scholasticall Divinity, with practical, in several tractates.* Wing J508. Oxford: H. Hall [and A. Lichfield] for Thomas Robinson, 1660.

Johnson, Thomas. *Johnson. 1604 an Almanack and Prognostication, for the Yere of our Lord God. MDCIIII.* STC 466.7. London: E. Allde for the Company of Stationers, [1604].

Joyce, James. *Ulysses.* London: The Bodley Head, 2008.

Junius, Hadrianus. *The Nomenclator, or Remembrancer of Adrianus Iunius Physician*, translated by John Higgins. STC 14860. London: for Ralph Newberie and Henrie Denham, 1585.

Juvenal. *The Satires of Decimus Junius Juvenalis*, translated by John Dryden. Wing J1288. London: for Jacob Tonson, 1693.

Juvenal and Persius, translated by Susanna Morton Braund. Cambridge, MA: Harvard University Press, 2004.

Lacey, William. *The Iudgment of an Vniuersity-Man Concerning M. William Chillingvvorth his Late Pamphlet, in ansvvere to Charity maintayned.* STC 15117. [Saint-Omer]: The English College Press, [1639].

Leland, John. *De uiris Illustribus/On Famous Men*, edited by James P. Carley with Caroline Brett. Oxford: Bodleian Library, 2010.

Levens, Peter. *Manipulus Vocabulorum. A Dictonarie of English and Latine Wordes, Set Forthe in Suche Order, as None Heretofore Hath Ben, the Englishe Going Before the Latine, Necessary not Onely for Scholers that Wa[N]T Varietis of Words, but also for such as Vse to Write in English Meetre.* STC 15532. London: Henrie Bynneman for Iohn Waley, 1570.

Lilly, William. *Monarchy or No Monarchy in England.* Wing L2228. London: for Humfrey Blunden, 1651.

Love, Christopher. *The Dejected Soules Cure: Tending to Support Poor Drooping Sinners.* Wing L3151. London, John Rothwell: 1657.

Markham, Gervase. *The English House-VVife. Containing the Inward and Outward Vertues Which Ought to be in a Compleate Woman.* STC 17353. London: Nicholas Okes for Iohn Harison, 1631.

Marprelate, Martin. *Oh Read Ouer D. Iohn Bridges, for it is Worthy Worke: or an Epitome of the Fyrste Booke, of that Right Worshipfull Volume, Written Against the Puritanes, in the Defence of the Noble Cleargie, by as Worshipfull a Prieste, Iohn Bridges, Presbyter, Priest or Elder, Doctor of Diuillitie, and Deane of Sarum.* STC 17453. [Fawsley, Northants.]: [Robert Waldegrave, 1588].

Marprelate, Martin. *The Martin Marprelate Tracts, A Modernized and Annotated Edition*, edited by Joseph L. Black. Cambridge: Cambridge University Press, 2008.

Master Broughtons Letters, Especially his Last Pamphlet to and Against the Lord Archbishop of Canterbury, About Sheol and Hades, for the Descent into Hell, Answered in their Kind. STC 3864. London: [F. Kingston for] John Wolfe, 1599.

Martial. *M. Val. Martialis epigrammaton libri*, edited by Thomas Farnaby. STC 17492. London: Felix Kingston for William Welby, 1615.

Martial. *Selected Epigrams of Martial*, translated by Thomas May. STC 17494. London: [Humphrey Lownes] for Thomas Walkley, 1629.

Martial. *Martial: Epigrams*, edited and translated by D. R. Shackleton Bailey, vols. 1–3. Cambridge, MA: Harvard University Press, 1993.

Martial. *Martial Book XIII, The Xenia*, edited by T. J. Leary. London: Bloomsbury, 2001.

Middleton, Thomas. *The Owles Almanacke. Prognosticating many strange accidents which shall happen to this kingdome of Great Britaine this yeere, 1618.* STC 6515. London: E.G. for Lawrence Lisle, 1619.

Middleton, Thomas. *No Wit, Help Like a Womans.* Wing M1985. London: [Thomas Newcombe] for Humphrey Moseley, 1657.

Middleton, Thomas. *No Wit, No Help Like a Woman's*, edited By Lowell E. Johnson. Lincoln: University of Nebraska Press, 1976.

Middleton, Thomas. *Thomas Middleton, The Collected Works*, edited by Gary Taylor and John Lavagnino. Oxford: Clarendon Press, 2007.

Middleton, Thomas. *An/The Old Law*, edited by Jeffrey Masten. In *Thomas Middleton, The Collected Works*, edited by Taylor and Lavagnino.

Middleton, Thomas. *The Masque of Heroes*, edited by James Knowles. In *Thomas Middleton, The Collected Works*, edited by Taylor and Lavagnino.

Middleton, Thomas. *No Wit/Help Like a Woman's; Or, The Almanac*, edited by John Jowett. In *Thomas Middleton, The Collected Works*, edited by Taylor and Lavagnino.

Middleton, Thomas. *The Owl's Almanac*, edited by Neil Rhodes. In *Thomas Middleton, The Collected Works*, edited by Taylor and Lavagnino.

Milton, John. *Paradise Lost*, edited by John Leonard. Harmondsworth: Penguin, 2000.

Montaigne, Michel de. *The Essayes or Morall, Politike and Millitarie Discourses of Lo: Michaell de Montaigne*, translated by John Florio. STC 18041. London: Val. Sims for Edward Blount, 1603.

Montulmo, Antonius de. *An Almanacke and Prognostication for the Yere of our Lord God D.CCCCC.LV. [sic]*. STC 483.14. London: Thomas Marche, [1555].

Moxon, Joseph. *Mechanick Exercises*, vol. 1. Wing M3013. London: for Joseph Moxon, 1677 [1678–1680].

Nash, Thomas. *Quaternio or a Fourefold Vvay to a Happie Life; Set Forth in a Dialogue Betweene a Countryman and a Citizen, a Divine and a Lawyer*. STC 18382. London: Iohn Davvson, 1633.

Nashe, Thomas. *The Anatomie of Absurditie*. STC 18364. London: I. Charlewood for Thomas Hacket, 1589.

Nashe, Thomas. *An Almond for a Parrat, or Cutbert Curry-knaues almes*. STC 534. [London?]: [Eliot's Court Press?], [1590].

Nashe, Thomas. *Pierce Penilesse his Supplication to the Diuell*. STC 18373. London: Abell Ieffes for Iohn Busbie, 1592.

Nashe, Thomas. *Strange Newes, of the Intercepting Certaine Letters, and a Conuoy of Verses, as They Were Going Priuilie to Victuall the Low Countries*. STC 18377. [London]: [J. Danter], 1592.

Nashe, Thomas. *Christs Teares Ouer Ierusalem*. STC 18366. London: Iames Roberts to be sold by Andrew Wise, 1593.

Nashe, Thomas. *Christs Teares Ouer Ierusalem*. STC 18367. London: [James Roberts and R. Field] for Andrew VVise, 1594.

Nashe, Thomas. *The Terrors of the Night or, A discourse of apparitions*. STC 18379. London: Iohn Danter for William Iones, 1594.

Nashe, Thomas. *The Vnfortunate Traueller, or the life of Iacke Wilton*. STC 18380. London: Thomas Scarlet for Cuthbert Burby, 1594.

Nashe, Thomas. *Haue vvith You to Saffron-VValden. or, Gabriell Harueys hunt is vp*. STC 18369. London: Iohn Danter, 1596.

Nashe, Thomas. *Nashes Lenten Stuffe, Containing, the Description and First Procreation and Increase of the Towne of Great Yarmouth in Norffolke: With a New Play Neuer Played Before, of the Praise of the Red Herring*. STC 18370. London: [Thomas Judson and Valentine Simmes] for N[icholas] L[ing] and C[uthbert] B[urby], 1599.

Nashe, Thomas. *A Pleasant Comedie, Called Summers Last Will and Testament*. STC 18376. London: Simon Stafford for Walter Burre, 1600.

Nashe, Thomas. *The Works of Thomas Nashe*, edited by Ronald B. McKerrow and revised by F.P. Wilson, 5 vols. Oxford: Blackwell, 1958.

Nashe, Thomas. *Thomas Nashe: The Unfortunate Traveller and Other Works*, edited by J. B. Steane. Harmondsworth: Penguin, 1972.

Ordinaunces Decreed for Reformation of Diuers Disorders in Pryntyng and Vtteryng of Bookes. STC 7754.8. [London: s.n., 1566].

Parker, Henry. *Observations Upon Some of his Majestie Late Ansvvers and Expresses*. Wing P413. [London: s.n., 1642].

Perceval, Richard. *A Dictionarie in Spanish and English*. STC 19620. London: Edm. Bolifant, 1599.

Persius. *Aulus Persius Flaccus his Satires*, translated by Barten Holyday. STC 1977.5. Oxford: Ioseph Barnes, 1616.

Phillips, Edward. *The New World of English Words: Or, a General Dictionary: Containing The Interpretations of Such Hard Words as are Derived From Other Languages*. Wing P2068. London: E. Tyler or Nath. Brooke, 1658.

Phillips, Thomas. *Booke of Lamentations; or, Geennēlogia a Treatise of Hell*. STC 19878.5. Cambridge: J. D[awson] for P[eter] Cole, 1639.

Plat[t], Hugh. *Sundrie Nevv and Artificiall Remedies Against Famine*. STC 19996. [London]: P[eter] S[hort], 1596.

Plat[t], Hugh. *The Nevv and Admirable Arte of Setting of Corne*. STC 19993. London: P. Short, 1601.

Plat[t], Hugh. *The Second Part of the Garden of Eden. Or An Accurate Description of All Flowers and Fruits Growing in England; With Partuicular [Sic] Rules How to Advance Their Nature and Growth, as Well in Seeds and Herbs, as the Secret Ordering of Trees and Plants*. Wing P2392. London: for William Leak, [1659].

Plattes, Gabriel. *The Profitable Intelligencer, Communicating his Knowledge for the Generall Good of the Common-Wealth and All Posterity*. Wing P2414 [London?]: for T.U., [1644].

Pliny the Elder, *The Historie of the VVorld: Commonly Called, the Naturall Historie of C. Plinius Secundus*, translated by Philemon Holland. STC 20030. London: Adam Islip, 1634.

Pond, Robert. *A Newe Treatise of the Right Reckoning of Yeares*. STC 20104. Edinburgh: Robert Waldegrave, 1599.

The Poetical Works of Alexander Pope. London: Macmillan & Co. Ltd, 1961.

Porter, Edmund. *Theos Anthropophoros. Or, God incarnate*. Wing P2985. London: printed for Humphrey Moseley, 1655.

Rabelais, François. *The First Book of the Workes of Mr. Francis Rabelais Doctor in Physick, Containing Five Books of the Lives, Heroick Deeds, and Sayings of Gargantua, and his Sonne Pantagruel*, translated by Thomas Urquhart. Wing R105. London: Richard Baddeley, 1653.

Rabelais, François. *Oeuvres Completes*, edited by Jacques Boulenger. Paris: Gallimard, 1955.

Roberts, Lewes. *The Merchants Mappe of Commerce: Wherein, the Universall Manner and Matter of Trade, is Compendiously Handled*. STC 21904. London: R. O[ulton, Eliot's Court Press?, Thomas Harper, and Felix Kingston] for Ralph Mabb, 1638.

Sandys, George. *A Paraphrase Upon the Divine Poems*. STC 21725. London: [John Legatt] to be sold by [Andrew Hebb], 1638.

Securis, John. *1574. An Almanacke and Prognostication, for the Yere of Our Lord God M.D.LXXIIII*. STC 512. London: Richarde Watkins and Iames Robertes, 1574.

Seyfridt, George. [*Almanake and Pronostication for the Yeare of Our Lorde MCCCCC, and XXXVII*]. STC 513. Antwerp: Widow of C. Ruremond?, [1537].

Shakespeare, William. *The Oxford Shakespeare: Complete Works*, 2nd edition, edited by John Jowett et al. Oxford: Clarendon Press, 2005.

Sidney, Philip. *Syr P.S. His Astrophel and Stella*. STC 22536. London: [John Charlewood] for Thomas Newman, 1591.

Sidney, Philip. *Sir Philip Sidney: The Major Works*, edited by Katherine Duncan-Jones. Oxford: Oxford University Press, 2008.

Sparke, Michael. *A Second Beacon Fired by Scintilla*. Wing S4818BA. London: for the author, 1652.

Spenser, Edmund. *The Faerie Queene*, edited by A. C. Hamilton. London: Routledge, 2013.

Statius Silvae, edited and translated D. R. Shackleton Bailey, with corrections by Christopher A. Parrot. Cambridge, MA: Harvard University Press, 2015.

Sterne, Lawrence. *A Sentimental Journey*. Harmondsworth: Penguin, 2005.

Swift, Jonathan. *The Complete Poems*. Harmondsworth: Penguin, 1983.

The Works of George Swinnock. Wing S6264. London: J.B. for Tho. Parkhurst, 1665.

All the Workes of Iohn Taylor the Water Poet. STC 23725. London: I[ohn] B[eale], Elizabeth Allde, Bernard Alsop, and Thomas Fawcet for Iames Boler, 1630.

Troughton, William. *Saints in England Under a Cloud: And Their Glory Eclipsed in this Life. or, the Case of Desertion Briefly Stated in a Few Considerations with Severall Symptomes of the Saints Decreasing and Declining in Spirituals*. Wing T2319. London: Mathew Simmons for Hannah Allen, 1648.

Urquhart, Thomas. *Ekskybalauron: or, The Discovery of a Most Exquisite Jewel*. Wing U134 London: Ja: Cottrel, to be sold by Rich. Baddeley, 1652.

Uthalmus, Levimos. *Fasciculus Florum: or, A Nosegay of Flovvers, Translated Out of the Gardens of Severall Poets, and Other Authors*. STC 24559. London: A[ugustine] M[athewes], 1636.

Valentine, Henry. *Private Devotions, Digested into Six Letanies*. Wing 24576.3. London: [M. Flesher] for John Marriot, 1635.

Valentinus, Basilius, *Of Natural & Supernatural Things*. Wing B1020. London: sold by Moses Pitt, 1671.

Vaughan, William. *The Golden Fleece diuided into three parts, vnder which are discouered the errours of religion, the vices and decayes of the kingdome, and lastly the wayes to get wealth, and to restore trading so much complayned of*. STC 24609. London: [William Stansby, Miles Flesher, and another] for Francis Williams, 1626.

Vernon, John. *The Compleat Comptinghouse: or, The Young Lad Taken From the Writing* School. Wing V249. London: J[ohn] D[arby] for Benj[amin] Billingsley, 1678.

Webster, John. *The White Diuel, or, The Tragedy of Paulo Giordano Vrsini, Duke of Brachiano, With the Life and Death of Vittoria Corombona the Famous Venetian Curtizan*. STC 25178. London: N[icholas] O[kes] for Thomas Archer, 1612.

White, John. *White 1638. A New Almanacke And Prognostication for the Yeere of Our Lord God 1634*. STC 527.28. London: William Stansby [and A. Mathewes], for the Company of Stationers, [1634].

Whitehead, George. *The He-Goats Horn Broken. Or, Innocency Elevated Against Insolency & Impudent Falshood. in Answer to Two Books Against the People of God Called Quakers*. Wing W1933. London: for Robert Wilson, 1660.

Wood, Anthony. *The History and Antiquities of the University of Oxford*, vol. 2, edited by John Gutch. Oxford: Printed for the Editor, 1796.

Younge, Richard. *Englands Unthankfulness Striving with Gods Goodness, for the Victory*. Wing Y152. [London: J. Bell for James Crump, 1643?].

Other Sources

Claesz, Pieter. *Still Life with a Turkey Pie.* 1627. Oil on panel, 75×132 cm. Amsterdam, Rijksmuseum.

Heda, Willem Claeszoon. *Still Life with Glasses and Tobacco.* 1633. Oil on panel, 508×756 mm. Boston, Museum of Fine Arts.

Heda, Willem Claeszoon. *Still Life with a Gilt Cup.* 1635. Oil on panel, 88×113 cm. Amsterdam., Rijksmuseum.

Heda, Willem Claeszoon. *Tobacco Still Life.* 1637. Oil on wood, 41.91×54.61 cm. Los Angeles County Museum of Art.

Heda, Willem Claeszoon. *Breakfast with a Lobster.* 1648. Oil on canvas, 118×118 cm. St Petersburg, The Hermitage.

Hogarth, William. *The Bad Taste of the Town.* 1724. Engraving on paper, 170×242 mm. London, Victoria and Albert Museum.

van Hoogstraten, Samuel. 1664. *Trompe-l'oeil Still-Life.* Oil on canvas, 460×580 mm. Dordrecht, Dordrechts Museum.

Teniers the Younger, David. *Boors Drinking.* 1635–1640. Oil on panel, 37.5×27 cm. Edinburgh, National Galleries of Scotland.

Secondary Sources

Adams, Brandi, K. 'Fair/Foul'. In *Shakespeare/Text: Contemporary Readings in Textual Studies, Editing, and Performance*, edited by Claire M. L. Bourne, 29–49. London: The Arden Shakespeare, 2021.

Anderson, Randall L. 'Metaphors of the Book as Garden in the English Renaissance'. *The Yearbook of English Studies* (2003): 248–261.

Appadurai, Arjun, ed. *The Social Life of Things: Commodities in Cultural Perspective.* Cambridge: Cambridge University Press, 1986.

Aston, Margaret. 'English Ruins and English History: The Dissolution and the Sense of the Past'. *Journal of the Warburg and Courtauld Institutes* 36 (1973): 231–255.

Aston, Margaret. *Lollards and Reformers: Images and Literacy in Late Medieval England.* London: Hambledon Press, 1984.

Aston, Margaret. *England's Iconoclasts: Laws Against Images*, vol. 1. Oxford: Clarendon Press, 1988.

Aston, Margaret. 'Public Worship and Iconoclasm'. In *The Archaeology of Reformation, 1480–1580*, edited by David Gaimster and Roberta Gilchrist, 9–28. Leeds: Maney, 2003.

Avrin, Leila. *Scribes, Scripts, and Books: The Book Arts from Antiquity to the Renaissance.* London: The British Library, 1991.

Baker, David J. *On Demand: Writing for the Market in Early Modern England.* Stanford: Stanford University Press, 2010.

Baker, Herschel C. *The Wars of Truth: Studies in the Decay of Christian Humanism in the Earlier Seventeenth Century.* Eugene: Wipf and Stock, 2006.

Barrett, Tim. 'Paper through Time: Non-Destructive Analysis of 14th through 19th Century Papers'. University of Iowa, 2012. Last accessed August 19, 2020. http://paper.lib.uiowa.edu/european.php#sizing.

BBC News. 'M6 Toll built with pulped fiction'. BBC News, December 18, 2003. Last accessed August 13, 2020. http://news.bbc.co.uk/1/hi/england/west_midlands/3330245.stm.

Beal, Peter. *A Dictionary of Manuscript Terminology, 1450–2000*. Oxford: Oxford University Press, 2011.
Bennett, Jane. *Vibrant Matter: A Political Ecology of Things*. Durham: Duke University Press, 2010.
Bennett, Kate. 'John Aubrey and the Printed Book'. *Huntington Library Quarterly* 76, no. 3 (2013): 393–411.
Bergeron, David M. *Textual Patronage in English Drama, 1570–1640*. Ashgate: Aldershot, 2006.
Bidwell, John. 'French Paper in English Books'. In *Cambridge History of the Book in Britain*, vol. 4, edited by John Barnard and D. F. McKenzie, 538–601. Cambridge: Cambridge University Press, 2002.
Blagden, Cyprian. 'The English Stock of the Stationers' Company: An Account of Its Origins'. *The Library*, series 5, 10:3 (1955): 163–185.
Blagden, Cyprian. 'The Distribution of Almanacks in the Second Half of the Seventeenth Century'. *Studies in Bibliography* 11 (1958): 107–116.
Blagden, Cyprian. *The Stationers' Company: A History, 1403–1959*. London: George Allen and Unwin, 1960.
Bland, Mark. *A Guide to Early Printed Books and Manuscripts*. Chichester: Wiley-Blackwell, 2010.
Boehm, Katharina. 'The Economies of Intellectual Labor in Thomas Middleton's City Comedies'. *Studies in English Literature, 1500–1900* 56, no. 2 (2016): 351–372.
Borlik, Todd A. *Ecocriticism and Early Modern English Literature*. Abingdon: Routledge, 2011.
Bosanquet, Eustace. *English Printed Almanacks and Prognostications, A Bibliographical History to the year 1600*. London: The Bibliographical Society at the Chiswick Press, 1917.
Bosanquet, Eustace. 'English Seventeenth-Century Almanacks'. *Library* 4, no. 10 (1930): 361–397.
Boscagli, Maurizia. *Stuff Theory: Everyday Objects, Radical Materialism*. London: Bloomsbury, 2014.
Bradshaw, Henry. *List of the Founts of Type and Woodcut Devices Used by Printers in Holland in the Fifteenth Century*. London: Macmillan, 1871.
Brayman Hackel, Heidi. *Reading Material in Early Modern England: Print, Gender, and Literacy*. Cambridge: Cambridge University Press, 2005.
Brooks, Douglas A., ed. *Printing and Parenting in Early Modern England*. Aldershot: Ashgate, 2005.
Brown, Bill. 'Thing Theory'. *Critical Inquiry* 28, no. 1 (2001): 1–22.
Brown, Georgia. *Redefining Elizabethan Literature*. Cambridge: Cambridge University Press, 2004.
Brown, Matthew P. *The Pilgrim and the Bee: Reading Rituals and Book Culture in Early New England*. Philadelphia: University of Pennsylvania Press, 2007.
Brown, Michelle P. 'A New Fragment of a Ninth-Century English Bible'. *Anglo-Saxon England* 18 (1989): 33–43.
Brown, Michelle P. 'Sir Robert Cotton, Collector and Connoisseur?' In *Illuminating the Book: Makers and Interpreters, Essays in Honour of Jane Backhouse*, edited by Michelle P. Brown and Scot McKendrick, 291–298. London: The British Library, 1998.
Butterworth, Emily. 'Apothecaries' Cornets: Books as Waste Paper in the Renaissance'. *MLN* 133, no. 4 (2018): 891–913.

Calhoun, Joshua. *The Nature of the Page: Poetry, Papermaking, and the Ecology of Texts in Renaissance England.* Philadelphia: Pennsylvania University Press, 2020.
Capp, Bernard. *Astrology and the Popular Press: English Almanacs, 1500–1800.* London: Faber and Faber, 1979.
Capp, Bernard. 'The Political Dimension of Apocalyptic Thought'. In *The Apocalypse in English Renaissance Thought and Literature: Patterns, Antecedents, and Repercussions*, edited by C. A. Patrides and Joseph Wittreich, 93–124. Manchester: Manchester University Press, 1984.
Carey, John. *John Donne: Life, Mind and Art.* London: Faber and Faber, 1981; rep. 1983.
Carley, James P. and Colin G. C. Tite. 'Sir Robert Cotton as Collector of Manuscripts and the Question of Dismemberment: British Library MSS Royal 13 D.I and Cotton Otho D. VIII'. *The Library* 6, no. 2 (1992): 94–99.
Carley, James P. 'The Royal Library under Henry VIII'. In *The Cambridge History of the Book in Britain*, vol. 3, edited by Lotte Hellinga and J. B. Trapp, 274–282. Cambridge: Cambridge University Press, 1999.
Carley, James P. 'Monastic collections and their dispersal'. In *The Cambridge History of the Book in Britain*, vol. 4, edited by John Barnard and D. F. McKenzie, 339–347. Cambridge: Cambridge University Press, 2002.
Chandler, John. *John Leland's Itinerary: Travels in Tudor England.* Stroud: Sutton Publishing, 1993.
Chapman, Alison A. 'Almanacs'. In *The Oxford Encyclopedia of British Literature*, vol. 1, edited by David Scott Kastan, 26–30. Oxford: Oxford University Press, 2006.
Chapman, Alison A. 'Marking Time: Astrology, Almanacs, and English Protestantism'. *Renaissance Quarterly* 60, no. 4 (2007): 1–25.
Clegg, Cyndia Susan. 'The Authority and Subversiveness of Print in Early-Modern Europe'. In *The Cambridge Companion to the History of the Book*, edited by Leslie Howsam, 125–142. Cambridge: Cambridge University Press, 2015.
Cockayne, Emily. *Rummage: A History of the Things We Have Reused, Recycled, and Refused to Let Go.* London: Profile Books, 2020.
Cohen, Jeffrey Jerome. *Stone: An Ecology of the Inhuman.* Minneapolis: University of Minnesota Press, 2015.
Coleman, D.C. *The British Paper Industry, 1495–1860.* Oxford: Clarendon Press, 1958.
Corcoran, Lorelei H. and Marie Svoboda, *Herakleides: A Portrait Mummy from Roman Egypt.* Los Angeles: Getty Publications, 2010.
Coughlan, Sean. '"Incredibly rare" William Caxton print discovered'. *BBC*, May 9, 2017. Last accessed May 9, 2017. https://www.bbc.co.uk/news/education-39846929.
Craig, Heidi. 'English Rag-Women and Early Modern Paper Production'. In *Women's Labour and the History of the Book in Early Modern England*, edited by Valerie Wayne, 29–46. London: Bloomsbury, 2020.
Craik, Katharine A. 'John Taylor's Pot-Poetry'. *The Seventeenth Century* 20, no. 2 (2005): 185–203.
Cressy, David. 'Books as Totems in Seventeenth-Century England and New England'. *The Journal of Library History* 21, no. 1 (1986): 92–106.
Crewe, Jonathan V. *Unredeemed Rhetoric: Thomas Nashe and the Scandal of Authorship.* Baltimore: John Hopkins University Press, 1982.
Crick, Julia. 'An Anglo-Saxon fragment of Justinus' Epitome'. *Anglo-Saxon England* 16 (1987): 181–196.
Crowther, Kathleen. 'The Scientific Revolution'. In *The Oxford Handbook of Early Modern European History, 1350–1750*, vol. 2, *Cultures & Power*, edited by Hamish Scott, 56–80. Oxford: Oxford University Press, 2015.

Curry, Patrick. *Prophecy and Power: Astrology in Early Modern England*. Oxford: Polity Press, 1989.

Dane, Joseph. *The Myth of Print Culture: Essays on Evidence, Textuality, and Bibliographical Method*. Toronto: Toronto University Press, 2003.

Da Rold, Orietta. *From Pulp to Fictions: Paper in Medieval England*. Cambridge: Cambridge University Press, 2020.

Day, Geoffrey and Amélie Junqua. 'Á la recherche du papier perdu—Some Possible Circulations of Waste Paper in Eighteenth-Century England'. In *Intermediality and the Circulation of Knowledge in the Eighteenth Century*, edited by Peter Wagner and Frédéric Ogée, 203–218. Trier: Wissenschaftlicher Verlag Trier, 2019.

Daybell, James. *The Material Letter in Early Modern England: Manuscript Letters and the Culture and Practices of Letter-Writing, 1512–1635*. Basingstoke: Palgrave Macmillan, 2012.

Diringer, David. *The Book before Printing: Ancient, Medieval and Oriental*. New York: Dover, 1982.

Dolan, Frances E. *Digging the Past: How and Why to Imagine Seventeenth-Century Agriculture*. Philadelphia: University of Pennsylvania Press, 2020.

Dolan, Frances E. 'Hester Pulter's Dunghill Poetics'. *Journal for Early Modern Cultural Studies* 20, no. 2 (2020): 16–42.

Donaldson, Ian. 'The Destruction of the Book'. *Book History* 1, no. 1 (1998): 1–10.

Duncan-Jones, Katherine. 'Christs Teares, Nashe's "Forsaken Extremities"'. *The Review of English Studies* 49, no. 194 (1998): 167–180.

Dunke, William and Falk Eisermann. 'Fortschritt oder Fidibus? Zur Bestimmung, Bewahrung und Bedeutung von Inkunabelfragmenten'. In *Fragment und Makulatur: Überlieferungsstörungen und Forschungsbedarf bei Kulturgut in Archiven und Bibliotheken*, edited by by Hanns Peter Neuheuser and Wolfgang Schmitz, 281–308. Wiesbaden: [Harrassowitz], 2015.

Dupont, Florence. 'The Corrupted Boy and the Crowned Poet: Or, the Material Reality and the Symbolic Status of the Literary Book at Rome', translated by Holt N. Parker. In *Ancient Literacies: The Culture of Reading in Greece and Rome*, edited by William A. Johnson and Holt N. Parker, 143–163. Oxford: Oxford University Press, 2009.

Eamon, William. 'Astrology and Society'. In *A Companion to Astrology in the Renaissance*, edited by Brendan Dooley, 141–192. Leiden: Brill, 2013.

Eisenstein, Elizabeth. *The Printing Press as an Agent of Change: Communications and Cultural Transformations in Early Modern Europe*, 2 vols. Cambridge: Cambridge University Press, 1979.

Ettenhuber, Katrin. *Donne's Augustine: Renaissance Cultures of Interpretation*. Oxford: Oxford University Press, 2011.

Fairfield, Lesley P. *John Bale: Mythmaker for the English Reformation*. West Eugene: Wipf and Stock Publishers, 1976.

Feeney, Dennis. 'Representation and the Materiality of the Book in the Polymetrics'. In *Catullus: Poems, Books, Readers*, edited by Ian Du Quesnay and Tony Woodman, 29–47. Cambridge: Cambridge University Press, 2012.

Findlay, Ronald. 'The Emergence of the World Economy'. In *Contemporary Economic Issues: Proceedings of the Eleventh World Congress of the International Economic Association, Tunis*, vol. 3, edited by Daniel Cohen, 82–123. Basingstoke: Macmillan Press, 1998.

Fleming, Juliet. *Graffiti and the Writing Arts of Early Modern England*. London: Reaktion Books, 2001.

Fleming, Juliet. 'Damask Papers'. In *The Elizabethan Top Ten: Defining Print Popularity in Early Modern England*, edited by Andy Kesson and Emma Smith, 179–191. Farnham: Ashgate, 2013.

Fleming, Juliet. *Cultural Graphology: Writing After Derrida*. Chicago: University of Chicago Press, 2016.

Floyd-Wilson, Mary and Garret A. Sullivan, eds. *Geographies of Embodiment in Early Modern England*. Oxford: Oxford University Press, 2020.

Folger Library. 'Private Libraries in Renaissance England'. Last accessed August 19, 2020. https://plre.folger.edu/books.php.

Folkerth, Wes. 'Pietro Aretino, Thomas Nashe, and Early Modern Rhetorics of Public Address'. In *Making Publics in Early Modern Europe: People, Things, Forms of Knowledge*, edited by Bronwen Wilson and Paul Yachnin, 68–79. New York: Routledge, 2010.

Fowler, Caroline. *The Art of Paper: From the Holy Land to the Americas*. New Haven: Yale University Press, 2019.

Franklin-Wallis, Oliver. 'Plastic recycling is a myth: What really happens to your rubbish'. *The Guardian*, August 17, 2019. Last accessed March 5, 2022. https://www.theguardian.com/environment/2019/aug/17/plastic-recycling-myth-what-really-happens-your-rubbish.

Fritze, Ronald Harold. '"Truth Hath Lacked Witnesse, Tyme Wanted Light": The Dispersal of the English Monastic Libraries and Protestant Efforts at Preservation, ca. 1535–1625'. *The Journal of Library History* 18, no. 3 (1983): 274–291.

Frow, John. 'Invidious Distinction: Waste, Difference, and Classy Stuff'. In Hawkins and Muecke, *Culture and Waste*, 25–38.

Gaisser, Julia Haig. *Catullus*. Oxford: Wiley-Blackwell, 2009.

Galinsky, Karl. *Augustan Culture: An Interpretive Introduction*. Princeton: Princeton University Press, 1996.

Galitz, Kathryn Calley. *The Metropolitan Museum of Art: Masterpiece Paintings*. New York: Skira Rizzoli, 2016.

Garrington, Abbie. *Haptic Modernism: Touch and the Tactile in Modernist Writing*. Edinburgh: Edinburgh University Press, 2013.

Gee, Sophie. *Making Waste: Leftovers and the Eighteenth Century Imagination*. Princeton: Princeton University Press, 2010.

Geneva, Ann. *Astrology and the Seventeenth-Century Mind: William Lilly and the Language of the Stars*. Manchester: Manchester University Press, 1995.

George, David. 'Weather-Wise's Almanac and the Date of Middleton's No Wit No Help Like a Woman's'. *Notes and Queries* 13, no. 8 (1966): 297–301.

Geyssen, John. 'Sending a book to the Palatine: Martial 1.70 and Ovid'. *Mnemosyne* 52, no. 6 (1999): 718–738.

Ghosh, Pallab. 'Scan technique reveals secret writing in mummy cases'. *BBC News*, December 31, 2017. Last accessed August 18, 2020. https://www.bbc.co.uk/news/science-environment-42357259.

Gillespie, Vincent. *Syon Abbey: Corpus of British Medieval Library Catalogues*, vol. 9. London: The British Library, 2001.

Gilman, Ernest B. *Plague Writing in Early Modern England*. Chicago: University of Chicago Press, 2009.

Graham, Timothy. 'Matthew Parker and His Manuscripts: A Study of an Elizabethan Library and Its Use'. In *The Cambridge History of Libraries in Britain and Ireland*, vol. 1, edited by Elisabeth Leedham-Green and Teresa Weber, 332–342. Cambridge: Cambridge University Press, 2006.

Grafton, Anthony, Glenn W. Most, and Salvatore Settis, eds. *The Classical Tradition*. Cambridge: The Belknap Press, 2010.

Gratwick, A. S. 'Vale, patrona virgo: The Text of Catullus 1.9'. *Classical Quarterly* 52, no. 1 (2002): 305–320.

Greenberg, Stephen. 'Plague, the Printing Press, and Public Health in Seventeenth-Century London'. *Huntington Library Quarterly* 67, no. 4 (2004): 508–527.

Greg. W.W. and E. Boswell, eds. *Records of the Court of the Stationers' Company, 1576–1602, from Register B*. London: The Bibliographical Society, 1930.

Grier, Miles P. 'Inkface: The Slave Stigma in the Early Imperial Imagination'. In *Scripturalizing the Human: The Written as the Political*, edited by Vincent L. Wimbush, 193–220. Abingdon: Routledge, 2015.

Guibbory, Achsah. *Returning to John Donne*. Farnham: Ashgate, 2015.

Hadfield, Andrew. *Literature, Travel, and Colonial Writing in the English Renaissance, 1545–1625*. Oxford: Oxford University Press, 1998.

Halasz, Alexandra. *The Marketplace of Print: Pamphlets and the Public Sphere in Early Modern England*. Cambridge: Cambridge University Press, 1997.

Halperin, David M. 'Out of Australia'. In Hawkins and Muecke, *Culture and Waste*, 1–7.

Hamel, Christopher de. *Syon Abbey: The Library of the Bridgettine Nuns and their Peregrinations after the Reformation*. London: Roxburghe Club, 1991.

Hamilton, A. C., ed. *The Spenser Encyclopedia*. Toronto: University of Toronto Press, 2003.

Hamling, Tara. *Decorating the 'Godly' Household: Religious Art in Post-Reformation Britain*. New Haven: Yale University Press, 2010.

Hanabusa, Chiaki. 'Notes on the Second Edition of Thomas Nashe's The Unfortunate Traveller'. *Notes and Queries* 56 (2009): 556–559.

Hand, Meredith Molly. '"More Lies Than True Tales": Scepticism in Middleton's Mock-Almanacs'. In the *Oxford Handbook of Thomas Middleton*, edited by Gary Taylor and Trish Thomas Henley, 312–328. Oxford: Oxford University Press, 2012.

Hanß, Stephan. 'Zooming into History: An interview with Cristina Balloffet Carr on Examining Early Modern Textiles Under the Microscope'. *Materialized Identities* (blog). July 4, 2017. Last accessed August 19, 2020. http://www.materializedidentities.com/single-post/2017/07/04/Zooming-into-History-An-Interview-with-Cristina-Balloffet-Carr-on-Microscoping-Early-Modern-Textiles.

Harman, Graham. *Tool-Being: Heidegger and the Metaphysics of Objects*. Chicago: Open Court, 2011.

Harris, Jonathan Gil. 'The New New Historicism's Wunderkammer of Objects'. *European Journal of English Studies* 4, no. 2 (2000): 111–123.

Harris, Jonathan Gil. *Untimely Matter in the Time of Shakespeare*. Philadelphia: University of Pennsylvania Press, 2009.

Harris, Jonathan Gil and Anna Neill. 'Hollywood's Pacific Junk: The Wreckage of Colonial History in Six Days and Seven Night and Rapa Nui'. In Hawkins and Muecke, *Culture and Waste*, 85–101.

Harris, Victor. *All Coherence Gone: A Study of the Seventeenth Century Controversy Over Disorder and Decay in the Universe*. London: Frank Cass & Co. Ltd., 1966.

Harvey, Barbara. *The Obedientiaries of Westminster Abbey and their Financial Records, c. 1275–1540*. Woodbridge: Boydell, Press, 2002.

Harvey, David J. *The Law Emprynted and Englysshed: The Printing Press as an Agent of Change in Law and Legal Culture 1475–1642*. London: Bloomsbury, 2015.

Hawkins, Gay and Stephen Muecke, eds. *Culture and Waste: The Creation and Destruction of Value*. Lanham: Rowman and Littlefield, 2003.

Head, Randall. *Making Archives in Early Modern Europe: Proof, Information, and Political Record-Keeping, 1400–1700*. Cambridge: Cambridge University Press, 2019.

Healy, Margaret. 'Alchemy, Magic, and the Sciences'. In *Ben Jonson in Context*, edited by Julie Sanders, 322–329. Cambridge: Cambridge University Press, 2010.

Heidegger, *Being and Time*, translated by Joan Stambaugh and revised by Dennis J. Schmidt. New York: State University of New York Press, 2010.

Helgerson, Richard. *The Elizabethan Prodigals*. Oakland: University of California Press, 1976.

Hellinga, Lotte. *Texts in Transit: Manuscript to Proof and Print in the Fifteenth Century*. Leiden: Brill, 2014.

Hellinga, Lotte. *Incunabula in Transit: People and Trade*. Leiden: Brill, 2018.

Hibbard, G. R. *Thomas Nashe: A Critical Introduction*. London: Routledge & Kegan Paul, 1962.

Higgins, Ben. '"The Book-sellars Shop": Browsing, Reading, and Buying in Early Modern England'. In *The Oxford Handbook to the History of the Early Modern Book*, edited by Adam Smyth. Oxford: Oxford University Press, forthcoming.

Hills, Richard Leslie. *Papermaking in Britain, 1488–1988: A Short History*. London: Bloomsbury, 1988.

Hirsch, David A. Hedrich. 'Donne's Atomies and Anatomies: Deconstructed Bodies and the Resurrection of Atomic Theory'. *Studies in English Literature, 1500–1900* 31, no. 1 (1991): 69–94.

Hui, Andrew. *The Poetics of Ruins in Renaissance Literature*. New York: Fordham University Press, 2016.

Hunt, Katherine. *The Brazen World*, forthcoming.

Hunter, Dard. *Papermaking: The History and Technique of an Ancient Craft*. New York: Dover Publications, 1978.

Hunter, Matthew C. 'Picture, Object, Puzzle, Prompter: Devilish Cleverness in Restoration London'. *Art History* 36, no. 3 (2013): 546–567.

Hutson, Lorna. *Nashe in Context*. Oxford: Clarendon Press, 1989.

Jackson, William A., ed. *Records of the Court of the Stationers' Company, 1602–1640*. London: The Bibliographical Society, 1957.

Jamie, Kathleen. *Findings*. London: Sort of Books, 2005.

Jenner, Mark. 'Plague on a Page: Lord Have Mercy upon Us in Early Modern London'. *The Seventeenth Century* 27, no. 3 (2012): 255–286.

Jensen, Phebe. *Astrology, Almanacs, and the Early Modern English Calendar*. Abingdon: Routledge, 2021.

Johns, Adrian. *The Nature of the Book: Print and Knowledge in the Making*. Chicago: University of Chicago Press, 1998.

Johnston, Warren. *Revelation Restored: The Apocalypse in Later Seventeenth-Century England*. Woodbridge: The Boydell Press, 2011.

Jotischky, Andrew. *The Perfection of Solitude: Hermits and Monks in the Crusaders States*. University Park: University of Pennsylvania Press, 1995.

Jowett, John. 'Middleton's *No Wit* at the Fortune'. *Renaissance Drama* 22 (1991): 191–208.

Kadue, Katie. *Domestic Georgic: Labors of Preservation from Rabelais to Milton*. Chicago: The University of Chicago Press, 2021.

Kearney, James. *The Incarnate Text: Imagining the Book in Reformation England*. Philadelphia: University of Pennsylvania Press, 2009.

Ker, N. R. *Medieval Libraries of Great Britain: A List of Surviving Books*. London: Royal Historical Society, 1964.

Ker, N. R. *Fragments of Medieval Manuscripts Used as Pastedowns in Oxford Bindings: with a Survey of Oxford Binding, c.1515–1620*. Oxford: The Oxford Bibliographical Society, 1964, rep. 2004.

Kiessling, Nicolas K. *The Library of Anthony Wood*. Oxford: Oxford Bibliographical Society, 2002.

King, John N. *English Reformation Literature: The Tudor Origins of the Protestant Tradition*. Princeton: Princeton University Press, 1982.

Knight, Jeffrey Todd. *Bound to Read: Compilations, Collections, and the Making of Renaissance Literature*. Philadelphia: University of Pennsylvania Press, 2013.

Knight, Leah. *Of Books and Botany in Early Modern England: Sixteenth Century Plants and Print Culture*. Farnham: Ashgate, 2009.

Knowles, David. *Bare Ruined Choirs: The Dissolution of the English Monasteries*. Cambridge: Cambridge University Press, 1976.

Kurlansky, Mark. *Paper: Paging through History*. New York: W. W. Norton & Company, 2016.

Kwakkel, Eric. 'Destroying Medieval Books—and Why That's Useful'. *Medieval Books* (blog). October 31, 2014. Last accessed August 19, 2020. https://medievalbooks.nl/2014/10/31/destroying-medieval-books-and-why-thats-useful/.

Kwakkel, Eric. 'X-Rays Expose a Hidden Medieval Library'. *Medieval Books* (blog). December 18, 2015. Last accessed August 19, 2020. https://medievalbooks.nl/category/fragments/.

Lakoff, George and Mark Jonson. *Metaphors We Live By*. Chicago: University of Chicago Press, 1980.

Lander, Jesse M. 'Martin Marprelate and the Fugitive Text'. *Reformation* 7 (2002): 135–185.

Landon, Richard. 'Collecting and the Antiquarian Book Trade'. In *The Cambridge History of the Book in Britain*, vol. 5, edited by Michael F. Suarez and Michael L. Turner, 711–722. Cambridge: Cambridge University Press, 2009.

Lapidge, Michael. 'Frithegod of Canterbury'. *Anglo-Saxon England* 17 (1988): 48–49.

Leedham-Green, E. S. *Books in Cambridge Inventories: Book-lists from Vice-Chancellor's Court Probate Inventories in the Tudor and Stuart Periods*, vol. 2. Cambridge: Cambridge University Press, 1986.

Leong, Elaine. 'Papering the Household'. In *Working With Paper: Gendered Practices in the History of Knowledge*, edited by Carla Bittell, Elaine Leong, Christine Von Oetzen, 32–45. Pittsburgh: University of Pittsburgh Press, 2019.

Levy-Navarro, Elena. *The Culture of Obesity in Early and Late Modernity: Body Image in Shakespeare, Jonson, Middleton, and Skelton*. Basingstoke: Palgrave Macmillan, 2008.

Lewis, C. S. *English Literature in the Sixteenth Century, Excluding Drama*. Oxford: The Clarendon Press, 1954.

Ligatus. 'Language of Bindings Thesaurus'. https://www.ligatus.org.uk/lob/.

Llewellyn, Nigel. *The Art of Death: Visual Culture in the English Death Ritual, c.1500–c.1800*. London: The Victoria and Albert Museum, 1997.

Loughlin, Marie H., ed. *Same-Sex Desire in Early Modern England: An Anthology of Literary Texts and Contexts*. Manchester: Manchester University Press, 2014.

Lowrie, Michele. *Writing, Performance, and Authority in Augustan Rome*. Oxford: Oxford University Press, 2010.

Lyon, Harriet. *Memory and the Dissolution of the Monasteries in Early Modern England*. Cambridge: Cambridge University Press, 2021.

Lytle, Guy Fitch and Stephen Orgel, eds. *Patronage in the Renaissance*. Princeton: Princeton University Press, 1981.

Mare, A. C. de la and Stanley Gillam. *Duke Humfrey's Library and the Divinity School, 1488–1988*. Oxford: Bodleian Library, 1988.

Marks, Laura. *Touch: Sensuous Theory and Multisensory Media*. Minneapolis: University of Minnesota Press, 2002.

Milner, Matthew. *The Senses and the English Reformation*. Farnham: Ashgate, 2011.

McCusker, Honor. 'Books and Manuscripts Formerly in the Possession of John Bale'. *The Library* 16, no. 2 (1935): 144–165.

McDowell, Paula. 'Of Grubs and Other Insects: Constructing the Categories of "Ephemera" and "Literature" in Eighteenth-Century British Writing'. *Book History* 15 (2012): 48–70.

McGillivray, Barbara and Gábor Mihály Tóth. *Applying Language Technology in Humanities Research: Design, Application and the Underlying Logic*. Basingstoke: Palgrave Macmillan, 2020.

McKenzie, D. F. and Maureen Bell, eds. *A Chronology and Calendar of Documents Relating to the London Book Trade, 1641–1700*, vol. 3. Oxford: Oxford University Press, 2005.

McKitterick, David. *A History of Cambridge University Press*, vol. 1, *Printing and the Book Trade in Cambridge, 1534–1698*. Cambridge: Cambridge University Press, 1992.

Mentz, Steve. 'Day Labor: Thomas Nashe and the Practice of Prose in Early Modern England'. In *Early Modern Prose Fiction: The Cultural Politics of Reading*, edited by Naomi Liebler, 18–32. London: Routledge, 2004.

Miller, Daniel. *Stuff*. Cambridge: Polity Press, 2009.

Monro, Alexander. *The Paper Trail: An Unexpected History of a Revolutionary Invention*. New York: Penguin Random House, 2014.

Morrison, Susan Signe. *The Literature of Waste: Material Ecopoetics and Ethical Matter*. New York: Palgrave Macmillan, 2015.

Mukherjee, Ayeesha. *Penury into Plenty: Dearth and the Making of Knowledge in Early Modern England*. London: Routledge, 2015.

Müller, Lothar. *White Magic: The Age of Paper*, translated by Jessica Spengler. Cambridge: Polity Press, 2014.

Nagel, Alexander and Christopher S. Wood. *Anachronic Renaissance*. New York: Zone Books, 2010.

Nicholl, Charles. *A Cup of News: The Life of Thomas Nashe*. London: Routledge, 1984.

Nixon, Howard M. *Catalogue of the Pepys Library at Magdalene College, Cambridge, vol. 6: Bindings*. Cambridge: Brewer, 1984.

Noble, Louise. *Medical Cannibalism in Early Modern English Literature and Culture*. Basingstoke: Palgrave Macmillan, 2011.

Noblett, William. 'Cheese, Stolen Paper, and the London Book Trade, 1750–99'. *Eighteenth-Century Life* 38, no. 3 (2014): 100–110.

O'Neill, Patrick P. 'Another Fragment of the Metrical Psalms in the Eadwine Psalter'. *Notes and Queries* 233 (1988): 434–436.

Palma, Vittoria di. *Wasteland: A History*. New Haven: Yale University Press, 2014.

Parry, Graham. *The Trophies of Time: English Antiquarians of the Seventeenth Century*. Oxford: Oxford University Press, 2007.

Paster, Gail Kern. *Humoring the Body: Emotions and the Shakespearean Stage*. Chicago: University of Chicago Press, 2004.

Pearson, David. *Oxford Bookbinding 1500–1640. Including A Supplement to Neil Ker's Pastedowns in Oxford Bindings*. Oxford: Oxford Bibliographical Society, 2000.

Pearson, David. *English Bookbinding Styles 1450–1800*. London: The British Library, 2005.

Pender, Patricia. *Early Modern Women's Writing and the Rhetoric of Modesty*. Basingstoke: Palgrave Macmillan, 2012.

Perkins, Maureen. *Visions of the Future: Almanacs, Time, and Cultural Change, 1775–1870*. Oxford: Clarendon Press, 1996.

Peters, Kate, Alexandra Walsham, and Liesbeth Corens, eds. *Archives and Information in the Early Modern World*. Oxford: Oxford University Press, 2018.

Pickwoad, Nicholas. 'Onward and Downward: How Binders Coped with the Printing Press before 1800'. In *A Millenium of the Book: Production, Design and Illustration in Manuscript and Print, 900–1900*, edited by Robin Myers and Michael Harris, 61–106. Winchester: St Paul's Bibliographies, 1994.

Pickwoad, Nicholas. 'The Use of Fragments of Medieval Manuscripts in the Construction and Covering of Bindings on Printed Books'. In *Interpreting and Collecting Fragments of Medieval Books*, edited by Linda L. Brownrigg and Margaret M. Smith, 1–20. London: The Red Gull Press, 2000.

Pigman III, George Wood. 'Versions of Imitation in the Renaissance'. *Renaissance Quarterly* 33, no. 1 (1980): 1–32.

Pollard, Tanya. 'Spelling the Body'. In *Environment and Embodiment in Early Modern England*, edited by Mary Floyd-Wilson and Garret A., Sullivan Jr., 171–186. Basingstoke: Palgrave Macmillan, 2007.

Poole, Kristen and Owen Williams, eds. *Early Modern Histories of Time: The Periodizations of Sixteenth- and Seventeenth-Century England*. Philadelphia: University of Pennsylvania Press, 2019.

Prendergast, Maria Teresa Micaela, *Railing, Reviling, and Invective in English Literary Culture, 1588–1617: The Anti-Poetics of Theater and Print*. New York: Routledge, 2012.

Prescott, Anne Lake. *Imagining Rabelais in Renaissance England*. New Haven: Yale University Press, 1998.

Price, Leah. *How to Do Things with Books in Victorian Britain*. Princeton: Princeton University Press, 2012.

Raitt, Suzanne. 'Psychic Waste: Freud, Fechner, and the Principle of Constancy'. In Hawkins and Muecke, *Culture and Waste*, 73–84.

Ramsay, Nigel. '"The Manuscripts flew about like Butterflies": The Break-Up of English Libraries in the Sixteenth Century'. In *Lost Libraries: The Destruction of Great Book Collections since Antiquity*, edited by James Raven, 125–144. Basingstoke: Palgrave Macmillan, 2004.

Raymond, Joad. *Pamphlets and Pamphleteering in Early Modern Britain*. Cambridge: Cambridge University Press, 2003.

Razzall, Lucy. *Boxes and Books in Early Modern England: Materiality, Metaphor, Containment*. Cambridge: Cambridge University Press, 2021.

Reckford, K. J. 'Studies in Persius'. In *Persius and Juvenal*, in *Persius and Juvenal*, edited by Maria Plaza, 51–56. Oxford: Oxford University Press, 2009.

Reid, Lindsay Ann. *Ovidian Bibliofictions and the Tudor Book: Metamorphosing Classical Heroines in Late Medieval and Renaissance England*. Farnham: Ashgate, 2014.

Reid, Pauline. *Reading by Design: The Visual Interfaces of the English Renaissance Book*. Toronto: University of Toronto Press, 2019.

Reynolds, Anna. '"Worthy to be Reserved": Bookbindings and the Waste Paper Trade in Early Modern England and Scotland'. In *The Paper Trade in Early Modern Europe: Practices, Materials, Networks*, edited by Daniel Bellingradt and Anna Reynolds, 342–368. Leiden: Brill, 2021.

Reynolds, Anna. 'What Do Texts and Insects Have in Common?; or Ephemerality before Ephemera'. In *Practices of Ephemera in Early Modern England*, edited by

Callan Davies, Hannah Lilley, and Catherine Richardson, 25–43. Abingdon: Routledge, 2023.

Reynolds, Anna. 'Waste, Offcuts, Remains, Reuse: The History of the Book in Pieces'. In *The Oxford Handbook to the History of the Early Modern Book*, edited by Adam Smyth, 365–384. Oxford: Oxford University Press, 2023.

Rhodes, Neil. *Elizabethan Grotesque*. London: Routledge, 1980.

Rhodes, Neil. 'On Speech, Print, and New Media: Thomas Nashe and Marshall McLuhan'. *Oral Tradition* 24, no. 2 (2009). doi:10.1353/ort.0.0036.

Rhodes, Neil. 'Time'. In *The Ashgate Research Companion to Popular Culture in Early Modern England*, edited by Andrew Hadfield, Matthew Dimmock, and Abigail Shinn, 283–294. Farnham: Ashgate, 2014.

Richards, Jennifer. *Voices and Books in the English Renaissance: A New History of Reading*. Oxford: Oxford University Press, 2019.

Roemer, Patricia. 'The Papyrus Roll in Egypt, Greece, and Rome'. In *A Companion to the History of the Book*, edited by Simon Eliot and Jonathan Rose, 84–94. Oxford: Wiley-Blackwell, 2009.

Roman, Luke. 'The Representation of Literary Materiality in Martial's Epigrams'. *Journal of Roman Studies* 91 (2001): 113–145.

Rosenberg, Jessica. 'A Digression to Hospitality: Thrift and Christmastime in Shakespeare and the Literature of Husbandry'. In *Shakespeare and Hospitality: Ethics, Politics, and Exchange*, edited by David B. Goldstein and Julia Reinhard Lupton, 39–66. London: Routledge, 2016.

Ross, Trevor. *The Making of the English Literary Canon: From the Middle Ages to the Late Eighteenth Century*. Montreal: McGill-Queen's University Press, 1998.

Rudy, Kathryn M. 'Dirty Books: Quantifying Patterns of Use in Medieval Manuscripts Using a Densitometer'. *Journal of Historians of Netherlandish Art* 2, nos. 1–2 (2010). http://www.jhna.org/index.php/past-issues/volume-2-issue-1-2/129-dirty-books.

Rundle, David. 'Habits of Manuscript Collecting: The Dispersals of the Library of Humfrey, Duke of Gloucester'. In *Lost Libraries: The Destruction of Great Book Collections Since Antiquity*, edited by James Raven, 106–124. Basingstoke: Palgrave Macmillan, 2004.

Russell, Gillian. *The Ephemeral Eighteenth Century: Print, Sociability, and the Cultures of Collecting*. Cambridge: Cambridge University Press, 2020.

Rutkin, Daniel H. 'Astrology'. In *The Cambridge History of Science*, edited by Katharine Park and Lorraine Daston, 548–552. Cambridge: Cambridge University Press, 2006.

Ruvoldt, Maria. *The Italian Renaissance Imagery of Inspiration: Metaphors of Sex, Sleep, and Dream*. Cambridge: Cambridge University Press, 2004.

Rycker, Kate De. 'Commodifying the Author: The Mediation of Aretino's Fame in the Harvey-Nashe Pamphlet War'. *English Literary Renaissance* 49, no. 2 (2019): 145–171.

Rycker, Kate De. 'Mapping Thomas Nashe, from Lowestoft to London'. *The Thomas Nashe Project*. Last accessed August 19, 2020. https://research.ncl.ac.uk/thethomasnasheproject/resources/mappingnashe/

Ryley, Hannah. 'Waste Not, Want Not: The Sustainability of Medieval Manuscripts'. *Green Letters* 19, no. 1 (2015): 63–73.

Ryley, Hannah. *Re-Using Manuscripts in Late Medieval England*. Woodbridge: Boydell and Brewer, 2022.

Saenger, Michael. *The Commodification of Textual Engagements in the English Renaissance*. Aldershot: Ashgate, 2006.

Sansom, Ian. *Paper: An Elegy*. London: Fourth Estate, 2012.

Saunders, Gill. '"Paper Tapestry" and "Wooden Pictures": Printed Decoration in the Domestic Interior before 1700'. In *Printed Images in Early Modern Britain: Essays in Interpretation*, edited by Michael Hunter, 317–336. Farnham: Ashgate, 2010.

Scanlan, John. *On Garbage*. London: Reaktion, 2005.

Schoenfeldt, Michael. *Bodies and Selves in Early Modern England: Physiology and Inwardness in Spenser, Shakespeare, Herbert, and Milton*. Cambridge: Cambridge University Press, 1999.

Schwyzer, Phillip. 'Summer Fruit and Autumn Leaves: Thomas Nashe in 1593'. *ELR* 24, no. 3 (1994): 583–619.

Schwyzer, Phillip. *Literature, Nationalism and Memory in Early Modern England and Wales*. Cambridge: Cambridge University Press, 2004.

Scott-Warren, Jason. 'Nashe's Stuff'. In *The Oxford Handbook of English Prose, 1500–1640*, edited by Andrew Hadfield, 204–218. Oxford: Oxford University Press, 2013.

Shanks, Michael and Christopher Tilley. *Re-Constructing Archaeology: Theory and Practice*. Cambridge: Cambridge University Press, 1992.

Shell, Hannah Rose. *Shoddy: From Devil's Dust to the Renaissance of Rags*. Chicago: University of Chicago Press, 2020.

Sherman, William H. *Used Books: Marking Readers in Renaissance England*. Philadelphia: University of Philadelphia Press, 2008.

Shinn, Abigail. '"Extraordinary discourses of vnnecessarie matter": Spenser's Shepheardes Calendar and the Almanac Tradition'. In *Literature and Popular Culture in Early Modern England*, edited by Matthew Dimmock and Andrew Hadfield, 137–149. Farnham: Ashgate, 2009.

Shorter, Alfred H. *Paper Mills and Paper Makers in England, 1495–1800*. Hilversum: The Paper Publications Society, 1954.

Simpson, James. *Reform and Cultural Revolution*. Oxford: Oxford University Press, 2002.

Smith, Bruce. *Phenomenal Shakespeare*. Oxford: Wiley-Blackwell, 2010.

Smith, Helen. *Grossly Material Things: Women and Book Production in Early Modern England*. Oxford: Oxford University Press, 2012.

Smith, Helen. '"A Unique Instance of Art": The Proliferating Surfaces of Early Modern Paper'. *Journal of the Northern Renaissance* 8 (2017). Last accessed August 19, 2020, http://northernrenaissance.org

Smith, Ian. 'Othello's Black Handkerchief'. *Shakespeare Quarterly* 64, no. 1 (2013): 1–25.

Smyth, Adam. 'Almanacs, Annotators, and Life-Writing in Early Modern England'. *ELR* 38, no. 2 (2008): 200–244.

Smyth, Adam. 'What We Talk about When We Talk about Scissors'. In *Ornamentalism: The Art of Renaissance Accessories*, edited by Bella Mirabella, 293–307. Ann Arbor: University of Michigan Press, 2011.

Smyth, Adam. '"Shreds of holinesse": George Herbert, Little Gidding, and Cutting Up Texts in Early Modern England'. *ELR* 42, no. 3 (2012): 452–481.

Smyth, Adam. 'Almanacs and Ideas of Popularity'. In *The Elizabethan Top Ten: Defining Print Popularity in Early Modern England*, edited by Andy Kesson and Emma Smith, 125–133. Farnham: Ashgate, 2013.

Smyth, Adam. *Material Texts in Early Modern England*. Cambridge: Cambridge University Press, 2018.

Solodow, Joseph B. 'On Catullus 95'. *Classical Philology* 82, no. 2 (1987): 141–145.

Spraggon, Julie. *Puritan Iconoclasm during the English Civil War*. Woodbridge: Boydell Press, 2003.

Spufford, Margaret. *Small Books and Pleasant Histories: Popular Fiction and Its Readership in Seventeenth-Century England*. Cambridge: Cambridge University Press, 1981.

Stallybrass, Peter. 'The Value of Culture and the Disavowal of Things'. In *The Culture of Capital: Property, Cities, and Knowledge in Early Modern England*, edited by Henry S. Turner, 275–292. London: Routledge, 2002.

Stauffer, *Book Traces: Nineteenth-Century Readers and the Future of the Library*. Philadelphia: University of Pennsylvania Press, 2021.

Stenner, Rachel. *The Typographic Imaginary in Early Modern Literature*. London: Routledge, 2019.

Stern, Tiffany. '"On each Wall and Corner Poast": Playbills, Title-Pages, and Advertising in Early Modern London'. *ELR* 36, no. 1 (2006): 57–89.

Stern, Tiffany. *Documents of Performance in Early Modern England*. Cambridge: Cambridge University Press, 2009.

Stocker, David. 'Rubbish Recycled: A Study of the Re-Use of Stone in Lincolnshire'. In *Stone Quarrying and Building in England AD 43–1525*, edited by David Parsons, 83–101. Chichester: Phillimore, 1990.

Sullivan, Erin. 'Physical and Spiritual Illness: Narrative Appropriations of the Bills of Mortality'. In *Representing the Plague in Early Modern England*, edited by Rebeccao Totaro and Ernest B. Gilman, 76–94. London: Routledge, 2011.

Summersgill, Travis L. 'The Influence of the Marprelate Controversy Upon the Style of Thomas Nashe'. *Studies in Philology* 48, no. 2 (1951): 145–160.

Summit, Jennifer. *Memory's Library: Medieval Books in Early Modern England*. Chicago: University of Chicago Press, 2008.

Tarlow, Sarah. 'Reformation and Transformation: What happened to Catholic Things in a Protestant World?' In *The Archaeology of Reformation, 1480–1580*, edited by David Gaimster and Roberta Gilchrist, 108–121. Leeds: Maney, 2003.

Taussig, Michael. 'Miasma'. In Hawkins and Muecke, *Culture and Waste*, 9–23.

Taylor, Gary. 'Thomas Middleton: Lives and Afterlives'. In *Thomas Middleton, The Collected Works*, edited by Taylor and Lavagnino, 25–58.

Taylor, Gary and John Lavagnino (eds). *Thomas Middleton and Early Modern Textual Culture: A Companion to the Collected Works*. Oxford: Oxford University Press, 2007.

Thill, Brian. *Waste*. New York: Bloomsbury, 2015.

Thirsk, Joan. *Economic Policy and Projects: The Development of a Consumer Society in Early Modern England*. Oxford: Clarendon Press, 1978.

Thirsk, Joan. *Food in Early Modern England: Phases, Fads, Fashions 1500–1760*. London: Hambledon Continuum, 2007.

Thomas, Emily (host). 'What happened to newspaper wrapped fish and chips?' *The Food Chain* (podcast). November 1, 2017. Last accessed August 13, 2020. https://www.bbc.co.uk/programmes/p05lq6yq.

Thomas, Keith. *Religion and the Decline of Magic: Studies in Popular Beliefs in Sixteenth and Seventeenth Century England*. London: Weidenfeld and Nicolson, 1971.

Thompson, Michael. *Rubbish Theory: The Creation and Destruction of Value*. London: Pluto Press, 2017.

Tilley, Christopher. *Metaphor and Material Culture*. Oxford: Blackwell, 1999.

Tilley, Heather. 'Waste Matters: Charles Dickens' Our Mutual Friend and Nineteenth-Century Book Recycling'. In *Book Destruction from the Medieval to the Contemporary*, edited by Adam Smyth and Gill Partington, 152–171. Basingstoke: Palgrave Macmillan, 2014.

Upper, Elizabeth. 'Red Frisket Sheets, c. 1490–1700: The Earliest Artefacts of Colour Printing in the West'. *Papers of the Bibliographical Society of America* 108, no. 4 (2014): 477–522.

Vine, Angus. *In Defiance of Time: Antiquarian Writing in Early Modern England*. Oxford: Oxford University Press, 2010.

Viney, William. *Waste: A Philosophy of Things*. London: Bloomsbury, 2015.

Vivier, Eric D. 'Thomas Nashe's Unprofitable Satire'. *Modern Philology* 117, no. 4 (2020): 423–444.

Vlieghe, Hans. *Flemish Art and Architecture, 1585–1700*. New Haven: Yale University Press, 1998.

Walker, Katherine. 'Early Modern Almanacs and the Witch of Edmonton'. *Early Modern Literary Studies* 18, no. 1 (2015): 1–25.

Walker, Katherine. '"Daring to Pry into the Privy Chamber of Heaven": Early Modern Mock-Almanacs and the Virtues of Ignorance'. *Studies in Philology* 115, no. 1 (2018): 129–153.

Wall, Wendy. *The Imprint of Gender: Authorship and Publication in the English Renaissance*. Ithaca: Cornell University Press, 1993.

Wall, Wendy. *Recipes for Thought: Knowledge and Taste in the Early Modern Kitchen*. Philadelphia: University of Pennsylvania Press, 2015.

Wall-Randell, Sarah. *The Immaterial Book: Reading and Romance in Early Modern England*. Ann Arbor: University of Michigan Press, 2013.

Wall-Randell, Sarah. 'Marlowe's Lucan: Winding-Sheets and Scattered Leaves'. In *Christopher Marlowe, Theatrical Commerce, and the Book Trade*, edited by Kirk Melnikoff and Roslyn L. Knutson, 11–25. Cambridge: Cambridge University Press, 2018.

Walsham, Alexandra. '"Like Fragments of a Shipwreck": Printed Images and Religious Antiquarianism in Early Modern England'. In *Printed Images in Early Modern Britain: Essays in Interpretation*, edited Michael Hunter, 87–109. Abingdon: Routledge, 2010.

Watson, Rowan. 'Medieval Manuscript Fragments'. *Archives* 13, no. 61 (1988): 61–73.

Watson, Rowan. 'Some Non-Textual Uses of Books'. In *A Companion to the History of the Book*, edited by Simon Eliot and Jonathan Rose, 480–492. Oxford: Blackwell, 2007.

Werrett, Simon. 'The Sociomateriality of Waste and Scrap Paper in Eighteenth-Century England'. In *Working with Paper: Gendered Practices in the History of Knowledge*, edited by Carla Bittell, Elaine Leong, and Christine Von Oetzen, 46–59. Pittsburgh: University of Pittsburgh Press, 2019.

Williams, Kelsey Jackson. *The Antiquary: John Aubrey's Historical Scholarship*. Oxford: Oxford University Press, 2016.

Wilson, F. P. 'Some English Mock-Prognostications'. *The Library* 4, no. 19 (1938): 6–43.

Winsbury, Rex. *The Roman Book: Books, Publishing and Performance in Classical Rome*. London: Bloomsbury, 2009.

Wolfe, Heather. 'An Example of Early Modern English Writing Paper'. *The Collation* (blog). February 4, 2014. Last accessed August 19, 2020. http://collation.folger.edu/2014/02/an-example-of-early-modern-english-writing-paper/.

Wolfe, Heather. 'Rethinking the Price, Quality, and Social Significance of Writing Paper in Early Modern England'. Paper presented at the Paper, Pen and Ink: Manuscript Cultures in Early Modern England Seminar Series, London, April 14, 2014.

Wolfe, Heather. 'Filing, Seventeenth-Century Style'. *The Collation* (blog). March 7, 2016. Last accessed August 19, 2020. http://collation.folger.edu/2013/03/filing-seventeenth-century-style/.

Wolfe, Heather. 'Was Early Modern Writing Paper Expensive?' *The Collation* (blog). February 12, 2018. Last accessed August 19, 2020. https://collation.folger.edu/2018/02/writing-paper-expensive/.

Wolfe, Richard J. *Marbled Paper: Its History, Techniques, and Patterns*. Philadelphia: University of Pennsylvania Press, 1991.

Woolf, Daniel. *The Social Circulation of the Past: English Historical Culture, 1500–1730*. Oxford: Oxford University Press, 2003.

Wright, C. E. 'The Dispersal of the Monastic Libraries and the Beginnings of Anglo-Saxon Studies. Matthew Parker and his Circle: A Preliminary Study'. *Transactions of the Cambridge Bibliographical Society* 1, no. 3 (1951): 208–237.

Yaeger, Patricia. 'Trash as Archive, Trash as Enlightenment'. In Hawkins and Muecke, *Culture and Waste*, 103–116.

Yale, Elizabeth. 'With Slips and Scraps: How Early Modern Naturalists Invented the Archive'. *Book History* 12, no. 1 (2009): 1–36.

Yale, Elizabeth. *Sociable Knowledge: Natural History and the Nation in Early Modern Britain*. Philadelphia: University of Pennsylvania Press, 2015.

Yates, Julian. *Error, Misuse, Failure: Object Lessons from the English Renaissance*. Minneapolis: University of Minnesota Press, 2003.

Yates, Julian. 'Thomas Middleton's Shelf Life'. In *The Oxford Handbook of Thomas Middleton*, edited by Gary Taylor and Trish Thomas Henley, 16–31. Oxford: Oxford University Press, 2012.

Youings, Joyce. *The Dissolution of the Monasteries*. Fakenham: Allen and Unwin, 1971.

Index

For the benefit of digital users, indexed terms that span two pages (e.g., 52–53) may, on occasion, appear on only one of those pages.
Figures are indicated by an italic f following the para. ID.